Contents

5·95

Television Principles and Practice

J. S. Zarach

Senior Lecturer,
North Staffordshire Polytechnic

Noel M. Morris

Principal Lecturer,
North Staffordshire Polytechnic

MACMILLAN

© J. S. Zarach and Noel M. Morris 1979

All rights reserved. No part of this publication may be reproduced or transmitted, in any form or by any means, without permission.

First edition 1979
Reprinted 1981, 1983, 1984

Published by
Higher and Further Education Division
MACMILLAN PUBLISHERS LTD
London and Basingstoke
Companies and representatives throughout the world

Typeset in 10/12 Times by
Reproduction Drawings Ltd, Sutton, Surrey
Printed in Hong Kong

British Library Cataloguing in Publication Data

Zarach, J S
Television principles and practice.
1. Television—Repairing
I. Title II. Morris, Noel Malcolm
621.3888′7 TK6642

ISBN 0-333-19220-6
ISBN 0-333-19221-4 pbk

Preface

The aim of this book is to present the reader with a comprehensive description of the principles involved in the operation and servicing of modern TV receivers. The treatment of the subject is such that is should appeal to students preparing for examinations of the City and Guilds Institute for **Radio, TV and Electronics Mechanics.** In this connection the syllabuses of the two options, Television (Colour and Monochrome) and Additional TV, receive extensive coverage.

Students taking CGLI **Radio, TV and Electronics Technicians** course, as well as the various 'television options' within the **Technician Education Council** framework, should also find the text of considerable assistance.

Many service engineers who wish to refresh their knowledge will find the subject presented in an easy-to-follow manner.

The contents of the book reflect the circuit techniques adopted in modern TV receivers; these include transistors, I.C.s, thyristors, valves (still used in many time base circuits) and other devices as appropriate.

The first three chapters cover the important aspects of colour perception and the formation of monochrome and colour signals (with special reference to the PAL system). A review of a complete TV receiver in block diagram form is presented in chapter 4; this provides the basis for detailed circuit descriptions in later chapters.

Chapters 5 to 7, inclusive, cover the main signal path from the tuner to the video (luminance) amplifiers and decoder. Practical circuits and principles of the inter-carrier sound reception are fully outlined in chapter 8. The field and line time base circuits (including e.h.t. generation) are featured in chapters 9 and 10. This is followed in chapter 11 by a description of picture tubes, including delta gun and in-line tubes, together with information about convergence and pincushion correction circuits.

Receiver setting-up procedures are outlined in chapter 12, including details of test cards F and G. Aerials and practical aspects of their installation are discussed in chapter 13; finally, in chapter 14 receiver power supplies are described, ranging from simple rectifier circuits to switched-mode arrangements.

Throughout the book, reference is made to the practical aspects of TV servicing and to the accompanying safety considerations.

The authors wish to thank the following organisations for permission to use their technical information in this book: BBC, Decca, GEC, ITT, Mullard, Philips, Rank Radio International and Thorn.

J. S. ZARACH
N. M. MORRIS

1 Principles of Colour and Colour Perception

1.1 Colour Perception of the Eye

In this chapter we want to concentrate on a subject that may appear at first glance to have more connection with physics than with TV engineering. To appreciate how the transmission and reception of colour TV signals has been made possible, we must examine the whole process of colour perception, or—as one may say—the illusion of colour.

As can be demonstrated, so-called 'white' light may be split into a large number of individual colours which form the whole light spectrum. If the source of light is sunlight, then the spectrum consists of all colours from dark violet to deep red. If the 'white' light may be split into a number of colours, is the process reversible? Is it possible to combine a number of colours, and produce 'white' light? This is indeed easily achieved; it has been found that it is not even necessary to mix all the colours of the rainbow to produce an impression of 'white' light. The human eye is constructed to be principally sensitive to the overall light intensity—that is, its brightness: only a small percentage of the receiving area of the eye is sensitive to the actual colour of the light (this phenomenon will be recalled later in conjunction with the composite TV signal).

It has been discovered that the impression of colour seems to be centred around three actual colours: blue, green and red. The eye, as it were, analyses the image it receives primarily in terms of these three *hues* and also in terms of the overall intensity of the light, and this is precisely what a colour TV camera does. The sensitivity of the eye to the three colours is not uniform—it is at its greatest to the yellow-green, followed by red and then blue. Further experiments show that if we combine red light, blue light and green light in appropriate intensities, all three being emitted from separate sources, the eye will perceive white light.

Light is a form of electromagnetic radiation, rather like radio waves, but it differs from the telecommunication signals in the wavelength or frequency spectrum involved. The actual colour one sees is only a sensation produced by the eye when it is struck by radiation of a given frequency. While frequencies of TV channels extend up to only 850 MHz, the light spectrum occupies a frequency range from approximately 400 million MHz to 800 million MHz. In radio and TV the wavelength is expressed in metres or centimetres; when we speak of light frequencies, the wavelength is so short that the unit used is the *nanometre* (1 nm $= 10^{-9}$ m; therefore there are 10 million nanometres in one centimetre). Figure 1.1 shows the distribution of the different forms of electromagnetic radiation, their frequencies and wavelength.

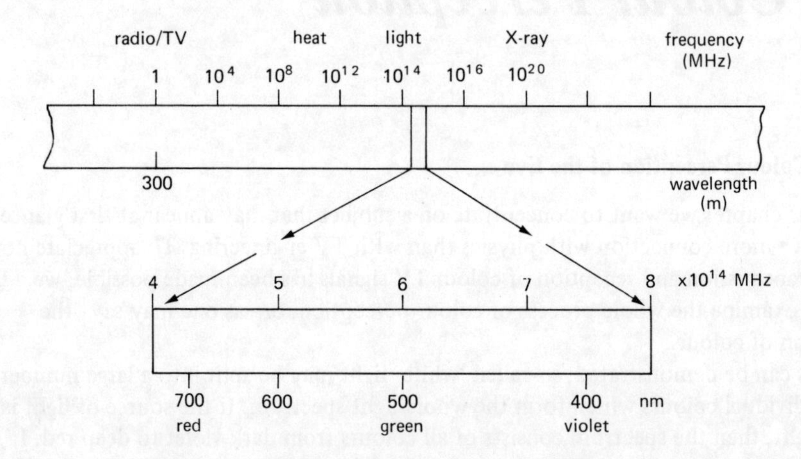

Figure 1.1 The spectrum of electromagnetic radiation

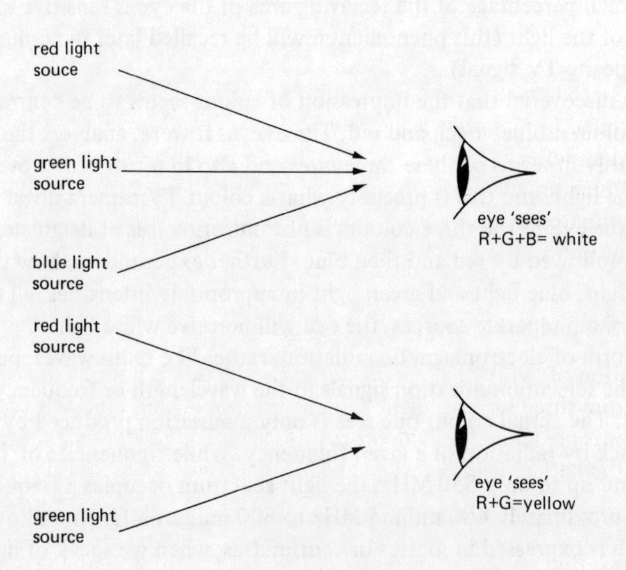

Figure 1.2 Additive mixing of colours

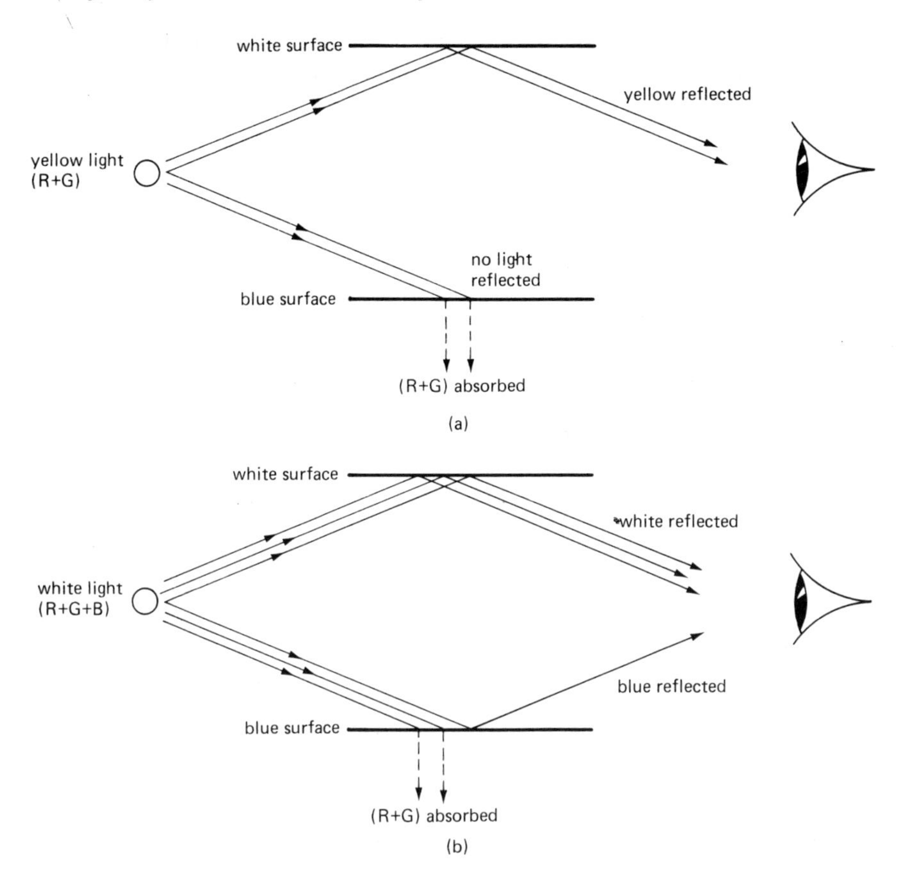

Figure 1.3 Subtractive mixing of colours using paints. (a) White paint reflects all colours; the blue paint can reflect blue light only; the original yellow light reaches the eye when reflected by the white surface, but light is not reflected off the blue surface, which now looks black (no blue in the yellow beam). (b) Both surfaces, when illuminated by white light, retain their intended colours (all three colours are present in the source of white light).

1.2 Additive and Subtractive Mixing of Colours

The sensation of 'white' can be produced by the eye as a result of combining the outputs from three separate light sources: red, green and blue. This implies that the three colours are added and the process is called *additive mixing.* (See figure 1.2.) Using mathematical shorthand the above statement can be repeated as follows

$$\text{white} = \text{red} + \text{green} + \text{blue}$$

It is obvious that only the correct intensities of the three colours must be used, for too much of any one of them will 'bias' our 'white' more towards that particular colour (producing, for example, the effect of 'warm white' or 'cool white').

The three colours which together produce white are called the *primary colours.* If only two primary colours are mixed at a time, the resultant hues are called the *complementary colours.* Provided the proportions of each primary are otherwise the same as those needed to produce white, the complementary colours are as follows

green + red = yellow

green + blue = cyan

red + blue = magenta

As will be shown later, the colour TV system operates on the principle of additive colour mixing. It is useful, therefore, to remember the above combinations because of their important practical implications in TV servicing.

Already, out of the three primaries, it is possible to obtain six different colours plus the white, and, of course, black on a TV screen simply means the absence of any light output. All other colours can be produced by altering the proportions of primary colours during the mixing process. For example, orange can be obtained by increasing the amount of red in the green–red mixture.

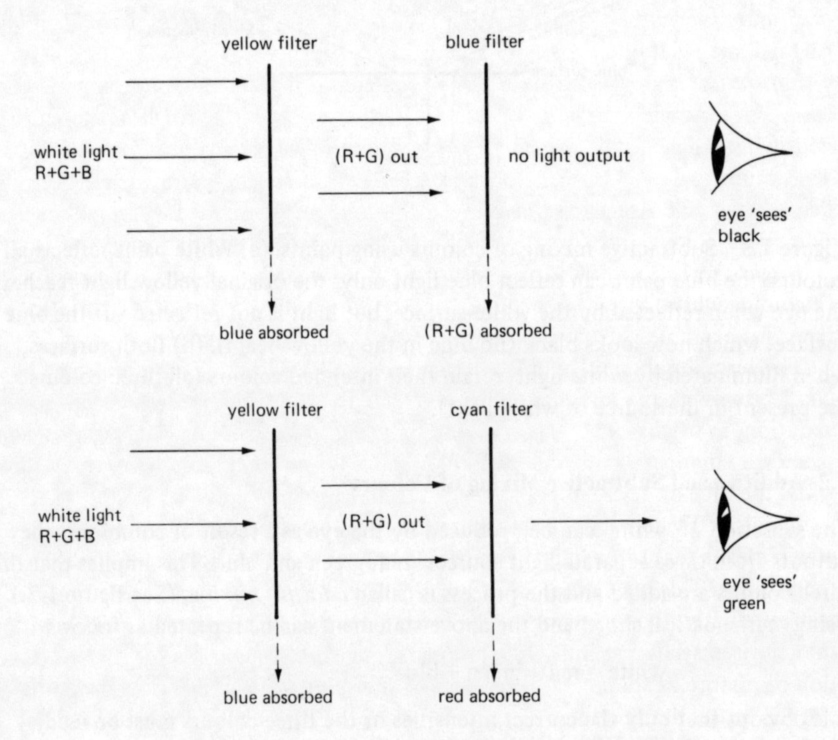

Figure 1.4 Subtractive mixing of colours using filters

There is another method of colour mixing which is used in connection with colour paints, colour photography, etc. This is known as *subtractive mixing* because the colour of an object is the result of certain colours being absorbed, or *subtracted* from the illuminating light, while other hues are then reflected towards the observer's eye. A number of examples of subtractive colour mixing based on the properties of different paints are given in figure 1.3. The diagrams in figure 1.4 illustrate this principle by means of colour filters. It must be stressed that the statement 'white = red + green + blue' is still valid. In additive colour mixing, however, the light reaches the eye *directly* from its source (lamps, TV screen), and in subtractive mixing the light has been reflected by another object which can alter the colour content of the original light source.

1.3 Hue, Saturation and Brightness

Later in the book we shall see how the principles of additive mixing are put into practice in a colour TV system. At this point certain terms have to be introduced which are used in connection with the TV signal.

Hue is the actual colour that the eye sees—i.e. we can have red, magenta, green, orange, etc., hues. Any of those colours may appear to be strong, rich—or, as it is called, *saturated*; alternatively they may appear to be light, pale—or referred to as *desaturated.* Desaturation of a colour is caused by the presence of white light in the original hue.

Saturation is an indication of how free a colour is from any dilution with white light. For example, in colour TV increasing the setting of the brightness control adds 'white' light to the display, which results in very pale colours (see section 7.6). That causes, say, a rich red colour to become pink—or a desaturated red. Of course, desaturated colours can also be part of the televised scene; for this reason information about colour saturation must also be transmitted in the TV signal.

Brightness is a measure of the intensity of light associated with a given colour (including white). Brightness is linked with the response of the eye to different hues; for example, yellow appears to be a brighter colour than green or red.

The term 'brightness' will be familiar to the TV engineer, as it is also used in connection with monochrome television—a black and white receiver displays the variations in brightness of the scene on its screen. The white or yellow parts of the scene would appear almost identically bright ('white') on a monochrome TV screen, while the blue or dark red content might look very dark grey.

In colour TV terms the brightness of the scene is described by its *luminance*, which means that some parts of the scene appear to be brighter, or more luminous, than others. This is a vital piece of information that must be transmitted for two reasons: firstly, monochrome receivers can only process the luminance content of the televised scene; secondly, the luminance signal is responsible for the reproduction of picture detail in colour receivers. It has already been mentioned that the eye recognises detail principally as a variation in light intensity; this applies especially to fine detail in the image.

2 Transmission of Monochrome and Colour Television Information

2.1 The TV Camera

The TV camera converts the image of the scene to be televised into an electrical waveform called the *video waveform.* The camera assembly can be considered as consisting of an optical and an electronic section.

The optical part of the camera follows the established photographic practice, but the picture is focused on to a 'target' layer of material whose resistance varies with the amount of light falling upon it.

The electronic section is formed by the camera tube and the associated amplifiers. The tube produces an electron beam by thermionic emission from a heated cathode. The electrons are accelerated by the anode, the electron beam being used to scan the target (see figure 2.1) so as to 'read' its illumination and generate a video waveform.

The scanning movement of the beam is in two directions–horizontally across the target and also in the downward direction; the image is being 'traced' as if by individual lines which are positioned below one another. Each time the beam reaches the end of one line, it is returned quickly to the beginning of the next one. The scanning is performed by the two sets of scan coils placed on the outside of the tube (figure 2.1). There is also a focusing coil to ensure that the electron beam is converged when it reaches the target.

The target part of the camera tube consists of a glass faceplate, on which a very thin, transparent layer of conducting tin oxide (signal plate) is deposited and then the target layer proper. The tin oxide allows the light to pass through it, but at the same time it conducts electric current to the target. A metal ring surrounds the signal plate to act as a seal for the tube and also as a connector to an external supply. The output signal voltage is developed across resistor R.

The target layer and the signal plate behave like a capacitor which is charged by the electron beam. When light falls upon the target, the resistance of the target layer drops and the capacitor loses some of its charge at that particular spot. The amount of charge lost depends upon the amount of light received. The design of the target is such that the loss of charge occurs only between the target and the

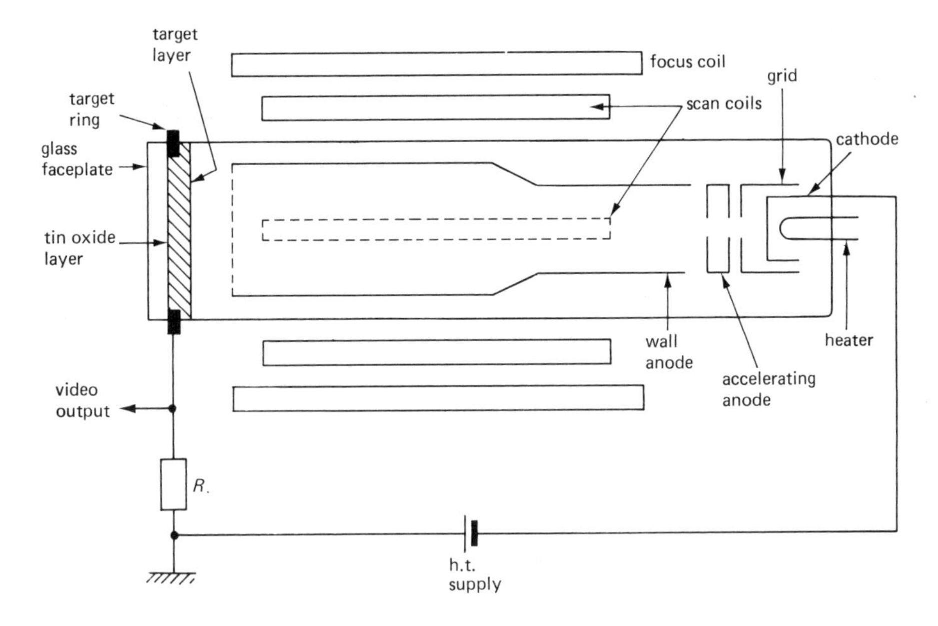

Figure 2.1 Simplified diagram of a camera tube

signal plate. The optical image consists of a large number of dark and light 'spots', and, therefore, what we have is not one capacitor but thousands of them, the number depending upon the detailed contents of the picture. As the beam scans the target, it encounters all these little capacitors in a varying state of discharge and it recharges them. As a result, we get a current flowing through the resistor R whose value, at any moment, depends upon the brightness of the scene. The magnitude of the current is only a few microamperes, and a suitable amplifier is needed to produce a usable signal. The voltage drop across R gives rise to the video waveform.

The most commonly used tubes are called the *Vidicon* and the *Plumbicon.* Their principle of operation is similar to the one described, but the target material is different in each case. Vidicon cameras are very compact and are used in closed circuit TV, industrial applications, portable outdoor colour TV cameras, etc.; Plumbicon tubes are mostly employed in studio colour TV cameras.

In the receiver the process is reversed—ultimately, a scanning beam is produced in the picture tube to 'trace' the same scene on the screen. It is necessary to keep the beam in synchronism with the camera beam; this is done by means of synchronising pulses which tell both the transmitter and receiver time bases when to start and when to finish the scan. Because the required deflection is in two directions, we have *line* (horizontal) *scan* and *field* (frame or vertical) *scan.* Thus the output from the camera will have the sync. pulses added to it before it can be modulated upon the carrier.

2.2 Block Diagram of Monochrome Transmitter

The diagram in figure 2.2 shows a monochrome transmitter arrangement in block form. A monochrome system has been chosen here since it provides an insight to the operation of the TV system without the complexity associated with colour television which will be described later.

The signal from the camera, together with the synchronising pulses [see figure 2.3(a)], modulate the vision carrier in the modulated r.f. (radio frequency) amplifier. Two methods of modulation are possible: *positive modulation* and *negative modulation.*

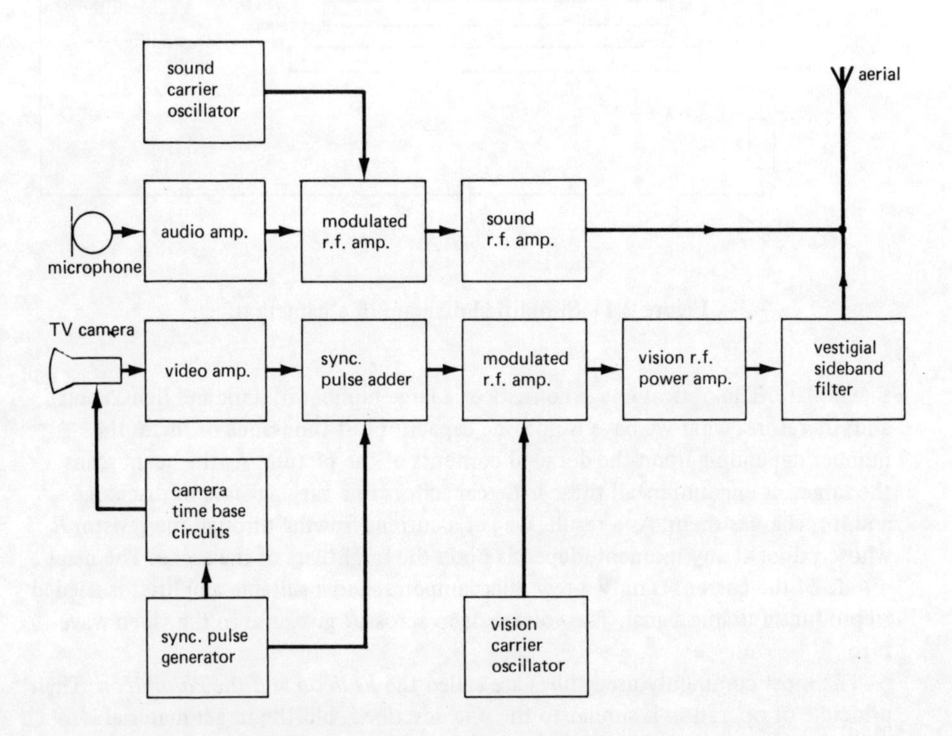

Figure 2.2 Simplified block diagram of a monochrome transmitter

With positive modulation, the amplitude of the carrier increases with the brightness of the scene [figure 2.3(b)]; maximum brightness, known as *peak white*, corresponds to 100 per cent amplitude of vision carrier. When negative modulation is used, the carrier amplitude diminishes with increasing brightness; peak white occurs at 20 per cent of the maximum amplitude [figure 2.3(c)]. Negative modulation is the most popular method, and is used in the UK for the 625-line standard. Positive modulation is used in a few countries and is also

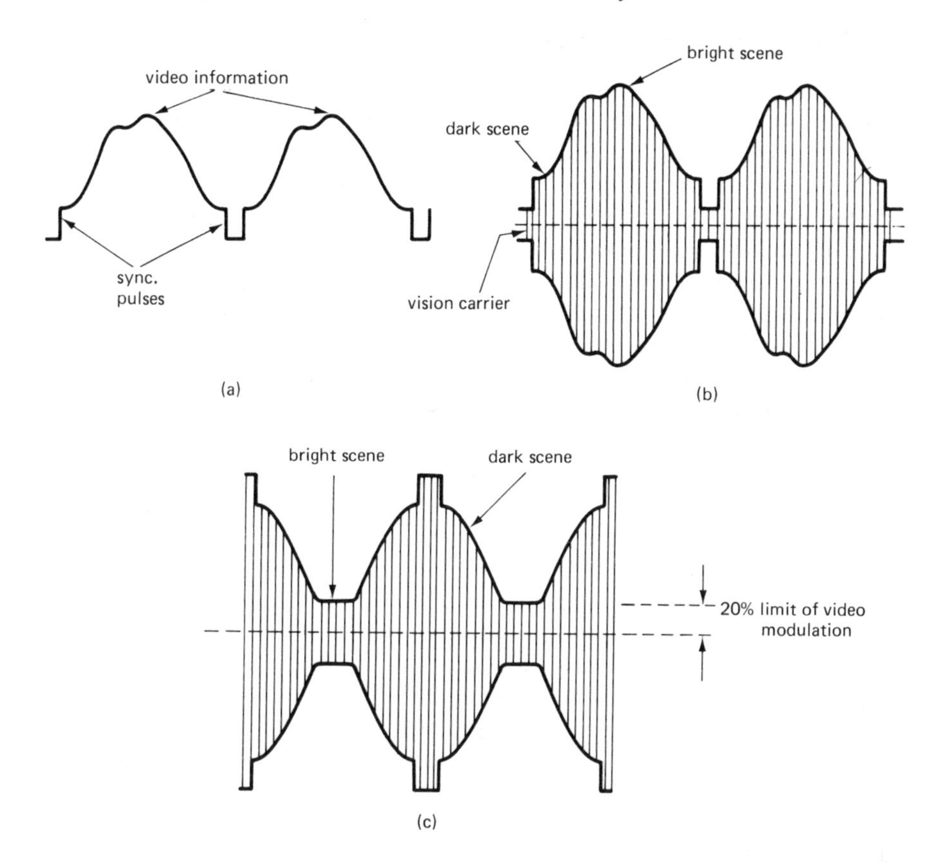

Figure 2.3 Examples of vision carrier modulation: (a) two lines of video waveform before modulation; (b) two lines of video waveform modulated on a carrier–positive modulation; (c) two lines of video waveform–negative modulation

employed in the UK for the 405-line standard. The chief advantage of negative over positive modulation is the improvement in picture quality. Any impulsive interference which normally affects the maximum amplitude of the signal will, in positive modulated systems, correspond to the 'whiter than white' region; this causes annoying white spots to appear on the screen. With negative modulation, such interference produces black spots which are less obvious to the viewer—unfortunately, this affects the synchronising pulses, and more elaborate circuitry is needed in the receiver line time base to ensure steady synchronisation.

The sound channel uses frequency modulation on the 625-line standard and amplitude modulation for the 405-line transmission. Before they are radiated from the aerial, both the vision modulated and the sound modulated carriers are amplified by their respective r.f. power amplifiers to raise the power levels of the signal.

2.3 Principles of Colour Transmission

As we saw in the previous chapter, it was possible to mix the three primary colours, red, green and blue, in their specified proportions to produce either white or other colours as required. The process is reversible—we can split the white and the various colours into their primary colour contents. Therefore, at the transmitter the televised scene can be analysed by the cameras into its red, green and blue content, while the receiver will mix the signal voltages to display the original colours on the screen. One of the main requirements of a colour transmission system is that it must be *compatible* with the existing black and white system. In other words, a monochrome receiver must reproduce the signal in black and white without any modification or loss of quality. Also, colour receivers must be able to display a monochrome picture; this is sometimes called *reverse compatibility*.

To satisfy these requirements, the transmitted information consists of the *luminance signal* and two so-called *colour difference signals* (*chrominance* or *chroma*). Black and white sets respond to the luminance content, since this provides the variations in brightness; monochrome receivers will be practically unaffected by the presence of the chrominance signal. Colour receivers require, apart from a different tube, a separate circuit called the *decoder* to detect the colour content of the scene.

The symbol used for the luminance signal is Y' (or E'_Y, and sometimes it is simply Y), and the two transmitted colour difference signals are known as $(R' - Y')$ or $(E'_R - E'_Y)$ or just $(R - Y)$ and $(B' - Y')$ or $(E'_B - E'_Y)$ or $(B - Y)$.

To produce a colour difference signal, we must first have a camera output voltage corresponding to, say, the red content of the scene, which after some additional processing we shall call R'. We also need a signal voltage which depends upon the luminance value of the *same* scene—that is, Y'. The latter signal is subtracted from the R' voltage in an electronic circuit to produce $(R' - Y')$. The circuit which performs the addition (or subtraction) of voltages is called a *matrix*. The second transmitted colour difference signal is obtained in a similar manner to yield $(B' - Y')$, where B' is related to the blue content of the scene.

The next step is to modulate the carrier with the signals ready for transmission. The process of modulation in a colour transmitter is fairly complex and it will be discussed separately.

We can now anticipate what will happen in the receiver, which, after detection and decoding, will have available Y', $(R' - Y')$ and $(B' - Y')$ information. The original colour signals, namely R' and B', are then recovered using a matrix circuit which adds together the detected voltages

$$(R' - Y') + Y' = R'$$

$$(B' - Y') + Y' = B'$$

The reader will have noticed that the third primary colour, green, appears to be missing from this system; one might expect to find $(G' - Y')$ somewhere. It can be shown, however, that the $(G' - Y')$ signal may be obtained by adding a proportion of the decoded $(R' - Y')$ voltage to a proportion of the $(B' - Y')$. The derivation of this relationship will not be carried out in this book, but it can be proved that $(G' - Y') = -0.51\,(R' - Y') - 0.186\,(B' - Y')$; the minus signs indicate that the signals used have to be of the opposite phase from those processed in the remainder of the circuit. Finally, the third primary colour, G', is obtained from another addition

$$(G' - Y') + Y' = G'$$

Why is $(G' - Y')$ not transmitted? As shown above, it is only necessary to transmit two of the three colour difference signals. It is preferable to exclude $(G' - Y')$ because it has been observed that, of all three, this one has the lowest magnitude; hence, it could be much more affected by noise, distortion, etc.

2.4 Colour TV Camera Arrangements

In view of the need for the three transmitted signals, it is possible to have a number of camera arrangements. At least three camera tubes must be used in a combined unit, all three looking at the scene through the same lens, but the image is first split into its colour content before it reaches the targets of the respective tubes. The task of splitting the light is performed by *dichroic* mirrors or prisms mounted between the lens and the tubes. The most important property of a dichroic mirror is that it reflects one colour only and allows the remaining colours to pass through it. An example of a much simplified three-tube arrangement is shown in figure 2.4(a).

The image from the lens falls on the dividing mirror (this is not dichroic), which allows some of the light to reach the luminance signal tube, the remainder being reflected towards the first dichroic mirror. Here the red content of the image is reflected and other colours pass through to the second dichroic mirror. This mirror reflects only the blue content and directs it to the tube. After some initial amplification and processing, the luminance and the 'colour' signals are fed to their respective matrix circuits, where the two colour difference signals are formed.

Yet another, and more compact, three-tube camera arrangement is shown in figure 2.4(b). Now the dichroic prisms supply the tubes with the contents of the scene corresponding to the three primary colours, red, green and blue. The prism P1 is provided with a blue reflecting dichroic surface so that the blue light is directed towards the 'blue' tube; similarly, the dichroic surface of P2 reflects the red contents of the scene towards the 'red' tube. The luminance signal is not available directly, and must be formed by the addition of the three colours in the Y-matrix. In other words we have to reverse the action of the dichroic prism by electronic means. Finally the 'red' and the 'blue' outputs are matrixed with the luminance to produce the colour difference signals.

Studio cameras sometimes have four tubes, so that the set-up appears to be a combination of the two just described. In this case three tubes are used to generate the colour signals; the fourth tube receives the complete image and produces the

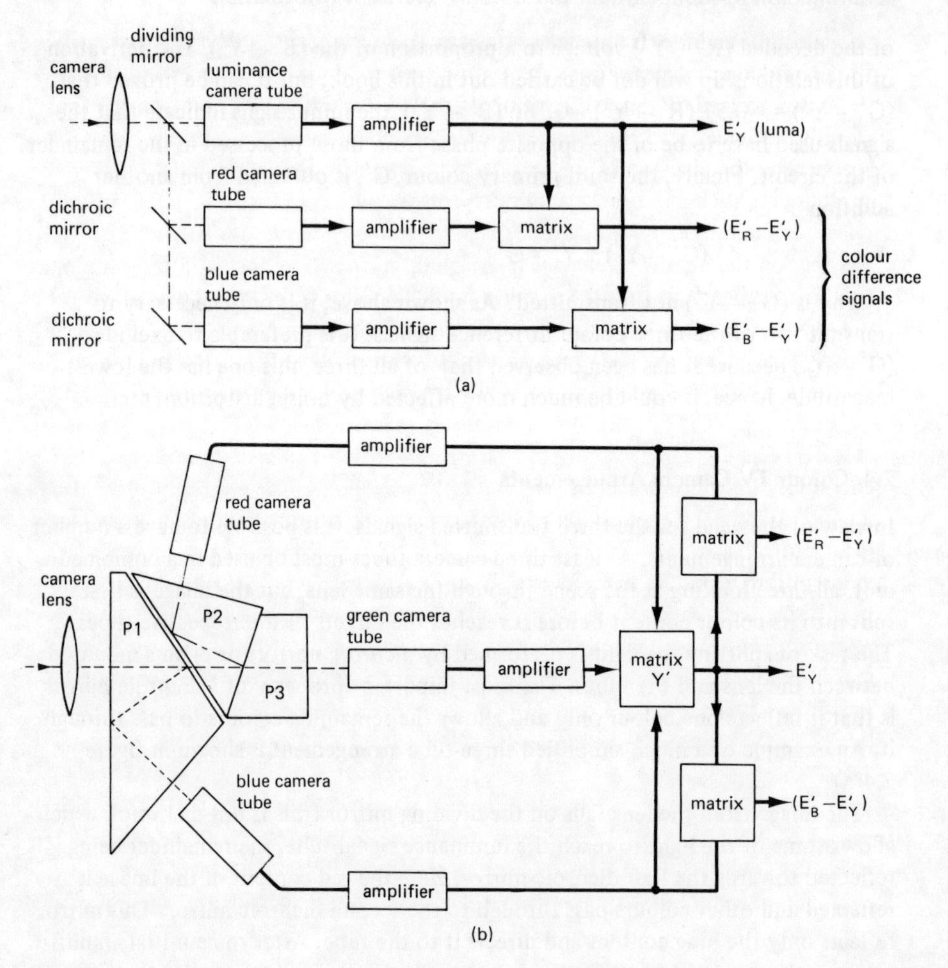

Figure 2.4 Three camera tube arrangements required for the production of colour TV signals: (a) simplified arrangement using dichroic mirrors; (b) practical arrangement with dichroic prisms

luminance signal. The latter output is used as a source of reference which is compared with that obtained by the addition of the red, green and blue camera tube outputs; any adjustments can then be made to equalise the two signal routes.

2.5 TV Signal Bandwidth

We now turn our attention to the range of frequencies occupied by the complete television signal, which we call the *bandwidth.* This is a very important property affecting not only the transmitter (it determines the operation of the transmitter, allocation of channel frequencies, etc.), but also the receiver design (principally its i.f. and video amplifier stages).

As we noticed from the operation of the TV camera, its output is in the form of pulses of electric current corresponding to the pattern of the image being scanned by the electron beam. In order to appreciate the frequencies involved in reproducing such a pattern, we have to consider the speed with which the beam traverses the target; as we shall see, the duration of one picture line is 52 μs. During that period the beam current has to rise and fall to follow the variations in picture brightness; the changes in the amplitude of the video waveform represent its frequency. If the image consists of fine detail, then the changes in brightness are very closely spaced, so that the repetition rate of the camera output voltage is high—it can be said that the video frequency of fine picture detail is high. For coarse picture detail the amplitude of the video waveform varies less and the resultant frequency is lower. The diagrams in figure 2.5 show parts of two picture patterns and the corresponding video waveforms along the scanning line X-X. In part (a) the display is of a relatively coarse pattern—when the beam current encounters the black area, it falls to zero and the output voltage is a maximum. The reader will recall that the video output from the camera tube circuit in figure 2.1 will be equal to the supply in the absence of a potential drop across the resistor R. When the beam scans the white area of our pattern, the current rises and the output voltage drops. Similar reasoning can be applied to the fine pattern in part (b). We have chosen a simple

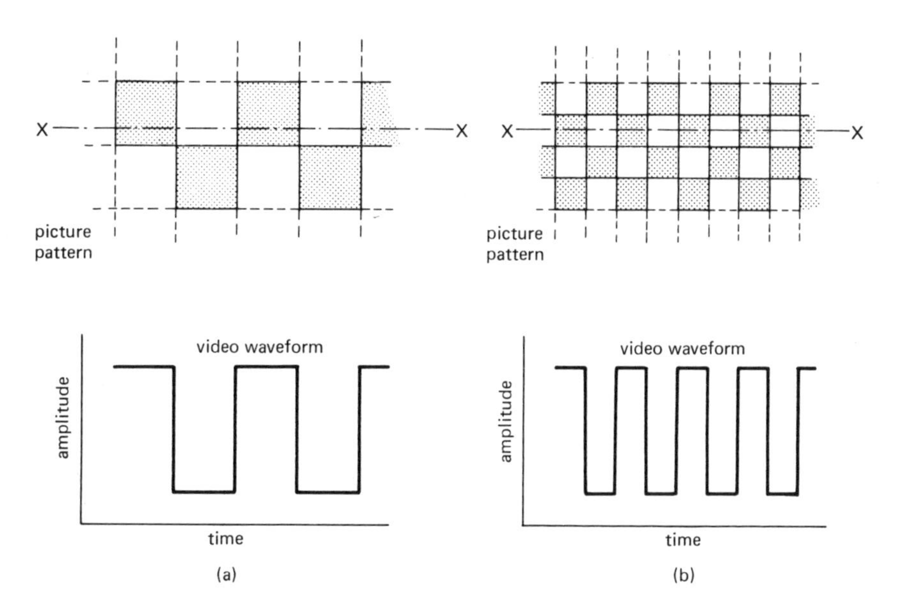

Figure 2.5 Effect of picture detail on video frequency. The video waveforms correspond to the portion of the pattern scanned along the line x-x. (a) Coarse detail—low frequency; (b) fine detail—high frequency

geometrical display, but the electrical effect would be the same if we analysed a more usual type of televised scene.

So far we have considered only the changes in the video waveform along one horizontal line; the display pattern also changes in the vertical direction. Here the complete picture is split into a fixed number of lines; if there are more lines per picture, the beam has to move much faster to complete the scan, and the time difference between the video peaks in figure 2.5(a) and (b) is shortened, resulting in a higher video frequency. The greater the number of lines, the better the quality of the displayed picture, because the changes in brightness in the downward direction are then likely to be dictated by the actual picture detail rather than by the line structure of the display. British readers can easily verify this by comparing a 405-line with a 625-line picture. The 405-line system originated in the days when electronic circuits could not easily cope with very high frequency signals.

Theoretically, the complete TV system should be able to transmit and receive satisfactorily a picture which could consist of the chequerboard pattern shown in figure 2.5. This display is divided into a number of elements, or small squares, whose sides correspond to the 'thickness' of one line. In a 625-line system the actual number of lines in one picture is 585 (as will be shown later, the remaining 40 lines are associated with other functions); therefore the dimensions of each picture element are equal to 1/585 of the overall height.

The total number of picture elements on a square screen would be in this case equal to 585 × 585. However, the transmitted display is a rectangle whose proportions are given by the *aspect ratio* of 4:3 (which means that the picture height is $\frac{3}{4}$ of its width). Since the number of elements is increased by the aspect ratio, the frequency of the corresponding video waveform must increase accordingly.

The speed of the downward movement of the beam is fixed by the *field frequency*. TV systems were initially designed to have the field frequency locked to the a.c. mains supply. In a colour TV system a very precise relationship exists between the various frequencies present and a drift in mains frequency would be unacceptable. Nevertheless, the field frequency is equal to the nominal supply frequency, which in the UK is 50 Hz.

If all the picture lines were scanned during each field, then the highest video frequency required to reproduce the fine chequerboard pattern would be given by

$$\text{video frequency} = \tfrac{1}{2} \times \text{maximum number of elements} \times \text{aspect ratio} \times \text{field frequency}$$

(As the picture elements consist of alternate black and white squares, two such squares correspond to one cycle of video frequency—hence the division by 2 in the above expression).

Substituting the values in the formula would give a video frequency of nearly 11.5 MHz. In fact, if the signal were to reproduce such rapid changes in brightness as the chequerboard pattern implies, the required frequency range would have to be extended still further. Finally, the process of modulating the video information upon a carrier gives rise to sidebands, which increase the bandwidth even more.

In order to reduce the overall bandwidth of a TV transmission, *interlaced scanning* is universally used. It has been assumed so far that all the 585 *active picture lines* are 'drawn' on the screen during each downward movement of the beam. In fact, only half that number is scanned during one field; the second half—which is produced during the next field—is placed in the spaces between the first group of lines. The principle of interlacing of the lines of two fields is illustrated in figure 2.6. With this method each picture (*frame*) consists of two fields. The field frequency is still 50 Hz, but there are only 25 pictures scanned per second. Despite such a low picture (frame) frequency, flicker is avoided, partly because of the persistence of the image on the screen and partly because of the 'inertia' of the eye, which retains the image for a fraction of a second.

Interlaced scanning halves the value of the highest video frequency as calculated above, because only 292½ active picture lines (312½ overall) are associated with every cycle of the field frequency. The calculations previously quoted were simplified to show the main factors involved; there are additional considerations which influence the final choice of video bandwidth, which is 5.5 MHz in the British 625-line standard. Since we have considered only the transmission of a black and white pattern, this bandwidth applies to the luminance or monochrome signal. The presence of other information will be discussed later.

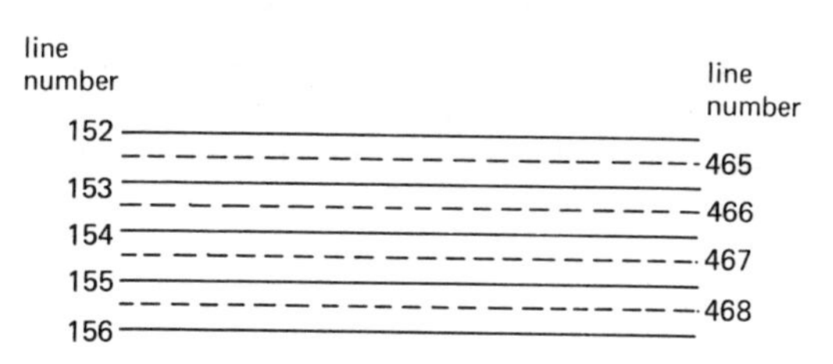

Figure 2.6 Principle of interlaced scanning. Lines 152, 153, etc., are scanned during one field; the interlaced lines 465, 466, etc., appear during the subsequent field. (Flyback not shown.)

2.6 Signal Sidebands

The problem of the transmitted signal bandwidth is further complicated by the process of modulating the carrier with the video information. It can be proved both experimentally and mathematically that when a carrier wave at a certain frequency, f_c, is amplitude modulated by a signal at another frequency, say f_s, the resultant output has been altered to such an extent that it consists of three signals at different frequencies, namely

(1) The carrier frequency, f_c.
(2) The carrier *plus* the modulating frequency, $f_c + f_s$; this is called the *upper sideband.*
(3) The carrier *minus* the modulating frequency, $f_c - f_s$, called the *lower sideband.*

The output from the transmitter now covers a very wide frequency range if it includes both the lower and the upper sidebands. For example, if the highest modulating video frequency is 5.5 MHz, then the transmitted signal would extend from 5.5 MHz below the carrier frequency to 5.5 MHz above the carrier, making the bandwidth 11 MHz. We can now see how much wider it would have been if we did not use interlaced scanning!

There would be two major practical consequences of using such a high bandwidth in the transmitted signal: firstly, each transmitter would occupy this frequency range and the available number of channels would have to be reduced to accommodate these wide spacings between them; secondly, the receiving circuits would have to cope with large bandwidths, adding to the complexity of the design (and servicing!).

Fortunately, it is possible to introduce economies which reduce the necessary bandwidth before transmission. It so happens that all the information required is available in *either* sideband, and the information in the upper sideband simply duplicates the contents of the signal in the lower sideband. If we could remove one of the sidebands, we would still broadcast and receive the original programme intelligence. The diagram in figure 2.7 shows the distribution of video frequencies and the sidebands associated with the carrier. We note that the output remains constant to include signals up to 5.5 MHz; afterwards it is attenuated, so that there ought to be no video information extending beyond 6 MHz. The slope of the diagram between 5.5 MHz and 6 MHz indicates that it is not possible to reject

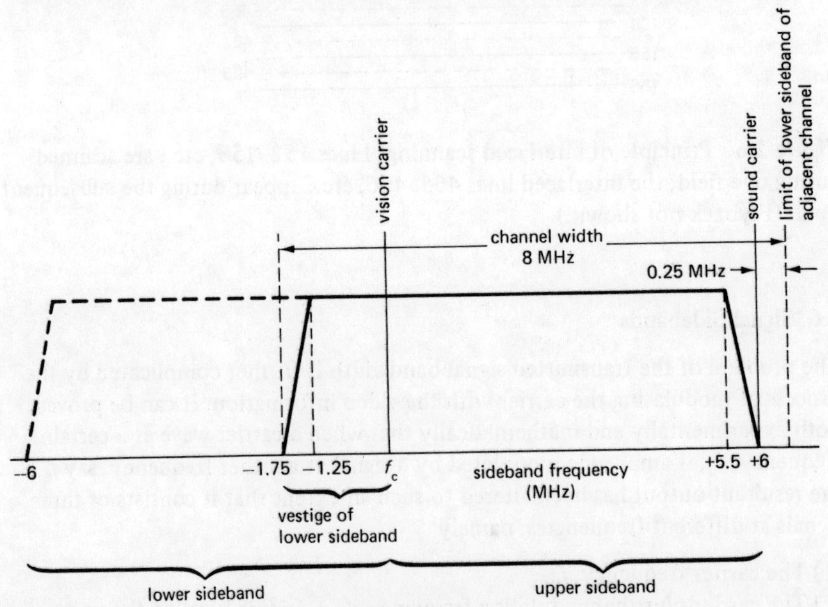

Figure 2.7 Diagram of the principle of vestigial double sideband transmission (transmitted frequency spectrum shown by the solid outline)

abruptly all frequencies beyond a certain limit; any filter used offers the relatively gradual reduction in output. For the same reason it is not possible to cut off one of the sidebands along the vertical line passing through the f_c; instead, it would be necessary to start filtering at some frequency before f_c so that the output becomes zero at the carrier frequency. In TV broadcasting this is unacceptable, because we would lose the very important lowest video frequencies which lie close to the carrier. The alternative is to start filtering after f_c, which leaves part of the other rejected sideband intact. In the UK the upper sideband is transmitted together with part of the lower sideband. This is called *vestigial sideband* transmission because the remaining part of the lower sideband is called vestige (meaning a trace). From figure 2.7 we see that the overall width of one channel is from −1.75 MHz (negative sign to denote the lower sideband) to 6 MHz, at which the sound carrier is placed. An additional 0.25 MHz is needed to accommodate the sound sidebands and also to offer a buffer space before the next channel—the total is now 8 MHz instead of about 12 MHz if both video sidebands were transmitted.

Incidentally, in sound broadcasting both sidebands are broadcast as they are narrower; the full audio spectrum extends only up to 20 kHz. It must be pointed out that frequency modulation produces the two types of sidebands, but the actual frequency distribution is somewhat different from that outlined here for an AM process.

2.7 Channel Allocation

From the above discussion the reader will appreciate why transmitting station frequencies have to be suitably spaced to prevent mutual interference which overlapping sidebands could produce. As TV standards may be different in other countries, the actual frequencies in use also tend to differ. At present, TV broadcasting in the UK is carried in five bands of frequencies. Broadly speaking, band I is from 40 MHz to 70 MHz and band III from 170 MHz to 220 MHz, and both are known as v.h.f.; they are still used for the 405-line standard and, to some extent, for the 625-line system fed from special relay stations and distributors. Band II, 88 MHz to 108 MHz, is reserved for f.m. sound broadcasting. TV transmission at u.h.f. falls into either band IV, 470–580 MHz, or band V, 620–855 MHz. Each station, or programme company, is given a channel number (letters are also used on v.h.f.) which automatically standardises its vision and sound carrier frequencies. Each channel on the 625-line standard occupies a bandwidth of 8 MHz which includes vision (luminance), sound and chrominance. In band I the channels are A, B, C, R (also from 1 to 5 on the 405-line standard); in band III they are from D to I (also from 6 to 13); in band IV from 21 to 34; in band V from 39 to 68.

2.8 Modulation of Colour Information

The introduction of colour transmission had to fulfil the requirement that the existing bandwidth would not be extended by the presence of the chrominance

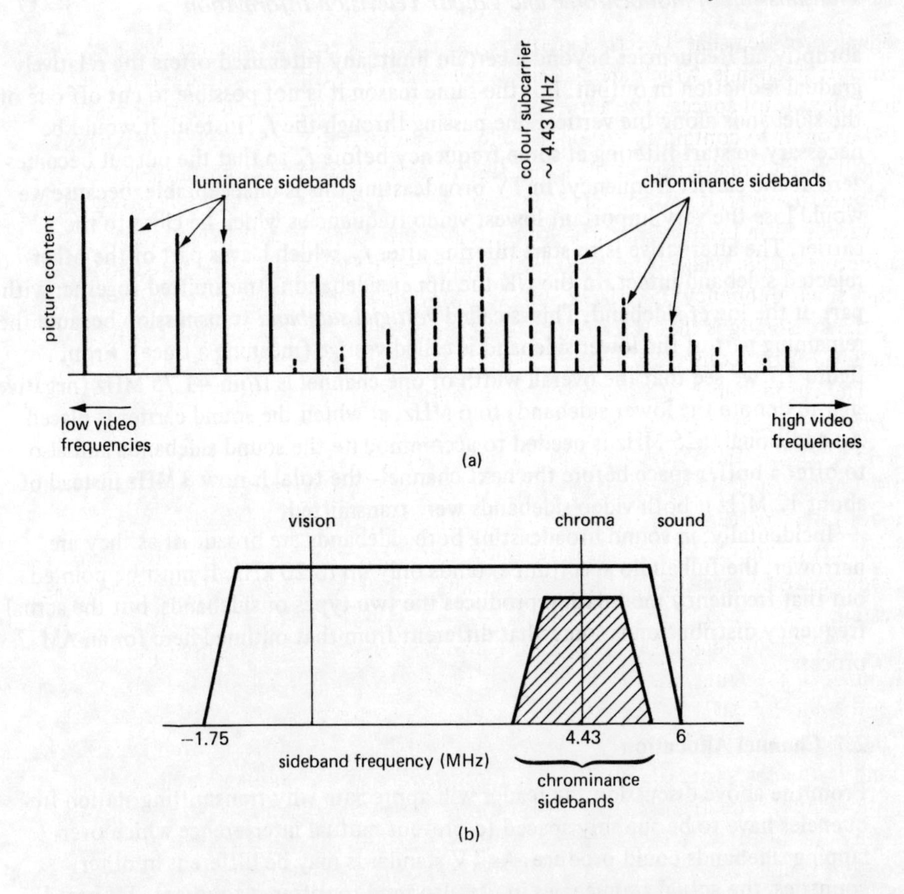

Figure 2.8 Principle of frequency interleaving: (a) distribution of signal energy in the luminance sidebands at high video frequencies; there are also frequency 'gaps' into which the chroma signal can be fitted; (b) position of the chroma signal within the existing video (luminance) bandwidth.

signal. At the same time, the colour signal and its inevitable sidebands should be accommodated without causing undue interference with the vision (luminance) and sound signals. Detailed study of the actual frequencies present in the video sidebands disclosed that they do not possess every conceivable value up to 5.5 MHz, as might be suspected at first. Instead, they are associated with multiples of the line time base frequency, so that there are gaps between those values which could be filled by another signal. This frequency distribution is given by the fact that the video waveform is not continuous but occurs at regular intervals caused by the line scan.

The available gaps, as shown in figure 2.8(a), can be filled by the chrominance signal, provided that its carrier frequency is very precise. This ensures that the chroma sidebands, which are also multiples of the line time base frequency, fall into the vacant spaces. The actual chrominance signal frequencies must be restricted in order to contain the resultant sidebands within the allocated channel bandwidth. Such a restriction implies that fine picture detail is not transmitted in colour. However, it was explained in chapter 1 that the eye notices fine detail only as variations in brightness; the colour effect is produced by somewhat coarser structure of the image. The reader will recall that fine picture detail corresponds to high video frequency; larger elements result in lower video frequency. The outcome of this reasoning is that the colour difference signals have a bandwidth of approximately 1 MHz; consequently their two sidebands are about 1 MHz each.

The chrominance signal is modulated on a carrier, usually referred to as a *subcarrier*, of a frequency normally quoted as 4.43 MHz, but whose precise value is 4.433 618 75 MHz. Such accuracy ensures the correct *frequency interleaving* of the luminance and chrominance signals. This value also minimises the creation of a dot interference pattern caused by the chroma signal on the screens of TV receivers (monochrome sets in particular). For the explanation of this process the reader must be referred to the more specialised books on the PAL colour television system.

As the chrominance information is centred around 4.43 MHz within the luminance spectrum, the upper sideband extends to (4.43 + 1) MHz, which approaches 5.5 MHz, while the lower sideband extends down to about 3.4 MHz. The diagram in figure 2.8(a) shows in a simplified form how the chrominance sideband frequencies (broken line) can be interleaved with the luminance signal (continuous line). The diagram also explains why the colour subcarrier lies towards the higher end of the video frequency spectrum. It appears that the average picture content is biased towards the low video frequencies, as indicated by the height of the verticals in the diagram. Therefore, by placing the chrominance in the region of low luminance content, any interaction between the two is kept to a minimum. In order to reduce interference between the different carriers comprising the composite TV signal (in particular, between sound and chrominance), and also to save transmitter power, the actual colour subcarrier is suppressed before transmission takes place. It was shown earlier that, as a result of the modulating process, there exist two sidebands and the carrier. The latter acts only as a 'vehicle' which simplifies demodulation in the receiver; if there were a simple alternative to the diode detector (see chapter 6), then in most applications the carrier would not have to be transmitted. The diagram in figure 2.9 shows what would happen if the video waveform of figure 2.3(a) were modulated and the carrier suppressed. In the receiver, a simple diode detector would give an output which corresponded to a rectified half of such a signal. As we see from figure 2.9(b), this would have very little resemblance to the original video signal because of the missing carrier reference. The reader is invited to compare the appearance of normal modulated wave in figure 2.3(c) with that in 2.9(a); the two envelopes, instead of being apart on either side of the zero line, are now shown crossing over.

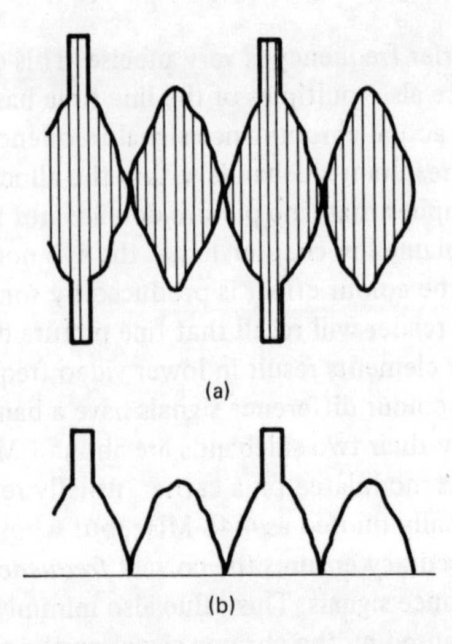

Figure 2.9 Possible effect on the normal video waveform if the vision carrier was suppressed: (a) modulated waveform (compare with figure 2.3c); (b) output from a simple diode detector [compare with the desired output in figure 2.3(a)]

To demodulate a suppressed carrier signal requires reinsertion of the missing carrier at the receiver itself. For correct demodulation, the regenerated colour subcarrier must be in perfect synchronism with that used at the transmitter. This can only be done if a colour synchronising signal is transmitted as part of the information. Such a signal is known as the *colour burst*, and it is considered in the next chapter.

There is also another problem in transmitting the chrominance signal. As described in section 2.3, there are two separate components, namely (R′ – Y′) and (B′ – Y′). Our discussion so far allowed for one signal to be interleaved which we called chrominance. To accommodate both components without interaction, a method known as *quadrature modulation* is used. This means that the same frequency subcarrier is used, but one of the modulators is supplied with the carrier via a 90° phase shift network—that is, the two modulated signals are in phase quadrature. As a result of this arrangement, when the subcarrier is passing through its maximum in one modulator, it is zero in the other, and vice versa during its next half-cycle. This system produces two separate signals which are added to form the composite chroma signal as shown in figure 2.10. The demodulating process in the receiver is the reverse of the modulation, as will be shown in chapter 7. There the reintroduced subcarrier will have two paths 90° out of phase with each other, so that the feeds to the demodulators will be in step with the transmitter and they will 'recognise' their respective signals as being the original (R′ – Y′) and (B′ – Y′).

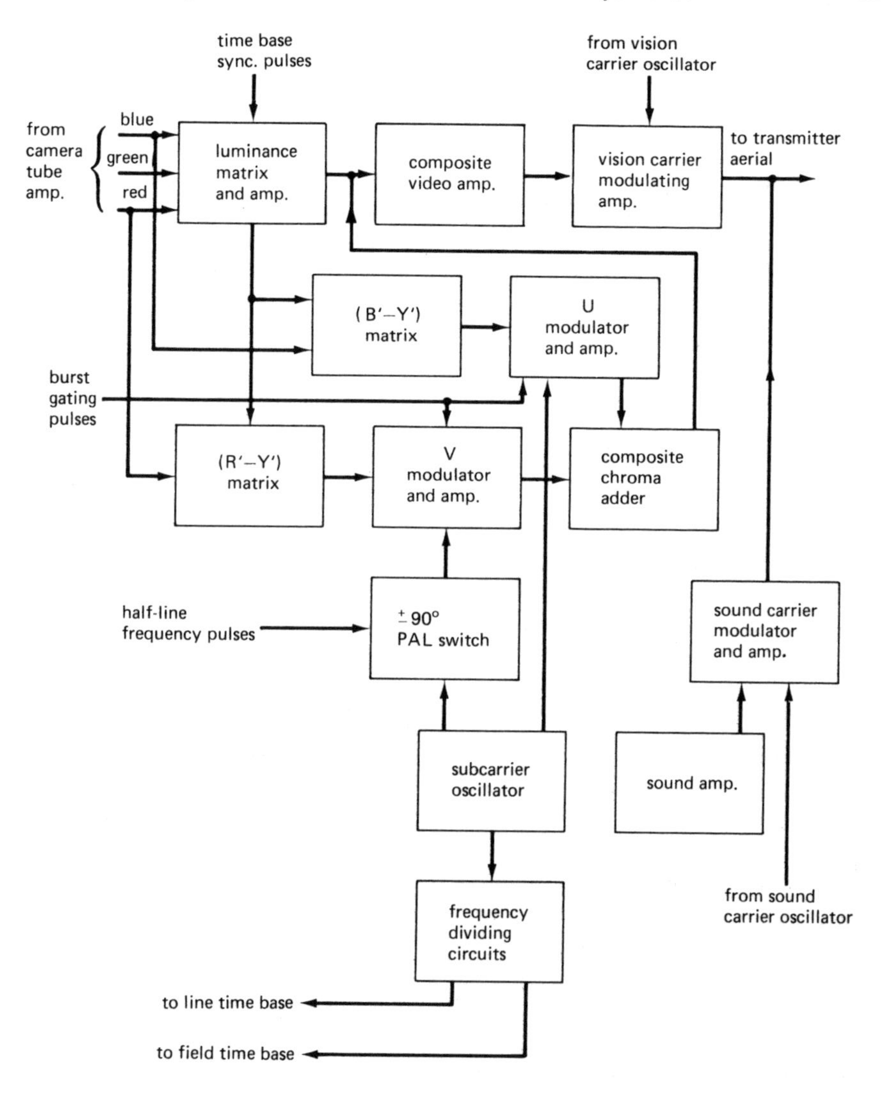

Figure 2.10 Simplified block diagram of PAL colour TV coder and transmitter arrangement

Readers will note that in the block diagram in figure 2.10 we have introduced new symbols, namely U to replace $(B' - Y')$ and V instead of $(R' - Y')$. This is to indicate the change in the character of the two initial colour difference signals, as will now be explained.

The composite modulated chroma is added to the luminance signal to produce composite video. The addition of the two could cause the amplitude of the output signal to exceed the permitted limits of modulation. It was mentioned in section 2.2 [and in figure 2.3(c)] that the amplitude of the video waveform must not drop below 20 per cent of the maximum. If chroma were superimposed upon the luminance signal, it would be possible to drive the composite video considerably below that limit. Similarly, at the other end of the video waveform the added chrominance could extend well beyond the synchronising pulse level; this point is discussed again in chapter 3. To prevent overmodulating, the two colour difference signals have their amplitudes reduced by what are known as *weighting factors.* Thus U and V are the weighted and modulated $(B' - Y')$ and $(R' - Y')$, respectively. As a result of the compression of their amplitude, the $(B' - Y')$ signal is reduced by a half and the $(R' - Y')$ by approximately 12 per cent [more exactly, $U = 0.493\ (B' - Y')$, $V = 0.877\ (R' - Y')$]. They are restored to their original proportions in the receiver amplifiers.

Referring to the block diagram in figure 2.10, the reader will observe that the time base frequencies are also derived from the colour subcarrier frequency to ensure luminance–chrominance interleaving, and to minimise the effect of subcarrier interference pattern on TV screens.

2.9 The PAL System

The colour encoding system used in the UK and in many other countries is known as the PAL system, and was invented in Germany by the Telefunken Co. The abbreviation PAL stands for *P*hase *A*lternation by *L*ine. Referring to figure 2.10, it will be seen that the subcarrier feed to the V modulator goes through the PAL 180° (±90°) switch. The purpose of the switch is to invert the phase of the subcarrier during alternate scanning lines, which, when fed to the V modulator, causes the V signal to reverse its phase every line. The reversal is brought about by the second input signal to the PAL switch, that is, a square waveform at half the line frequency. The output from the composite chroma adding circuit is (U + V) during one line and (U – V) during the next, to revert back to (U + V) again, and so on. The applied switching waveform is at half the line frequency, because one half-cycle will maintain the original phase of the subcarrier while the next half-cycle will cause the reversal.

This method of colour transmission makes the system relatively free from changes in actual colour (hue), which could otherwise occur owing to imperfections in the transmission–reception chain or imperfections in the performance or alignment of the receiver. Briefly, the above imperfections can cause a change in the mutual phase relationships between the U signal, the V signal and the subcarrier used in the modulation process (and a sample of which is also sent out as a separate signal in the form of a colour burst). These changes, unless corrected, would produce ‘confusion’ in the receiver demodulator, whose regenerated subcarrier relies on the various phase relationships being correct in order to ‘recognise’ the original values of $(R' - Y')$

and (B′ – Y′). As will be shown in the next chapter, relative alterations in the values of the colour difference signals mean changes in the hues reproduced on the screen. The phase reversal of the V signal causes the phase errors to be cancelled out (the error is positive on one line and negative on the next); however, there is some deterioration in the detected output in the form of reduced saturation. It is considered to be less important though, because the viewer might notice this as a reduction in colour intensity, easily counteracted by the colour control in the receiver. An explanation of the effect of PAL switching will be given in chapter 7.

The reader will recall that there is a colour synchronising signal, the colour burst, transmitted for the purpose of maintaining the phase of the receiver subcarrier oscillator with that of the transmitter. PAL switching introduces a new problem, because it causes the subcarrier phase to alternate through 180° every line. To maintain perfect synchronism, the same process must be repeated in the receiver. The phase of the burst is changed every line in step with the operation of the PAL switch, which allows the receiver to identify which line contains a reversed V component. The diagram in figure 2.10 shows how the colour burst is formed. During a short interval of time, no video information is transmitted (details in the next chapter) and a burst gating pulse is fed to the two modulators, which allows the subcarrier to pass through. (Normally the type of modulator used in this position automatically suppresses the subcarrier.) The gating pulse is of a very short duration and only 10 cycles at 4.43 MHz will be included in the composite chroma. From the block diagram, readers will note that the burst will consist of two components of the subcarrier: the one fed to the U modulator (U component) and that fed to the V modulator via the PAL switch (V component). Since the two components of the burst are fed to the composite chroma adder circuit, the resultant signal also undergoes a phase change every line. This signal is sometimes called a 'swinging' burst because of this alternation of its V component.

2.10 Other Colour Encoding Systems

In the United States a system is used known as *NTSC* (National Television System Committee, formed by a number of TV manufacturers). Historically this was introduced before the PAL system and, unfortunately, suffers from defects which the PAL arrangement managed to reduce. Indeed, in many respects the two systems have much in common—namely there are two chrominance signals quadrature modulated. However, instead of the two colour difference signals being kept separate, the American system uses a combination of (R′ – Y′) and (B′ – Y′) in each component. The system is so arranged that each chroma signal is associated with certain colours according to the ability of the eye to resolve detail in different colour. Because of this the bandwidth of one signal is almost three times the bandwidth of the other. This was the inventors' solution to the problem of accommodating the required colour information within the available TV bandwidth. American broadcasting standards use 4.5 MHz video bandwidth, which offers less space for frequency interleaving when compared with the British 5.5 MHz standard. The

NTSC system also uses suppressed subcarrier modulation and a colour burst, but there is no equivalent of the 180° switching. Perhaps the main criticism of the US colour TV system is that the various imperfections in the transmitting-receiving chain mentioned in the previous section do cause a change in reproduced colours. The viewer has to use a tint control (not to be confused with the tint control of PAL receivers—see chapter 11) to counteract this.

In France and in several other countries a system known as *SECAM* (*Séq*uential *à M*emoire—line sequential transmission with memory) is used. The system uses separate colour difference signals (R′ – Y′) and (B′ – Y′), but, instead of the two being transmitted simultaneously as quadrature modulated signals, they are broadcast sequentially. During one line the composite video consists of the luminance and modulated (R′ – Y′); during the next line we have the luminance and modulated (B′ – Y′). Since both signals are needed together to produce the colour picture, the receiver must store (memorise) the alternate colour signal for the duration of one line until the second chrominance component is available. As we shall see later, the modern PAL receiver also uses a 'memory' (PAL delay line), but for different reasons. In the French system the chroma signal is frequency modulated and there is no need for colour synchronising.

In conclusion, it must be mentioned that there are also slight variants of the PAL system dictated by the existing TV standards in any given country.

3 *The Composite Video Signal*

3.1 The Video Signal Waveform

In chapter 2 we discussed the formation of the various signals which together make up the composite signal broadcast by the transmitter. Now we shall look at these signals in some detail and study their waveforms. The design of a TV receiver is based on the existence of standardised video signals; knowledge of their waveforms helps the servicing technician in fault-finding or setting-up.

The luminance (or monochrome) video waveform is produced by the camera tube (or tubes) and the associated amplifiers as described in chapter 2. Time base synchronising pulses are added to the initial waveform in order to synchronise the scanning of the picture tube screen with the scanning of the camera tube target layer.

The diagram in figure 3.1 shows in detail the time and amplitude relationship

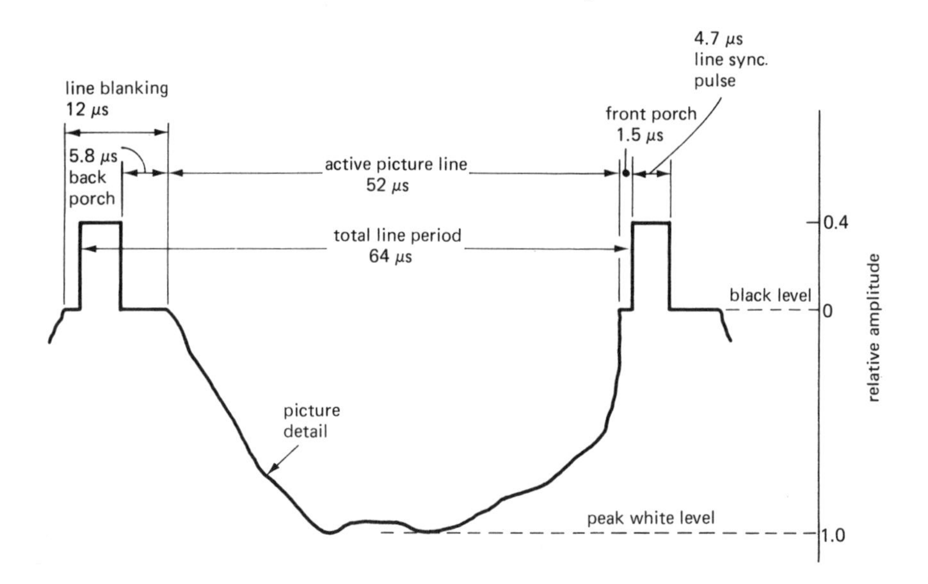

Figure 3.1 Time and amplitude relationships in a video waveform

between the various parts of a typical video waveform. Two line synchronising pulses have been added to define the limits of the actual scan period. The amplitudes are measured with respect to the reference point given by the *black level.* The black level is fixed at the transmitter and, ideally, it should also be maintained at the receiver, as will be shown in chapter 6. Picture modulation must always start from and end on 'black', irrespective of the actual content of the scene. To ensure that this occurs, there are two short 'resting' intervals on either side of the line synchronising pulse. The one preceding the pulse is called the *front porch*; this period allows receiver circuits to change their voltage levels to that corresponding to the black level. This is particularly important in the circuits leading to the synchronising pulse separator, as it is necessary to provide a strictly defined reference point, from which the separation of the pulses from the video can take place. It is especially needed when the line ends on white—a large change in signal level is required and that, in turn, takes time (for example, to charge or discharge various capacitors). Following the synchronising pulse, the picture modulation is still maintained at black level for a short period of time known as the *back porch*; this allows the receiver line time base to complete the flyback. During a colour transmission, the back porch is also a convenient part of the video waveform on which to place the *colour burst.*

The tips of the synchronising pulses associated with the waveform in figure 3.1 correspond to the maximum amplitude of the composite video. This is, in fact, how it appears at the output from the camera tube, where, for a very bright picture called *peak white*, the beam current is a maximum hence the output voltage becomes a minimum (section 2.1). Of course, the presence of various amplifiers can easily invert the appearance of the waveform as it happens, for example, in the receiver, as will be shown in chapter 6.

One line of the video signal consists of one synchronising pulse, the two porches and the actual picture information. The time duration of one line is 64 μs, out of which only 52 μs is used to transmit picture detail (this is known as the *active line*), while the remainder is occupied by the synchronising pulse and the maintenance of the black level. Therefore the 12 μs interval between the beginning of the front porch and the commencement of the next picture waveform is called the *line blanking period.*

The line time base frequency can be calculated from the total number of lines scanned per second, which is the number of lines per frame (625) multiplied by the number of frames per second (25); the resultant gives 15 625 Hz. The duration of one line, or its periodic time, is 1/15 625 = 64 μs. The time division between the active line and the blanking period is simply an accepted standard.

3.2 The Luminance Waveform of a Colour Bar Signal

The waveform in figure 3.1 represented one line of an arbitrary televised scene. If an oscilloscope were used to display a video waveform which corresponds to an average TV picture, the only clearly recognisable part of the waveform would be the synchronising pulses. The picture detail of a large number of lines (there are 15 625 lines per second) would be superimposed to such an extent that the scope display would become relatively meaningless.

With the advent of colour television, the use of precise waveform examination during servicing became much more important and the *colour bar waveform* was introduced. In a correctly adjusted colour receiver this type of signal produces a display of eight vertical stripes of equal width in the following order (from left to right): *white, yellow, cyan, green, magenta, red, blue* and *black.* This arrangement is from the brightest bar (white) to the darkest (black), or, in other words, the bars are in the order of decreasing luminance (see definitions in chapter 1). The reader will note that, in addition to the black and white, the display has the three primary and the three complementary colours. On the screen of a monochrome receiver this waveform will produce eight bars of decreasing brightness—from white, through different greys, to black. This type of pattern can be obtained either from a signal generator or from a colour bar transmission (including the test cards).

In order to produce the actual colour display, the waveform must contain both

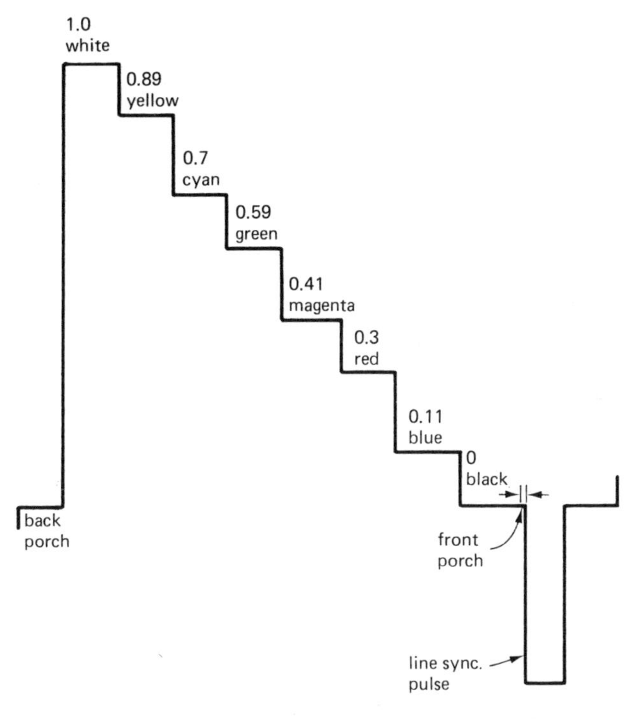

Figure 3.2 Video waveform—luminance part of a colour bar signal (in monochrome this waveform produces 'grey scale' display). Individual amplitudes are shown with respect to peak white

the luminance and the chrominance information. The required luminance waveform is given in figure 3.2.; it resembles a 'staircase', and it is often called the luminance steps. The amplitude of each step can be worked out from simple calculations, which we shall now consider. The reader will recall from chapter 2 that the luminance signal is formed from the three primary colour signals mixed in specified proportions. These amounts are given by the following equation

$$Y' = 0.59G' + 0.3R' + 0.11B'$$

That is, the luminance signal voltage is made up from 59 per cent of the green camera output voltage plus 30 per cent of the red and 11 per cent of the blue. The amounts used are due to the response of the eye to different colours (chapter 1). The camera tube amplifier outputs are adjusted so as to make them all equal when white is viewed (the impression of white is created when all three colours are suitably mixed). The actual voltages do not really matter at this point, so let us call that output 1 unit. Therefore, for the *white* bar $Y' = 0.59 \times 1 + 0.3 \times 1 + 0.11 \times 1 = 1.0$. This means that the luminance output voltage, either from a matrix or from a separate luminance camera, would also have to be 1 unit. The next bar is *yellow*; as shown in chapter 1, this is produced by mixing red and green; their proportions are given in the above equation—that is, $Y' = 0.59 \times 1.0 + 0.3 \times 1.0 = 0.89$. For example, for the *red* bar, only the red camera tube would produce its full output of one unit (the rest would be off) hence, $Y' = 0.3 \times 1.0 = 0.3$. Finally, for the *black* bar the output is zero from all camera tubes, and $Y' = 0$. The reader is invited to calculate the individual 'steps' and compare his answers with the values given in figure 3.2. The polarity of the waveform in figure 3.2 is opposite to that shown in figure 3.1, but it has been mentioned that the video signal may be inverted several times during signal processing, both at the transmitter and at the receiver; therefore either polarity may be encountered.

3.3 Chrominance Waveforms of the Colour Bar Signal

As shown in the previous chapter, the signal used to modulate the colour subcarrier is made up from two colour difference signals: $U(B' - Y')$ and $V(R' - Y')$. These are also found in the decoder section of a colour receiver; the service technician needs to be able to identify their waveforms in the process of fault-finding. The colour bar signal provides a stationary, easy-to-recognise set of waveforms which can be displayed on the screen of an oscilloscope.

By use of a technique similar to the method employed in calculating the luminance steps, the shape of the chrominance waveforms can be determined.

Let us consider the $(R' - Y')$ waveform. The expression $(R' - Y')$ means that from the output of the *red* camera tube for any given colour we must subtract the luminance value for that colour. For the *white* bar, as shown in section 3.2, $R' = 1.0$ and $Y' = 1.0$; hence, $(R' - Y') = 1.0 - 1.0 = 0$. For the *yellow* bar $R' = 1.0$, $Y' = 0.89$; therefore $(R' - Y') = 1.0 - 0.89 = 0.11$. For the *blue* bar $R' = 0$ (since there is no red in saturated blue), $Y' = 0.11$ (see figure 3.2); thus

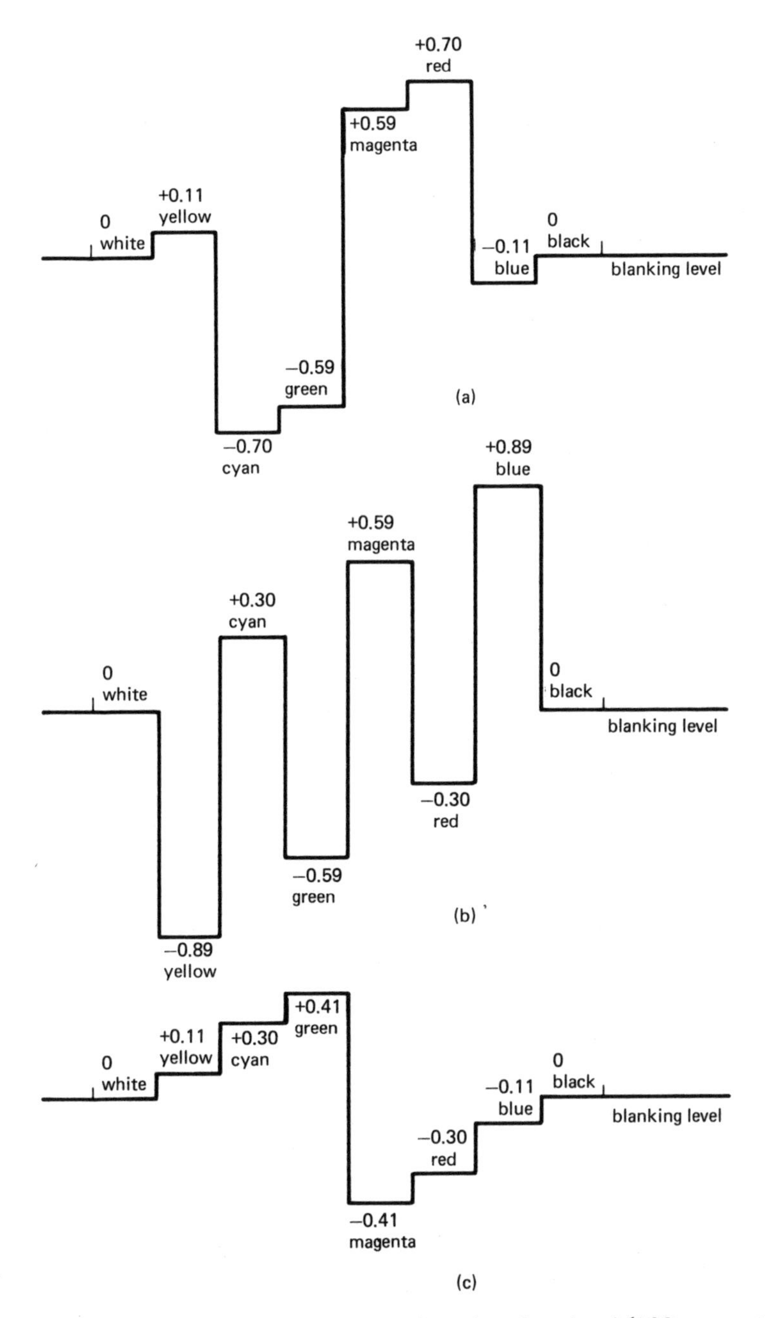

Figure 3.3 Colour difference waveforms of a colour bar signal (100 per cent saturation): (a) (R′ – Y′); (b) (B′ – Y′); (c) (G′ – Y′). Peak amplitudes for each bar are given in relation to the maximum value of the luminance signal (see also figure 3.2)

$(R' - Y') = 0 - 0.11 = -0.11$. The reader can check the calculations for the remaining colours and compare them with figure 3.3(a).

The second colour difference signal, $(B' - Y')$, is shown in figure 3.3(b). For the *white* bar $B' = 1.0$; hence, $(B' - Y') = 1.0 - 1.0 = 0$. For the *yellow* bar $B' = 0$ (no blue in fully saturated yellow); therefore $(B' - Y') = 0 - 0.89 = -0.11$; and so on for the other bars.

The third colour difference signal, $(G' - Y')$, is not transmitted, but it has to be formed in the receiver as outlined in section 2.3. It is instructive at this stage to construct the $(G' - Y')$ waveform, since the method is similar to that for the other two. For the *white* bar $G' = 1.0$; hence $(G' - Y') = 1.0 - 1.0 = 0$. For the *yellow* bar $G' = 1.0$, and $(G' - Y') = 1.0 - 0.89 = 0.11$. For the *red* bar $G' = 0$; hence $(G' - Y') = 0 - 0.3 = -0.3$. The complete waveform is given in figure 3.3(c).

One very important conclusion from the above calculations is that for both white and black (or any shade of grey, for that matter) all three colour difference signals are zero. This means that when a monochrome picture is being transmitted, the chrominance signal automatically disappears. This makes the colour decoder in the receiver temporarily 'redundant'; indeed, that section of the receiver is then automatically shut down to prevent unwanted coloured pattern appearing on the screen (see also chapter 7).

It will be noted from the colour difference waveforms that the synchronising pulses are absent; these are part of the low frequency luminance information and they would not appear in the 4.43 MHz chrominance channel.

The values of the waveforms given in figure 3.3 correspond to fully saturated colours—that is, with no white light added. For example, for the *yellow* bar it was assumed that only the red and the green outputs were present. If the colour were *desaturated*, then the additional white would consist of red, green and blue; therefore desaturation means that there is always an output, however small, from each camera tube. It has been argued that the use of fully saturated colours in test patterns is unfair, since the colours in a normal programme material will always be desaturated to some degree. For this reason the colour bar signal transmitted by the broadcasting authorities can be specified as being, for example, 95 per cent saturated.

The presence of desaturation and the weighting factors complicate the arithmetic involved in the calculation of the amplitudes of the above waveforms. However, from the practical point of view the shape of the signal waveforms remains basically similar to that shown in figure 3.3.

The modulated colour difference waveforms can now be considered—that is, when the weighted $(R' - Y')$ and $(B' - Y')$ signals become V and U, respectively. If the usual amplitude modulation technique were adopted, each colour difference signal would form the upper and the lower envelopes of the subcarrier in a manner similar to the video waveform in figure 2.3. As the colour subcarrier is suppressed, the two envelopes cross over the zero line [compare the possible effect on the video waveform in figure 2.9(a)], and the resultant waveforms are as shown in figure 3.4. The 'shading' inside the waveform denotes the high frequency sidebands which are, of course, present in this signal.

The precise method of mathematical construction of the composite chroma waveform is somewhat lengthy, and it is left to the reader to obtain the values shown in figure 3.5(b) with the help of the following explanation. Waveforms can be represented by a rotating phasor or, where the resulting wave is the sum of two other waveforms, by the sum of two phasors. Since the chrominance signal is pro-

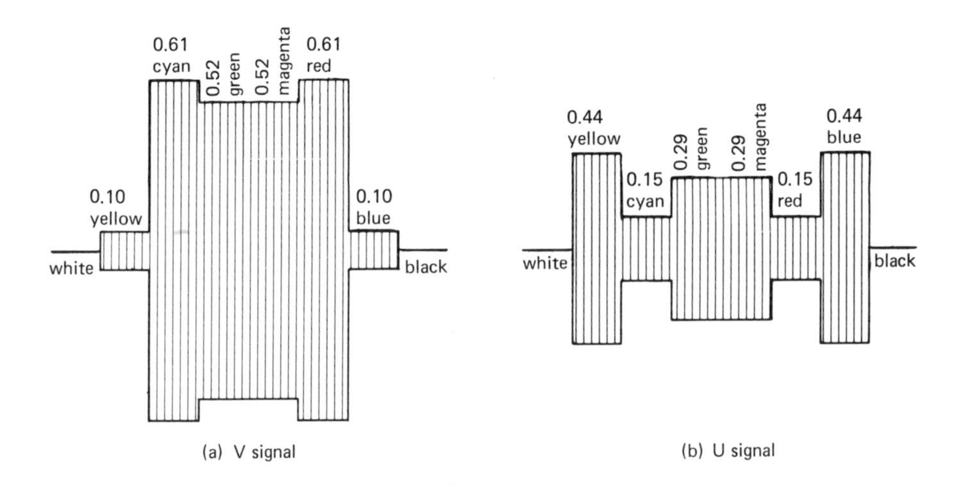

Figure 3.4 Modulated and weighted colour difference signals (100 per cent saturation) corresponding to a colour bar display. Amplitudes referred to the maximum value of the luminance signal (see also figure 3.2)

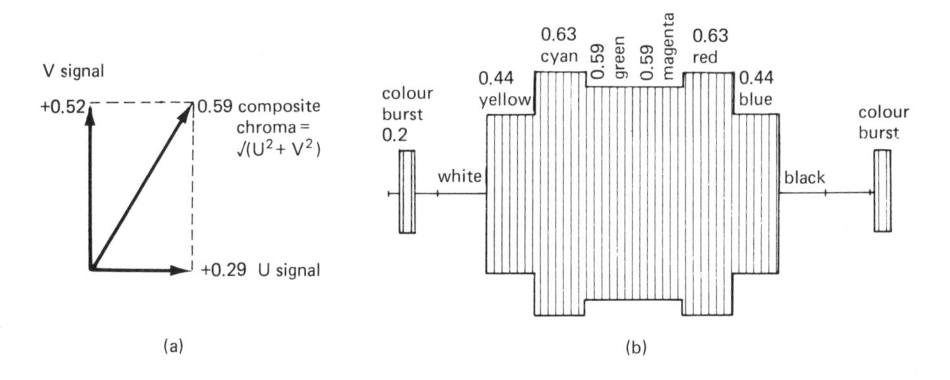

Figure 3.5 Composite chroma waveform of a colour bar signal (100 per cent saturation): (a) phasor addition of quadrature modulated signals (U + V); amplitudes correspond to the magenta bar (see also figure 3.4); (b) composite chroma waveform with colour burst

duced by the addition of the U and V signals, the resultant chroma can be represented by the U and V phasors [see figure 3.5(a)]. The process needs to be repeated for every colour bar except, of course, black and white. The values of U and V are obtained from (B′ – Y′) and (R′ – Y′) multiplied by their respective weighting factors (see section 2.8 and figure 3.3). The resultant chroma can be calculated by applying Pythagoras' theorem to the U/V phasor diagram, so that

$$\text{chroma} = \sqrt{(U^2 + V^2)}$$

The waveforms shown in this chapter include the colour burst, since at the transmitter both the U and V signals are accompanied by the burst, as implied in the block diagram in figure 2.10. In the receiver the colour burst is deleted before the separation of U and V takes place; therefore the reader would come across suitably modified waveforms, as shown in chapter 7. It will be noted that the amplitude of the burst is 20 per cent of the peak white value of the luminance signal.

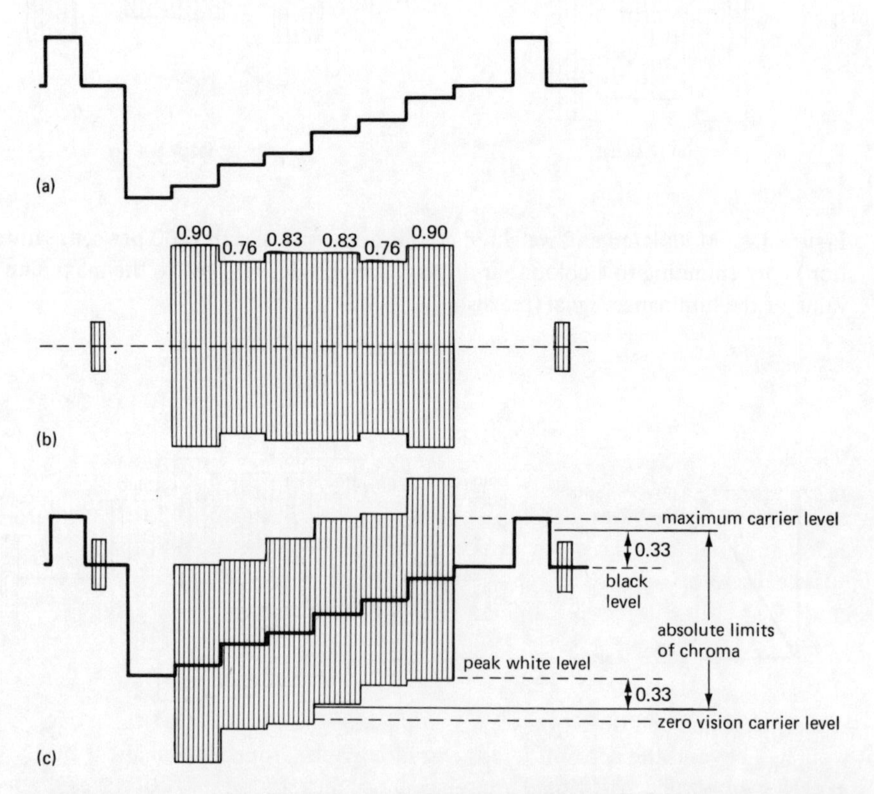

Figure 3.6 Formation of the composite video waveform of a colour bar signal: (a) luminance waveform; (b) unweighted composite chroma = $\sqrt{[(R' - Y')^2 + (B' - Y')^2]}$ with burst added; (c) composite video obtained by adding waveforms (a) and (b); the resultant exceeds the carrier levels (unacceptable overmodulation). Weighting factors are applied to maintain the composite waveform within the absolute limits of chroma

3.4 The Composite Video Waveform of the Colour Bar Signal

Composite video denotes that both the luminance and the composite chroma signals are present after the final addition in the encoder (figure 2.10). At this stage it is of interest to recall why weighting factors have to be used. Figure 3.6(a) represents the luminance waveform similar to that in figure 3.2, but in this case the polarity has been changed to show what happens when negative modulation of the vision carrier takes place. Figure 3.6(b) shows a composite chroma waveform made up from the modulated (R′ – Y′) and (B′ – Y′) without the weighting factors. The addition of the individual values of the composite chroma to the corresponding values of the luminance gives the composite video in figure 3.6(c). If we now include the vision carrier reference levels (zero and maximum), we note that the resultant video will considerably over-modulate the carrier. For the yellow, cyan and green, the presence of the chroma signal would drive the vision carrier below its zero level. If this were allowed, the vision detector would clip these excursions, which would result in distortion.

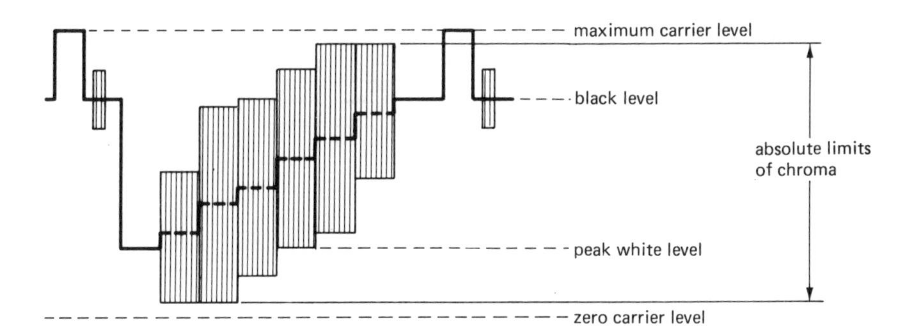

Figure 3.7 Composite video waveform of a colour bar signal (100 per cent saturation). Weighted chroma (figure 3.5) added to the luminance waveform (figure 3.6)—modulation limits no longer exceeded. If colours are less than 100 per cent saturated, the chroma amplitude is reduced and the resultant is well within the set limits

At the right-hand end of the waveform on the blue bar figure 3.6(c), the resultant would go beyond the peak of the synchronising pulse, which is well above the vision carrier amplitude; this would again be clipped and further distortion would occur. The values of the weighting factors were chosen to restrict the composite video to within 33 per cent of the peak white amplitude beyond either the white level or the black level. The modified composite video waveform, together with the 33 per cent limits, are shown in figure 3.7. As can be judged from figure 3.3, it is the (B′ – Y′) signal which has the highest amplitude; therefore the weighting factor is almost 0.5, compared with approximately 0.9 for the (R′ – Y′).

3.5 The Field Synchronising Pulse Sequence

Field synchronising pulses are part of the composite video signal; they ensure that the receiver field time base operates in step with the vertical scanning of the camera tube. Line time base synchronisation is achieved by a single pulse transmitted at the end of each picture line. Field synchronising information consists of a sequence of pulses which are transmitted at the end of each vertical scan. There are 15 pulses associated with this sequence: 5 narrow pulses (2.3 μs each) called *equalising pulses* (see figure 3.8), followed by 5 broad pulses (27.3 μs each) and, finally, there is another set of 5 equalising pulses. Firstly, this combination ensures correctly interlaced vertical scanning; secondly, it facilitates the separation of the line and field synchronising pulses by the respective time base circuits in the receiver; thirdly, it maintains line time base synchronisation during field flyback.

The entire sequence is shown in figure 3.8, from which it will be seen that

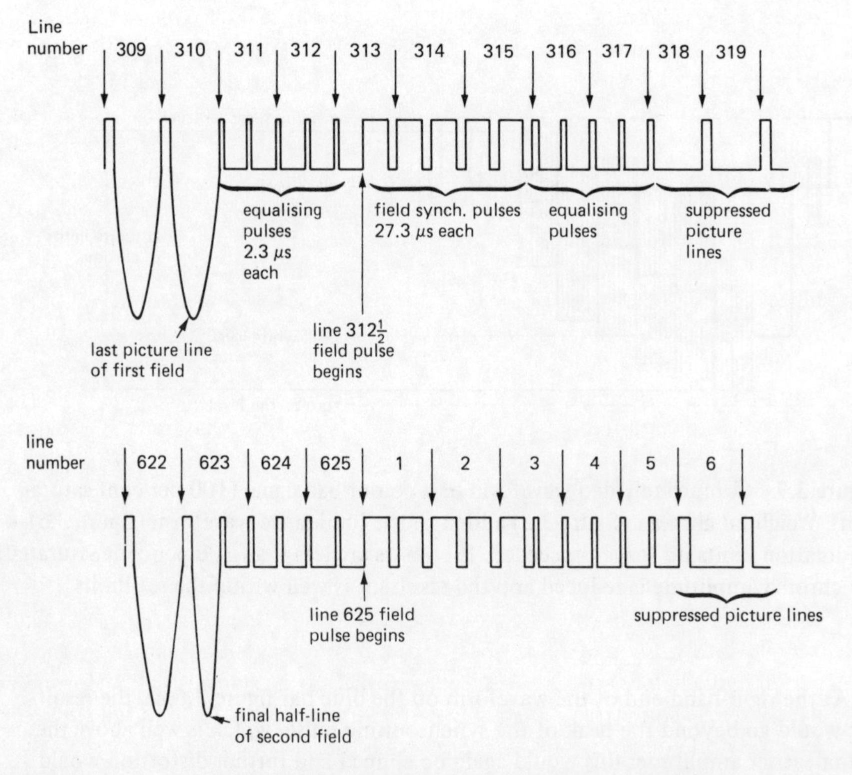

Figure 3.8 Field synchronising sequence for two successive fields. Vertical arrows show the instant of line time base synchronisation during the field flyback period. The complete sequence, together with the suppressed lines (only a few shown), reduces the number of picture lines to 585

following the last set of equalising pulses there are a number of 'blank' lines without any picture information. This is done to allow the field flyback to be completed before the scanning of the actual picture begins. The pulses in the field synchronising sequence occur at twice line frequency; therefore there is one pulse at what would have been the end of a line and one in the middle of a line. The necessity for the 'mid-line' pulses arises from the fact that there are 625 lines which must be divided between two successive fields. Therefore, the first field ends half-way through line 313, while the second field of the interlaced pair ends with a complete line 625. There are also differences at the beginning of each field; the first field starts at the beginning of line 1, and the second field commences from the middle of line 313. The 'mid-line' pulses are needed to initiate the field flyback sequence after the first $312\frac{1}{2}$ lines. Those pulses which coincide with the end of a line are necessary to maintain horizontal synchronisation. The two sequences shown in figure 3.8 demonstrate how the different pulses are utilised to synchronise the two time base circuits at the end of the first and the second field, respectively. The line time base triggering intervals have been marked by arrows; it can be seen that they occur at a regular rate irrespective of the nature of the pulse involved. The details of synchronisation in the receiver will be described in chapters 9 and 10, where it will be shown that the pulse processing circuits in the line time base respond to pulses of any width. The field trigger circuit is designed to respond to the broad 27.3 μs pulses in the sequence.

The differences between consecutive fields could lead to slight timing differences in the triggering of the field time base at the end of each vertical scan. Such inaccuracies would cause poor interlace—that is, the lines associated with the first field would not be scanned exactly half-way between those of the second field. The timing error would be too small to cause an obvious loss of synchronisation, but it would be sufficient to produce a slight displacement of one vertical scan with respect to the next. Equalising pulses are transmitted in the synchronising sequence to ensure perfect interlace in the manner to be described in chapter 9.

It has to be pointed out that while equalising pulses are desirable, they are not included in some transmitting standards; for example, the British 405-line system does not use them.

It was mentioned in section 2.5 that the TV picture consists of 585 lines instead of the accepted standard of 625. The apparent discrepancy can now be explained. The line time base does indeed provide 625 lines during a complete 'two-field' period, but the latter must include two synchronising sequences, shown in figure 3.8. Obviously, no picture information can be transmitted in that time, or in the period immediately after each sequence. Broadcasting standards allow for the field flyback to be completed during an interval equivalent to $12\frac{1}{2}$ lines following the second set of equalising pulses. Therefore the number of lines which are not available for picture modulation is: $2\frac{1}{2}$ for the first group of equalising pulses, $2\frac{1}{2}$ for the broad field pulses, $2\frac{1}{2}$ for the second set of equalising pulses, and, finally, $12\frac{1}{2}$ lines for the completion of field flyback. A total of 20 lines is 'lost' for every field, so that 40 lines are subtracted from each frame. The 20 line interval associated with the synchronising sequence is called a *field blanking period.* Two such periods can be observed on the screen of a receiver in the form of a broad black horizontal band whenever correct vertical synchronisation is lost.

3.6 Teletext Information

Teletext is a form of coded transmission of information by the broadcasting organisations which is carried out in addition to the normal picture information already described. In the UK the BBC use CEEFAX and the IBA system is known as ORACLE.

Some of the lines during the field blanking intervals are used to transmit *binary coded information*, which is in the form of black and white dots modulating the vision carrier. A special decoder is needed in the receiver to store the incoming signals and translate them into a display on the screen. The received information is grouped into sections, known as *pages* (for example, weather reports, news items, etc.), which the viewer can recall from the electronic store by operating appropriate controls. The teletext decoder involves complex circuitry, which is most conveniently provided by integrated circuits. Without a decoder, the information is meaningless, although the modulating dots can be seen if the field blanking bar is displayed on the screen.

3.7 Intercarrier Sound Transmission

TV sound is a frequency modulated signal placed on its own carrier whose frequency is exactly 6 MHz higher than that of the vision carrier. The term *intercarrier* refers to the method of separating the sound information from the video in the receiver. The two carriers, converted into their respective intermediate frequencies in the tuner, are amplified together in the receiver circuits and then fed into a mixer circuit to produce a beat frequency. The difference frequency between the two signals is 6 MHz, which becomes the intercarrier frequency. The actual audio information must still be detected from the 6 MHz frequency modulated signal.

The conversion into the intercarrier frequency can only be achieved if the amplitude of the modulated vision carrier is not allowed to drop to zero. Some vision carrier must remain to beat with the sound carrier; for this reason the peak white of the luminance signal does not fall below 20 per cent of the maximum amplitude of the vision carrier. Further discussion of the intercarrier sound system is deferred to chapter 8.

4 *The Essential Features of a TV Receiver*

4.1 Introduction

The purpose of this chapter is to introduce a complete TV receiver in a block diagram form. The reader will be able to see at a glance how the composite signal is processed before the individual outputs are obtained. A colour receiver has a number of features which are common to its monochrome counterpart. The practical circuits of the two types may differ in detail because of the more stringent requirements of a colour set, but the basic principles are similar. The block diagram approach is most helpful when one is dealing with unfamiliar receivers, whose circuit diagrams can then be divided into their various functions. In modern designs a number of integrated circuits are used to replace the blocks in our diagram. Alternatively, the layout may be in the form of modules or panels which carry the components associated with essential functions so as to simplify initial fault-finding. In field servicing it is then a simple matter of replacing a faulty module with a new one, since efficient repairs may require either a high degree of skill or complex test equipment.

Our discussion follows the diagram shown in figure 4.1, beginning with the signal input from the aerial. Detailed treatment of each section is given in subsequent chapters.

4.2 The Tuner

The tuner receives the signal of a selected channel from the aerial; the amplitude of the signal could be in the order of microvolts at the frequency up to several hundred MHz. Such a small input could be easily swamped by electrical 'noise' signals, which are very low voltages generated at random by the various components used in the circuit—transistors, resistors, etc. In audio amplifiers noise voltages cause a 'hissing' noise through the loudspeaker; in video amplifiers noise shows as a very 'grainy' background to the picture. To prevent the effect of noise destroying the entertainment value of the programme, it is essential to start with the largest possible signal very early in the receiving chain. This requires an efficient aerial installation, which is then followed by the amplifier in the tuner.

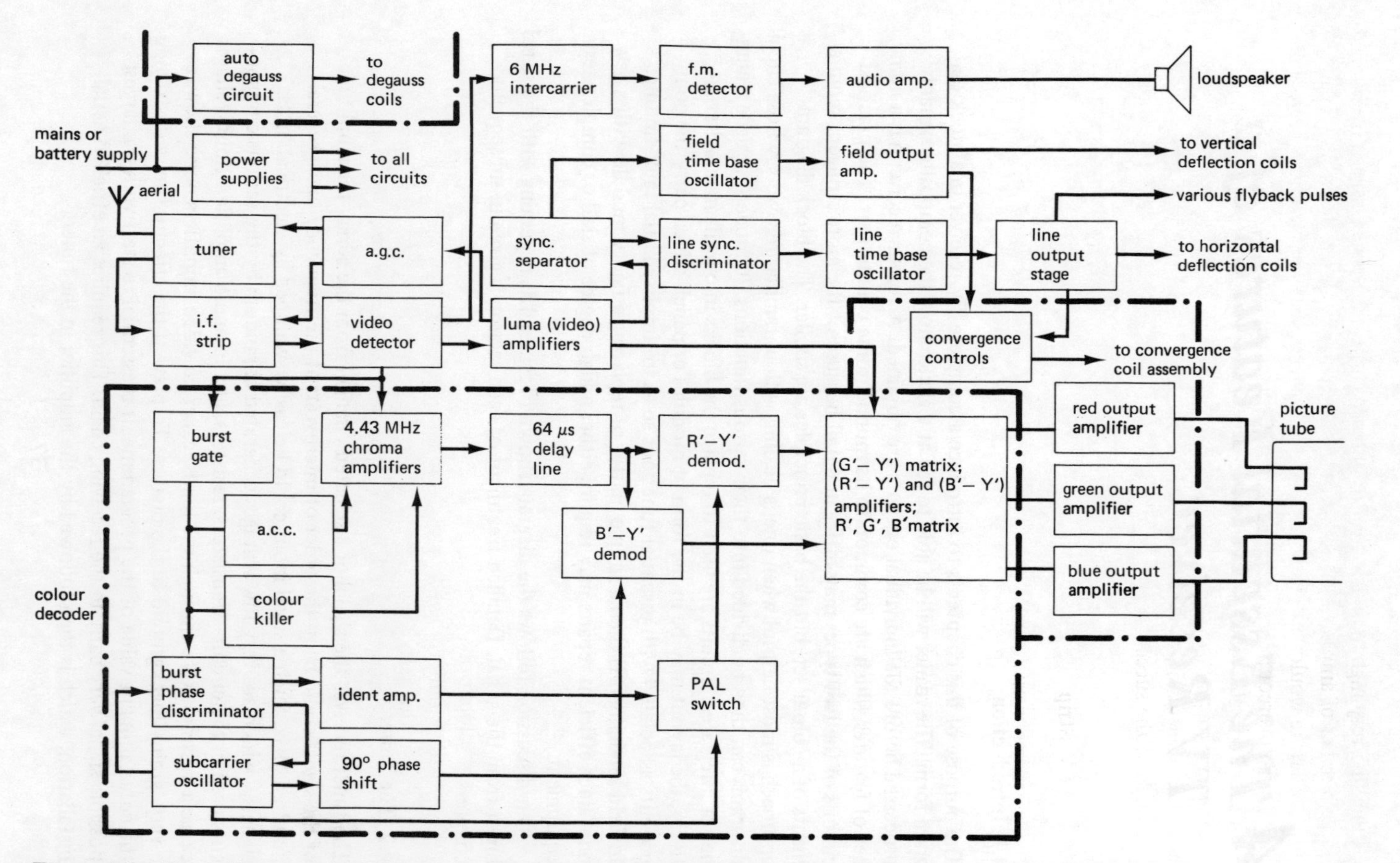

Figure 4.1 Block diagram of a TV receiver. Parts enclosed within broken line apply to colour receiver. In a black and white receiver the output from the video amp. feeds a monochrome tube directly

Handling very high and ultra high frequency signals is not easy, especially where a large number of amplifying stages are needed to produce a usable output. In addition, each stage would have to be retuned whenever a different channel were to be selected. To overcome these problems the incoming *radio frequency* (r.f.) signals are converted in the tuner mixer stage to a new, lower frequency known as the *intermediate frequency* (i.f.). The value of the i.f. remains constant irrespective of the channel selection, so that continuous retuning is not necessary, which simplifies the design of the succeeding i.f. amplifiers.

4.3 The I.F. Strip

The i.f. strip consists usually of three amplifying stages, which, between them, provide most of the amplification of the received signal. They are also responsible for ensuring a good quality picture which is free from mutual interference between the carriers comprising the composite video signal. In a colour receiver the design of the i.f. circuit is more critical than in monochrome sets, since it has to provide sufficient amplification to both the chrominance and the luminance signals.

In order to maintain a relatively constant picture quality under varying reception conditions, *automatic gain control* (a.g.c.) is used to alter the gain of either one or two of the i.f. amplifiers.

4.4 The Vision Detector and Video (Luminance) Amplifiers

When the magnitude of the i.f. signal has reached about 2 V, the composite video can be detected or separated from its intermediate frequency carrier. This is most frequently done by a diode detector, which rectifies the modulated i.f. to recover the original envelope. This is followed by a relatively simple filter which removes the residual i.f. Alternatively, an integrated circuit could be used; this operates as a synchronous detector which, by sampling the modulated carrier at a very fast rate, 'sees' the envelope voltages during each sampling interval. In this way the original envelope can be built up from the individual voltage levels.

It is also at the detector where the separation of chrominance and sound intercarrier from the composite video is possible so that the colour decoder and the sound section can be fed with their respective signals. In some receivers these take-off points can be found later in the video amplifiers.

Monochrome receivers can only make use of the luminance signal; they are relatively unaffected by the presence of the chrominance information. The signal is processed in *video amplifiers*, in which its voltage level is raised from the detected 1 V peak-to-peak to perhaps 150 V peak-to-peak; this is required to fully drive the picture tube from black to peak white. In a colour receiver the arrangement is similar, but the counterpart of the video amplifier is now called the *luminance amplifier.* To obtain a colour picture, the luminance signal has to be added (*matrixed*) to the colour difference signals, which are processed separately in the decoder. This addition either may take place in a circuit prior to feeding the tube or may be done using the electrodes (cathodes and control grids) of the picture tube itself; both methods can be found in practice. Our diagram shows a matrix block (within the colour decoder) external to the cathode ray tube; this is a very popular arrangement in modern receivers.

Line and field synchronising pulses are obtained from the video waveform; hence the need for the feed to the synchronising pulse separator.

The automatic gain control (a.g.c.) voltage may be derived from the amplitude of the video signal, and there is an a.g.c. stage which could be switched by line flyback pulses in some designs. Line flyback pulses are also fed to the luminance stages (or the matrix) to provide black level clamping. This ensures that every line of scan begins from the correct black level as determined at the transmitter.

The luminance and chrominance signals in a colour receiver follow separate paths, at the end of which both signals are combined again. The propagation time of signals through the colour decoder is longer than that in the luminance amplifier; a luminance delay line is incorporated to delay the signal, which ensures perfect registration of the luminance and chrominance information.

4.5 The Colour Decoder

This part of the circuit is applicable to every colour receiver. Its purpose is to decode the colour difference signals from the composite chrominance signal whose carrier has been suppressed. The decoder has a main path which processes the chroma itself, and a number of auxiliary circuits to provide the services necessary for the decoding. In many respects the colour decoder resembles that of the encoder at the transmitter because of the need for the reinsertion of the missing subcarrier; without it, demodulation of the colour difference signals would not be possible.

In the main path of the chroma signal there are a number of amplifiers which are tuned to the subcarrier frequency of 4.43 MHz. Here the signal level is raised from a few tens of millivolts to several volts before being fed to the demodulator. In modern PAL receivers a delay line is incorporated which stores one line of composite chroma information for the period of one picture line (64 μs). This system helps to take advantage of the ability of the PAL colour TV system to minimise the effects of changes in the chroma signal in the transmitter–receiver chain. It is done by averaging the received signal over a period of two lines: i.e. it averages the stored or delayed line with the current line. As a by-product of this process, separation of the U component from the V component takes place, which helps the process of demodulation. Following the $(R' - Y')$ and $(B' - Y')$ demodulators, the $(G' - Y')$ is produced by adding proportions of $(R' - Y')$ and $(B' - Y')$, and all three are then amplified and added to the luminance signal. From the matrix the outputs are in terms of the primary colours, R′, G′, B′, as generated by the camera amplifiers at the transmitter.

In the auxiliary decoder circuits the colour subcarrier is regenerated; its frequency and phase are identical with those at the transmitter. The receiver crystal oscillator is synchronised with the transmitted colour burst. The extraction of the subcarrier burst from the composite chroma is achieved by a *gated amplifier*, which is normally switched on only during the 'back porch' period. The necessary gating is done by means of suitably delayed line flyback pulses.

The regenerated subcarrier is then fed to the two synchronous demodulators; the $(B' - Y')$ subcarrier is fed via a 90° phase shift network, while the $(R' - Y')$ feed passes through the PAL switch.

The PAL switch operates line by line, but it must be synchronised with the transmitter PAL switch. A suitable synchronising signal, called the *ident*, is derived from the changing, or swinging, phase of the burst.

The amplitude of the colour burst is also detected, and is used to control the gain of one of the chroma amplifiers in the manner similar to the a.g.c. in the i.f. strip; this system is called *automatic chroma control*, or a.c.c.

During monochrome transmissions the chroma path is shut down; this avoids spurious colours and 'coloured noise' appearing on the screen. A circuit known as the *colour killer* detects, either directly or indirectly, the presence of the burst. Whenever the burst is absent, the colour decoder is 'killed'.

4.6 Sound Stages

The beat frequency between the vision and sound i.f. gives rise to a 6 MHz sound signal at intercarrier frequency. The audio information frequency modulates the sound carrier, but because of being mixed with the video signal, there may additionally be a high percentage of amplitude modulation. The latter is removed in *limiting amplifiers* which are usually associated with the 6 MHz tuned circuits or ceramic filters. The demodulation of the f.m. signal follows established techniques of detection—e.g. quadrature detection either inside an I.C. or in a ratio detector in which two diodes are used. The resultant audio signal drives an amplifier and, in turn, the loudspeaker. Integrated circuits are very common here, because of the identical circuit requirements found in both radio and TV receivers.

4.7 The Synchronising Pulse Separator

By detecting and amplifying the various pulses present in the video waveform, this circuit provides synchronising pulses to the line and the field time bases. In principle, the circuit is a level detector, which conducts only when the video waveform has reached a predetermined level, namely that of the pulses themselves.

4.8 The Line Time Base

The chief purpose of this circuit is to produce the necessary horizontal scanning of the screen by an electron beam. This deflection must be in synchronism with the transmitter time base, which is governed by the synchronising pulses. In conjunction with the field time base, the operation of the line scan circuit produces a display on the screen called *raster* which gives the picture its line structure.

The frequency of the line time base oscillator is controlled by means of a signal which is derived from the comparison of the frequency and phase of the synchronising pulses with those of the flyback pulses.

The output from the oscillator is used to drive the line output stage, which, via a transformer, provides the current waveform for the scan coils. When the current changes rapidly, a large back-e.m.f. is induced in the coils and in the associated output transformer, and is used to generate an e.h.t. (*extra high tension*). Its value could be 25 kV in colour receivers, and is required for the final anode supply of the picture tube. In addition, a somewhat lower voltage level, of between 600 V and 5 kV, is also required for other electrodes of the tube.

Modern line output stages can provide a number of power supplies to a TV receiver, both high voltage (h.t.), say between 100 and 200 V, and low voltage (l.t.) of between 15 and 40 V.

A variety of flyback pulses are obtained from the output stage and are used to provide a number of services in other parts of the receiver, some of which have already been mentioned.

In colour receivers, correction or *convergence* circuits are necessary to keep the three beams close together. Suitable waveforms are derived from the line output stage and fed to a separate convergence coil assembly mounted on the neck of the tube.

4.9 The Field Time Base

The field time base provides the vertical deflection for the scanning beam. The circuit consists of a synchronised oscillator, followed by an amplifier and the output stage feeding the scan coils mounted on the neck of the tube. As the electrical requirements of both the field and line time bases differ, the practical circuits of each also differ from one another. Synchronisation of the field time base is achieved by producing a single trigger signal for the oscillator from the transmitted field synchronising pulse sequence. The output stage requires a relatively complex arrangement of linearity correction circuitry.

In colour receivers, convergence coils are also provided which are fed with suitably shaped waveforms derived from the field time base output stage.

4.10 The Picture Tube

The tube provides the display of the transmitted picture. The cathode ray tube resembles, in many respects, a thermionic valve. Electrons are emitted from a heated cathode, and, by means of strong electric fields produced by other electrodes, a narrow beam of electrons is focused on to the face of the tube. The inside of the screen is coated with a material known as a *phosphor*, which emits light when the electron beam strikes it. The colour of the emitted light depends upon the composition of the phosphor; white is needed for monochrome tubes, while red, green and blue are required for colour TV tubes.

The brightness of the display is controlled by the amount of current reaching the screen, which, in turn, is affected by the voltages applied to the various electrodes of the tube. The potential difference between the cathode and the control grid is the video signal which modulates the beam current; hence, the picture is produced by the variation in the brightness level. The voltages applied to the remaining electrodes are normally constant, although they may be preset as part of the initial installation procedure.

The tube in a colour receiver has three sets of electrodes—each associated with its own colour signal (red, green or blue). The screen has three phosphor materials deposited in the form of either dots or stripes. The spacing between them is so close that, from a normal viewing distance, the emitted colours mix into a resultant depending upon the amounts of current associated with each of the three beams. It is necessary to ensure that the respective beams strike only their own phosphors. This is done by means of a special mask close behind the screen which prevents the 'wrong' electron beam reaching any given phosphor. There are small magnets on the neck of the tube, which also ensure the 'purity' of the displayed colour. Another problem with colour tubes is a tendency for the three beams to separate as they scan the screen both in the horizontal and in the vertical directions. If this occurs, the beams strike phosphors which are so far apart that the impression of colour merging is destroyed, and very unpleasant coloured fringes result. A convergence coil and magnet assembly is mounted on the neck of the tube to allow adjustments which reduce the above-mentioned effect to a minimum.

The overall dimensions of the tube dictate the size of a TV receiver. Monochrome receivers use a very wide angle of deflection approaching 120°; this makes the flared part of the tube relatively short.

Some colour tubes use 110° deflection, while others use only 90°. The wider angle requires a higher power consumption from the already hard-working line output stage. Also, the picture shape tends to depart from a rectangle and develop bowed horizontal and vertical edges. Such non-linearity (*pincushion* effect) is relatively easy to correct in monochrome tubes, but correction is more involved in colour receivers.

A colour tube has a coil mounted round its flared part, which is required to remove any residual magnetism in the surrounding external and internal steel parts of the tube. This is called a *degaussing coil*, and is fed from the mains supply via a special network which automatically reduces the operating current after a very short period of time.

4.11 Power Supplies

A TV receiver requires several d.c. power supplies, the value of the voltage depending on the magnitude of the signal being handled. For example, the tuner and the i.f. stages need a relatively low voltage, while the video and the line output stages operate at a relatively high voltage.

In a colour receiver a number of circuits require fully stabilised supplies. Often elaborate regulator arrangements are employed in addition to the more conventional rectifier and smoothing circuits.

An all-transistor monochrome receiver also has stabilised power supplies, primarily to feed the line time base circuits. A portable, battery-operated receiver must have a supply regulator, since the battery voltage varies over a wide range.

There is also a tendency to either eliminate or reduce the size of the mains transformer in TV receivers; this is achieved either by using the line output transformer to supply most of the power requirements of the TV set or by employing a low frequency to high frequency converter. In the latter the mains 50 Hz supply is rectified and smoothed to be fed to a switching circuit which can change the d.c. back into a.c.—for example, at line time base frequency. The resultant is easily transformed to any required voltage level and is then rectified and smoothed again.

All receivers whose power supplies are derived from the mains, especially where there is no isolating transformer, could present a safety risk to both the viewer and the servicing technician. Therefore the design and the components used must, by law, comply with a number of standard recommendations.

5 *Tuners and I.F. Amplifiers*

5.1 High Frequency Performance of Electronic Amplifiers

The frequency of signals processed by the tuner can extend from v.h.f. to u.h.f.— that is, from a few tens of MHz to several hundred MHz. A high frequency or *radio frequency* (r.f.) signal is converted in the tuner unit to an *intermediate frequency* (i.f.). The value of the i.f. is also relatively high—in the region of tens of MHz. Before we consider any detailed circuits, it is necessary to appreciate the effect of high frequencies on the performance of electronic amplifiers.

It is inevitable that every amplifier has limitations on the highest signal frequency which it can amplify. These limits are determined by two factors—namely the high frequency capability of the transistor (or I.C., or valve) and the overall design of the circuit. Generally, the amplifying device is chosen so that it has a gain which is maintained throughout the required frequency range. A replacement transistor or valve must therefore have a similar specification to that of the original device. When the circuit design is a limiting factor, the values of certain components, their construction and physical layout are usually critical. For example, a pair of conductors separated by an insulator behave like a capacitor; therefore a resistor or a coil will have what is known as *stray capacitance.* This capacitance exists between the component in question and adjacent components or wiring, including the chassis connections. The value of stray capacitance is usually in the region of a few picofarads, but its effects at high frequencies can be considerable because of the resultant low reactance. This can cause the signal to be shunted out of the required path, or it may couple the signal to a circuit where the link is not intended.

Similarly, a length of wire or such components as capacitors and resistors will have inductance in the order of a few microhenrys. This is again important at very high frequencies, where the 'stray' inductance may, perhaps combined with the stray capacitance, produce unwanted tuned (resonant) circuits.

The presence of such 'strays' is taken into consideration in the circuit design and construction. When testing or repairing high frequency circuits, it is important to appreciate the above factors. The connection of test equipment, probes and leads can easily upset the performance of the circuit. Component replacements must be of a similar type and fitted in a similar manner to the original (that is, length of connections, position with respect to other components, etc.).

It is unlikely that a particular circuit will have to amplify signals over a very wide frequency range; this applies especially to tuners and i.f. amplifiers. In fact, the range of signals to be amplified is limited by the bandwidth of the amplifier. Any frequencies falling outside this range are rejected—that is, the circuit must offer satisfactory *selectivity* as well as gain. The basis of such an amplifier is a *tuned circuit*, which consists of either a series or a parallel combination of a capacitor and a coil. Often the stray capacitance and inductance of wiring are incorporated into the resonant circuit. The response of a simple tuned circuit is centred around the frequency given by

$$\text{resonant frequency}, f_0 = \frac{1}{2\pi\sqrt{(LC)}}$$

If the inductance L is in μH and the capacitance C is in μF, then the frequency is in MHz.

5.2 The U.H.F. Tuner

The tuner is used firstly to select, or tune to, the desired TV channel, and secondly to convert the r.f. signal to an i.f. signal. A limited amount of amplification is also produced in the tuner, although for practical reasons most of the receiver gain is obtained from the i.f. section. The conversion of the r.f. vision carrier of any channel into a single frequency simplifies the receiver design; since each i.f. amplifier is tuned to a fixed frequency, no retuning is necessary when channels are changed by the user. Additionally, since the ultra high frequency carrier is converted to a much lower frequency signal, the design of subsequent circuits is easier.

The block diagram in figure 5.1 shows the arrangement of circuits in a u.h.f. tuner. The *aerial isolating circuit* is an important safety feature found in mains-operated receivers; it isolates the aerial installation from the receiver chassis, which could be 'live' (see chapter 14). Practical aerial isolating circuits are shown in figure 5.2; capacitors C_1 and C_2 in diagram (a) and C_1 in diagram (b) offer a reactance of

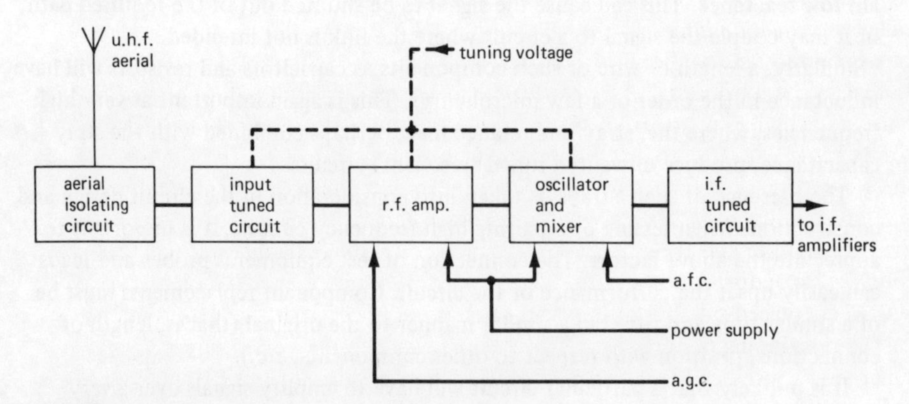

Figure 5.1 Block diagram of a u.h.f. tuner (tuning voltage is used in varicap tuners)

several MΩ at mains frequency. For example, a 470 pF capacitor has a reactance of 6.8 MΩ at 50 Hz, effectively reducing the danger of electric shock, but at the radio frequencies involved its reactance is practically zero. The large-value resistor, R_1 [figure 5.2(a)], prevents the accumulation of a static charge across C_1, which can

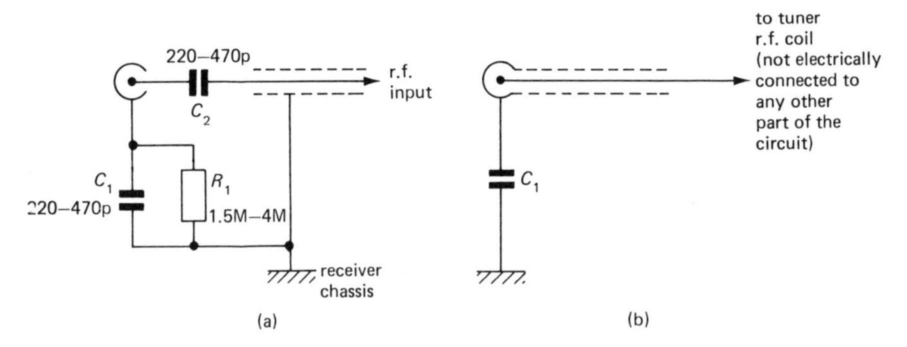

Figure 5.2 Aerial isolating circuits

arise from atmospheric conditions. In many modern receivers the entire circuit is an encapsulated unit, and should it develop a fault, the whole assembly would have to be replaced. For safety reasons the type of capacitors used in it have to conform to strict electrical specifications; therefore the isolating circuit is not a repairable item.

Tuned circuits used at u.h.f. are different in appearance from their v.h.f. counterparts. At ultra high frequencies it is difficult to accommodate the inevitable stray capacitance and inductance associated with conventional coils and their interconnections. Instead, the coils are replaced by straight lengths of a heavy gauge conductor, one end of which is usually joined to chassis, the other end connecting to a tuning capacitor or perhaps to one of the transistors. These conductors are called *transmission lines* or *Lecher lines*, which, in combination with the tuning capacitor, resonate at the required frequency (their behaviour is somewhat similar to that of the u.h.f. aerial elements described in chapter 13). The tuner unit contains individual compartments which house the Lecher lines together with the components associated with a particular section of the arrangement. The walls of these compartments form a part of the tuned circuit, and they also prevent unwanted coupling between individual sections. Where coupling is needed, say between the r.f. amplifier and the mixer, a small slot is cut in the dividing wall to act as a coupling capacitor. Alternatively, two or more lines, which could be of different length, are placed side by side and the arrangement behaves like a *bandpass filter.*

The positioning of components and the lengths of interconnections inside the tuner are very critical. Decoupling capacitors are of the *feed through* type in the walls of the tuner unit, which eliminates undesirable connecting leads necessary with conventional capacitors (see C_6, C_{12}, etc., in figure 5.4).

The tuning on u.h.f. is continuous over the entire channel range. Since a number of tuned circuits have to be adjusted together, a capacitor gang is used. The unit can be pretuned as required and channel selection is by means of push buttons. The push buttons actuate a mechanical lever, and the movement of the capacitor rotor is then governed by the position of adjustable stops which are preset during the initial tuning procedure. This type of tuner tends to be bulky, and the mechanical linkage can be troublesome.

5.3 The Varicap Diode Tuner

A *varicap diode*, also known as a *varactor diode*, is a silicon diode which is used in its reverse biased mode—the reverse bias varies its capacitance. A reverse biased p-n junction develops a *depletion layer* between the p- and n-regions, which acts as the dielectric of a capacitor, the p- and n-regions forming the plates of the capacitor. Since the width of this layer *increases* with the increasing voltage, the capacitance *decreases*, and vice versa. The available capacitance is small, but at v.h.f. and u.h.f. only a few pF are needed for tuning purposes.

The reverse bias potential to the diodes in the tuner is the *tuning voltage*; it is varied between approximately 0.5 V (lower channel numbers) and 28 V (upper channel numbers). The supply rail which provides this potential must be stabilised.

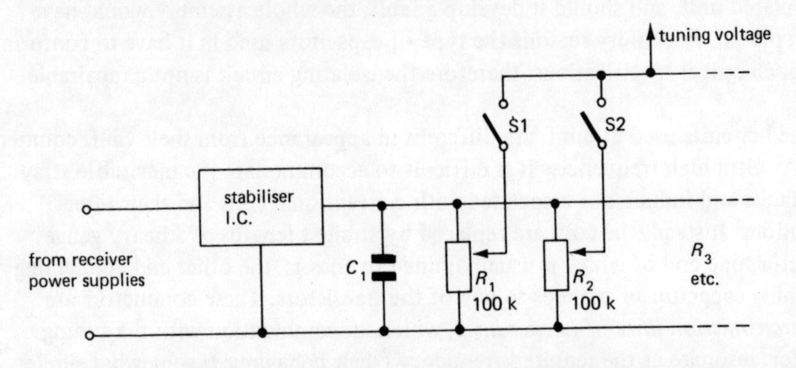

Figure 5.3 Varicap tuner channel selector

Any voltage fluctuations would cause a severe tuning drift, or even an inability to tune the channels at the extremes of their entire range. Supply voltage stabilisation is discussed fully in chapter 14. The arrangement shown in figure 5.3 uses an integrated circuit stabiliser whose output is decoupled by C_1. The resultant stabilised voltage is applied across the tuning potentiometers R_1, R_2, etc. (depending on the number of available channel selectors); associated switches S1, S2, etc., are operated by their respective push buttons. The channel selector circuit is fixed to the cabinet, the tuner itself being placed at a point close to the i.f. board.

A complete circuit of a u.h.f. varicap diode tuner is shown in figure 5.4. The radio frequency amplifier uses two transistors—TR1 and TR2; the biasing of TR1 is governed by the a.g.c. control voltage. Other tuner designs use only one r.f.

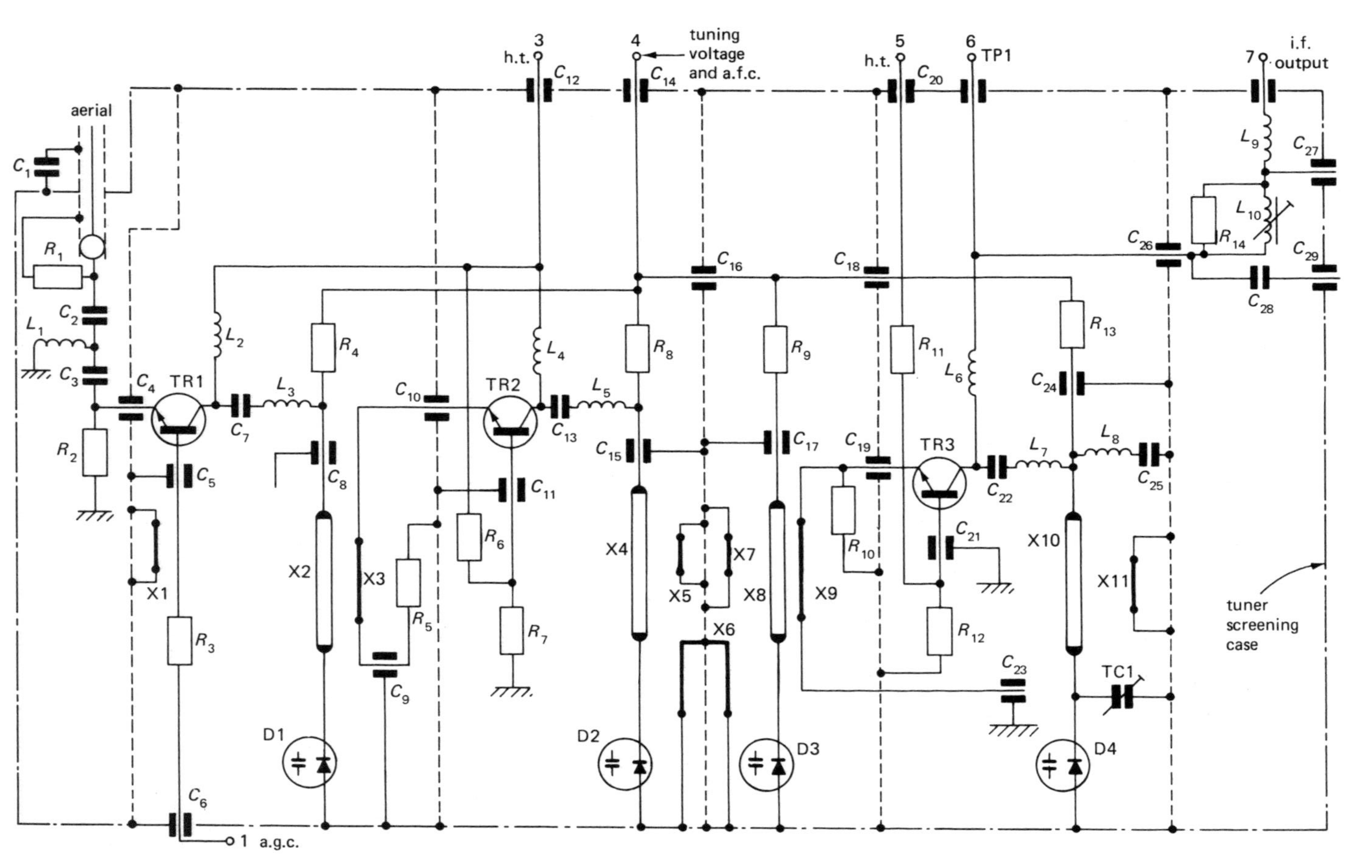

Figure 5.4 U.H.F. varicap diode tuner (Rank)

transistor whose gain may be controlled either automatically or manually (*local/ distant* reception switch). At the input to TR1 a filter consisting of L_1, C_2, C_3 rejects frequencies in the v.h.f. bands.

The transistors in the tuner are connected in *common base* mode—the input signal being applied to the emitter and the output taken from the collector, the base being decoupled to chassis. This method of connection offers a better high frequency performance than that of a common emitter amplifier. The base of TR1 is decoupled via C_5, while the d.c. connections to the transistor are provided by the emitter resistor R_2, the a.g.c. line via R_3 to the base, and the r.f. choke L_2 to the collector. The signal from the collector is coupled to TR2 via C_7 and the bandpass tuned circuit formed by the Lecher lines X1, X2, X3 and varicap diode D1. The d.c. tuning voltage from the channel selector (figure 5.3) is fed to the diode via R_4 and X2. The circuit of the second r.f. amplifier is similar to that of TR1; from TR2 the signal is coupled (through the tuned circuits X4, D2, X5, X6, X8, D3 and X9) to the *mixer oscillator* stage TR3. U.H.F. tuners use one transistor to perform both mixer and oscillator functions; the oscillator tuned circuit consists of X10, X11, varicap diode D4 and trimmer TC1. Feedback from the output is via X9 to the emitter of TR3, where the oscillator signal is added to the r.f. signal. The mixer produces a number of beat frequencies, out of which the desired i.f. is picked out by L_{10} and external capacitors, while L_9 is a choke rejecting the high oscillator frequency. The *automatic frequency control* (*a.f.c.*) voltage can be applied to the common tuning line at a point external to the tuner.

There are tuners which use a *diode mixer* stage. In these designs a transistor oscillator circuit is followed by a u.h.f. diode, which is sometimes described as the *first i.f. detector.* A diode mixer generates less electrical noise than a transistor stage and it produces a better quality output.

Where a tuning capacitor gang is used instead of a varicap diodes, the basic circuit arrangement is similar to the diagram in figure 5.4. Individual sections of the capacitor are then connected to the respective Lecher bars. If a.f.c. is used in a 'mechanical' tuner, the control voltage is applied separately to a single varicap diode incorporated in the oscillator tuned circuit. A more detailed discussion of a.f.c. appears later in this chapter.

5.4 V.H.F. and Integrated V.H.F./U.H.F. Tuners

In the UK the demand for v.h.f. tuners is restricted to the receivers which operate from special signal distribution networks and to receivers which were designed for the 405-line standard. In many other countries v.h.f. transmissions are used extensively in addition to u.h.f. programmes. Instead of two separate units, a combined v.h.f./u.h.f. tuner, called an *integrated tuner*, is used in modern receivers.

V.H.F. tuners of early design used fixed coils, each being pretuned to a given channel number. A suitable set of coils for each bandpass circuit, including the oscillator, was carried on a rotary switch assembly (turret tuners, wafer tuners). Final trimming (fine tuning) was then performed by adjusting the oscillator fre-

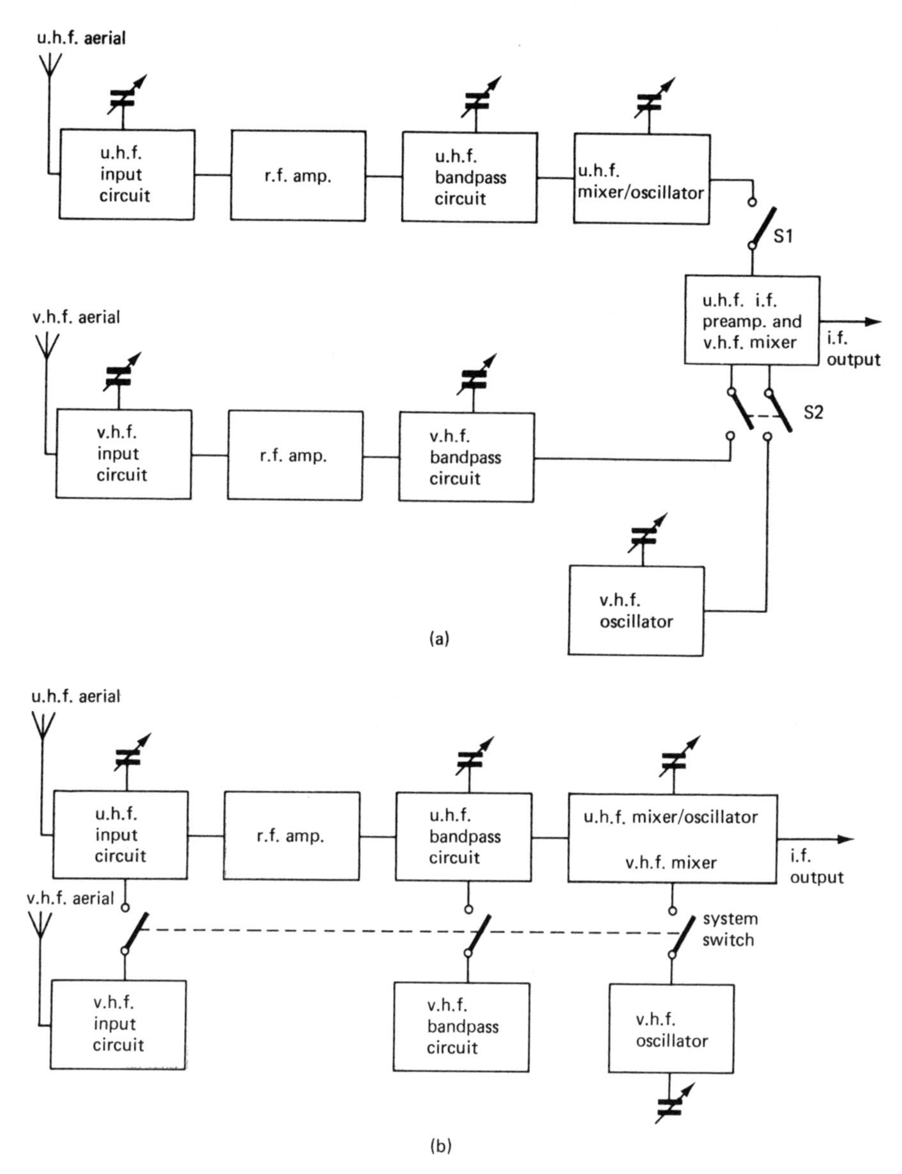

Figure 5.5 Block diagrams of integrated tuners: (a) u.h.f. and v.h.f. signal paths are separate; (b) compact tuner design–circuit separation kept to a minimum

quency for best reproduction of picture and sound. Such arrangements were bulky and the large number of moving contacts were a source of trouble.

Modern v.h.f. tuners are continuously tunable over the entire range of channels in Bands I and III. Because of the frequency gap caused by Band II (f.m. radio) a selector switch is normally used offering either Band I or Band III.

Integrated tuners follow one of the two basic arrangements shown in figure 5.5. In block diagram (a) the u.h.f. and v.h.f. signal paths are separated by the function switches S1 and S2 (the system is not applicable to the British 405/625 dual standard receivers with their different i.f. on each standard). The final transistor in the tuner acts as an i.f. preamplifier when switched to u.h.f. or as a mixer on v.h.f. The power supply to the sections not used at any time is also disconnected by a function switch. Figure 5.5(b) shows a more economical arrangement; the transistors perform dual functions depending on the position of the system switch. In the v.h.f. mode an additional oscillator transistor is required, followed by a v.h.f.

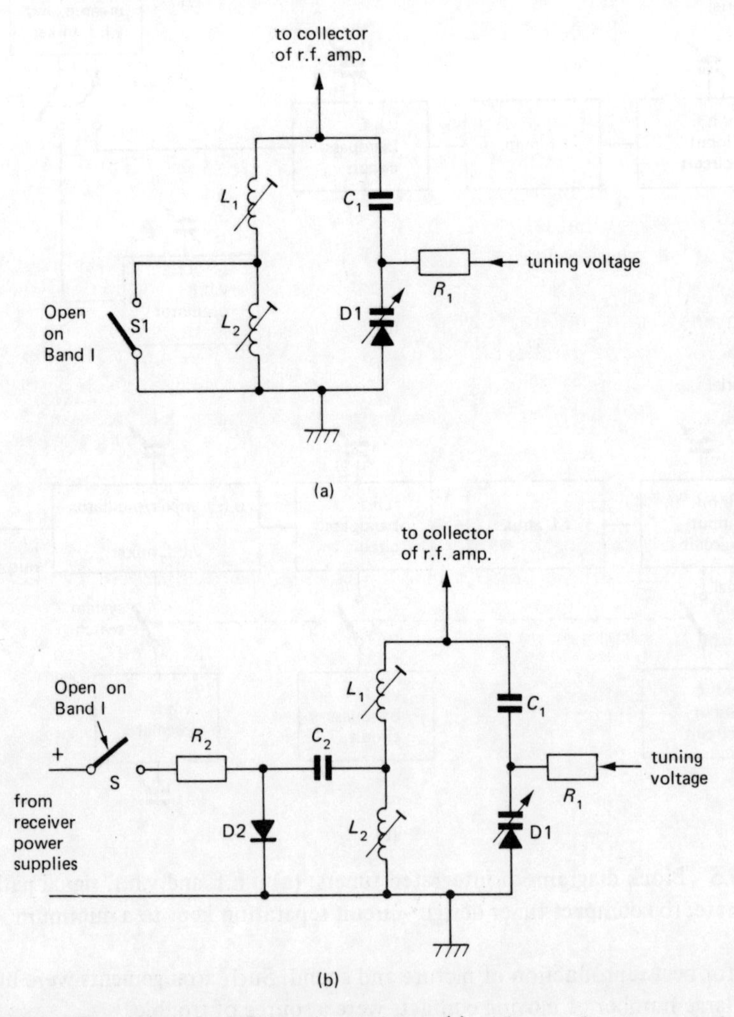

Figure 5.6 Examples of v.h.f. band switching: (a) ordinary shorting switch; (b) v.h.f. tuned circuit as in (a), but the switch is replaced by diode D2 with its own control circuit

mixer; on u.h.f. the latter serves as a mixer/oscillator in the manner described in section 5.3.

The tuned circuits used for v.h.f. reception utilise conventional coil design. The value of inductance needed at frequencies in Band I is higher than that for Band III, so that the inductors are tapped via the Band I/III changeover switch as shown in figure 5.6(a). Lecher lines are necessary for u.h.f. reception, but when the tuner is switched to v.h.f., they simply act as interconnectors which couple the signal from the v.h.f. coils to the amplifying transistors and to the tuning devices. Either varicap diodes or ganged capacitors can be used, their functions being shared on both u.h.f. and v.h.f.

Very low amplitude r.f. signals can be seriously affected by bad contacts of the system switch. To overcome this, *diode switching* can be employed; mechanical contacts are then used to apply a control bias voltage to a diode. This type of circuit is shown in figure 5.6(b). The two circuits (a) and (b) are identical as far as r.f. signals are concerned, but in (b) the switching diode D2 together with the switch S replace S1 shown in diagram (a). When switch S–diagram (b)–is open, D2 is not forward biased, so that it behaves as an open circuited switch, and L_1 is in series with L_2. (The magnitude of the signal across L_2 is by itself insufficient to turn the diode on.) Closing switch S applies a positive potential to the anode of D2 through the current limiting resistor R_2; D2 is now biased ON, which, in turn, acts as an r.f. short across L_2 (via the d.c. blocking capacitor C_2). Similar switching diodes can be used elsewhere in the tuner to steer the signal between parts of the circuit or to provide the switching of complete sections of an integrated tuner.

5.5 Touch Tuners

Touch tuners eliminate the need for mechanical methods of channel changing. Since it is relatively easy to control the tuning voltage of varicap tuners, they lend themselves to this form of 'automation'. The control circuit may contain either individual transistors or integrated circuits.

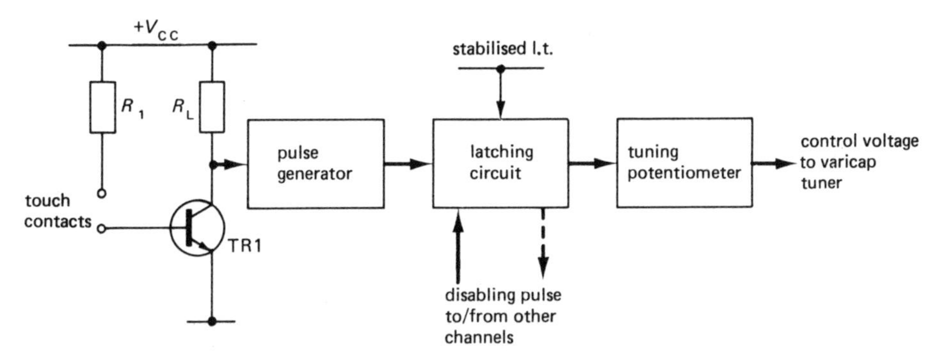

Figure 5.7 Principle of touch tuning. When energised, the circuit transfers the stabilised l.t. to the appropriate channel potentiometer

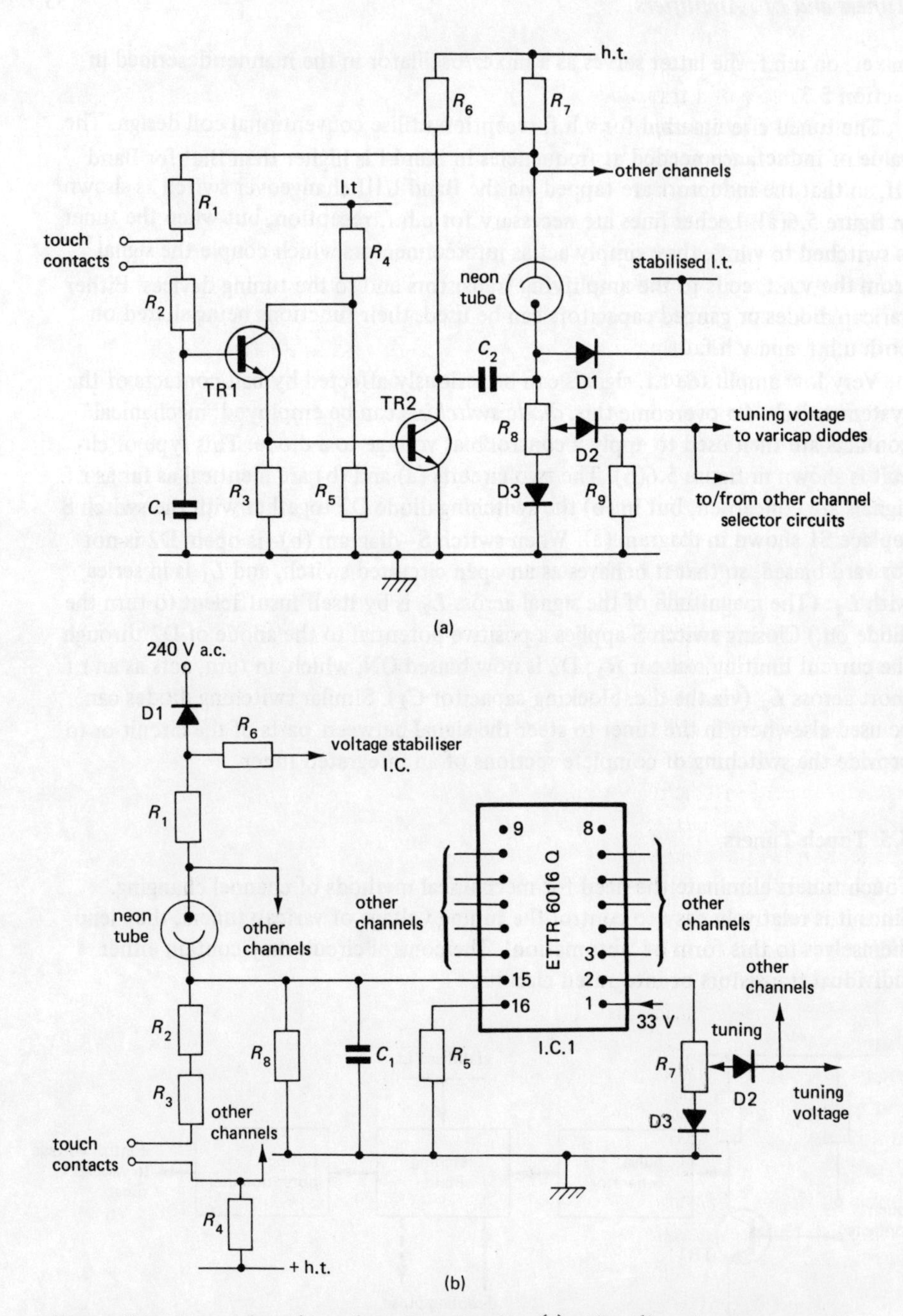

Figure 5.8 Examples of touch tuning circuits: (a) using discrete components–compare with figure 5.7 (Rank); (b) using an I.C. (GEC). Connections marked 'other channels' are made to identical circuits associated with the remaining touch contacts and tuning potentiometers

The principle of a transistorised control circuit of touch tuning is demonstrated in figure 5.7. The base bias circuit of the sensing transistor TR1 is broken by the touch contacts. When the resistance of the skin completes this circuit, TR1 turns on and its collector voltage falls. The resultant change in output is fed to a pulse generator; this produces a 'clean' switching pulse, which is used in turn to apply a d.c. potential to the tuning potentiometer. A latching circuit is needed to ensure that the tuning voltage remains after the removal of the finger from the touch contacts. At the same time, a 'disabling' pulse is sent to the other channel control circuits to ensure that they remain switched off; this action also 'unlatches' the unwanted channel. Often arrangements are made to 'bias' the selector in favour of one particular channel whenever the receiver is first switched on.

A similar method of touch sensing and subsequent switching can be built into an I.C. Alternatively, the touchplate forms part of a capacitor which couples the signal induced by the finger to a suitable integrated circuit.

In many cases lamps or miniature neon tubes are also connected to indicate the actual channel selection. Where neon tubes are used, they can be part of the switching circuit, as shown in figure 5.8(a). Touching the contacts biases both TR1 and TR2 ON; this, in turn, connects the left-hand plate of the charged capacitor C_2 to chassis. The resultant momentary high potential difference across the neon causes it to strike, which, in turn, switches the isolating diode D1 ON. The conducting diode transfers the stabilised l.t. supply to the channel selector potentiometer R_8. The tuning voltage is passed on to the tuner via another isolating diode D2 (D3 compensates for the effects of temperature changes on D2). The entire circuit shown in figure 5.8(a) is repeated for each set of touch contacts, as indicated by the connections to the identical points in the other channel selectors. When another set of touch contacts is used, the corresponding neon will strike, momentarily increasing the current through R_7. Consequently, the voltage across the previously operated tube falls and the lamp switches off; D1 and D2 remove their tuning voltage and the new channel is selected.

Figure 5.8(b) shows an arrangement using an I.C. Bridging the touch contacts connects the neon to a large potential difference—positive h.t. on one side and negative rectified a.c. on the other. When the tube strikes, the resultant pulse is communicated to the I.C. (pin 15) whose internal circuitry transfers the 33 V tuning supply voltage (pin 1) to the required channel selector potentiometer (for example, R_7 at pin 2). The output to the tuner is via the isolating diode D2 in the manner shown in figure 5.8(a), described above. Similarly, tuning to another channel will extinguish the previous neon, and a new 'switch-on' pulse will be initiated.

5.6 Remote Control

Remote control is a natural progression from touch tuning—again a suitable pulse is needed to initiate channel switching. A control pulse is sent by a hand-held transmitter to a 'pick-up' built into the TV receiver. The transmitted signal is in the form of ultrasonic vibrations which are generated by a special transducer driven by an electronic oscillator. The ultrasonic frequency lies in the range 30–45 kHz; one controller can employ a number of frequencies not only for channel switching, but also to control the brightness, volume, etc. If only one frequency is used, each successive burst of signal from the transmitter will cause sequential channel change.

The diagram in figure 5.9(a) shows the circuit of a battery-operated hand-held ultrasonic transmitter. A Hartley oscillator is used, its preset frequency being governed by L_1, and the circuit is activated by the control switch S1. The transducer requires a high amplitude a.c. signal superimposed on a *polarising d.c. voltage.* Therefore the oscillator output is stepped up to a voltage in the range 150–300 V; the a.c. is applied to the transducer via C_6, while the d.c. polarising potential is derived from the voltage doubler, D1, D2, C_4, C_5.

The 'pick-up' transducer in the receiver is similar to that in the transmitter, but

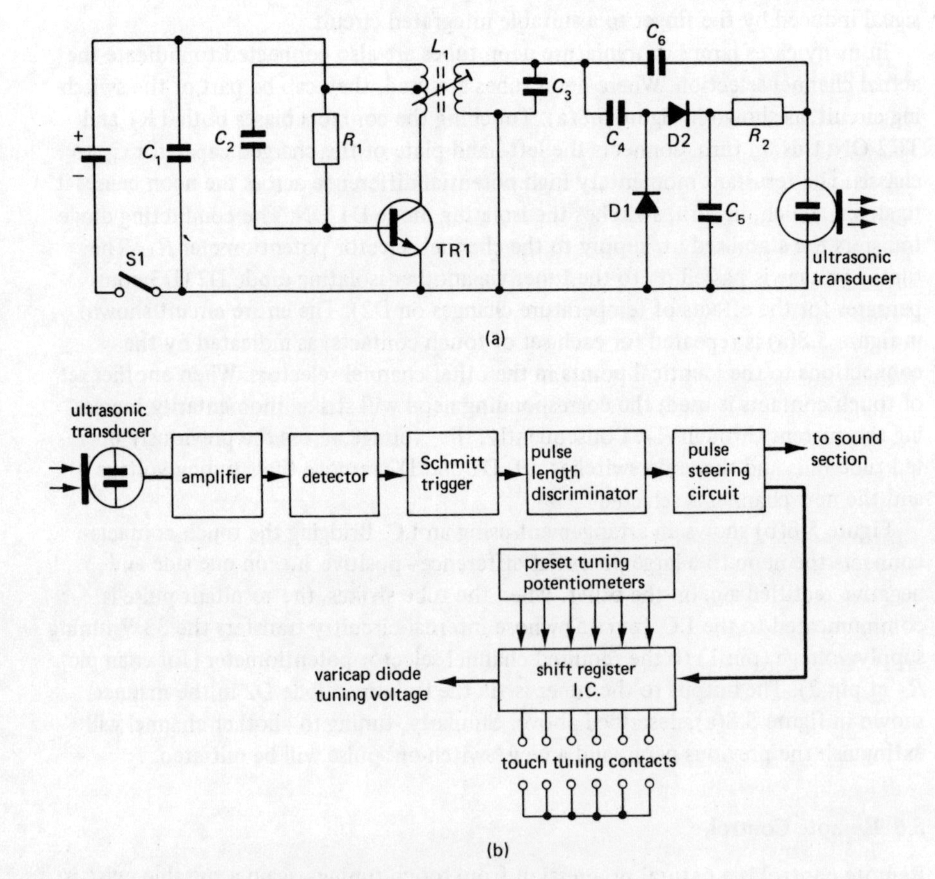

Figure 5.9 Remote tuning control (Rank): (a) transmitter unit; (b) block diagram of receiver unit. The receiver unit provides both channel selection and sound muting

the subsequent decoding circuit is more involved. A simplified block diagram of a decoder is shown in figure 5.9(b). A multistage amplifier feeds a detector, where the high frequency signal bursts are converted into individual pulses. A pulse shaping circuit (*Schmitt trigger*) applies 'clean' pulses to the *shift register* I.C. Whenever the I.C. accepts a pulse, the internal circuitry connects its output pin to the next tuning potentiometer in the sequence. The remote control system can be switched out and the tuner reverts to manual operation (touch tuning).

The circuit in figure 5.9(b) has an additional feature–namely that of *sound muting*. The viewer can transmit a longer pulse; this is noticed by the pulse length discriminator and is steered to the audio section. Shorter pulses are, of course, sent to the channel selector i.c.

The remote control system uses mechanical signals instead of electromagnetic waves; therefore transmission-reception at these ultrasonic frequencies follows a relatively straight line. For reliable operation the transmitting transducer should face the receiving transducer.

5.7 Tuner Servicing

The tuner unit is perhaps the most reliable part of the receiver, since it not only handles very small signals, but also dissipates very little power. Sometimes the tuner circuit diagram is not included in the service manual, because the manufacturers recommend that the unit be returned to them for servicing and realignment. Workshop repairs can easily disturb the critical layout of the circuit and, since replacement components can have different characteristics from the original parts, replacements may result in a degree of misalignment. Accurate tuner realignment is only possible with the aid of specialised equipment which is outside the scope of the average workshop.

Before the tuner unit itself becomes suspect, a number of points have to be checked first, because they can produce fault symptoms similar to those of a tuner failure. The following list of checks can be derived from the block diagram in figure 4.1: efficient aerial installation (especially plugs and sockets), d.c. power supplies to the tuner, a.f.c., a.g.c., i.f. amplifiers, etc. In addition, it may be necessary to inspect mechanical linkages in the tuner channel selector and the condition of contacts and switches.

D.C. readings can help to isolate the faulty stage, but it must be realised that any instrument connected inside the tuner is likely to disturb the signal conditions.

Examples of possible tuner faults: very noisy picture–suspect r.f. amplifier(s); no sound or picture, usually with some noise present–oscillator circuit; difficulties in tuning on all or certain channels only–oscillator circuit, misalignment, tuning voltage drift in varicap tuners (the latter normally due to circuits external to the tuner).

Should any components be replaced, then their positioning, the length of connections, etc., must follow the makers' original design.

5.8 Frequency Response of I.F. Amplifiers

I. F. amplifiers are responsible for most of the signal gain in the receiver and they also affect the overall quality of the displayed picture (and sound). This is done by rejecting unwanted channels and maintaining a correct balance between the signals which make up the composite video waveform.

The performance of an i.f. strip is represented by means of a frequency response graph. The graph shows the change in the output voltage, expressed in dB, to a base of the intermediate frequency (see figure 5.10). This range of frequencies arises from both the video signal bandwidth and the presence of sound and chrominance information.

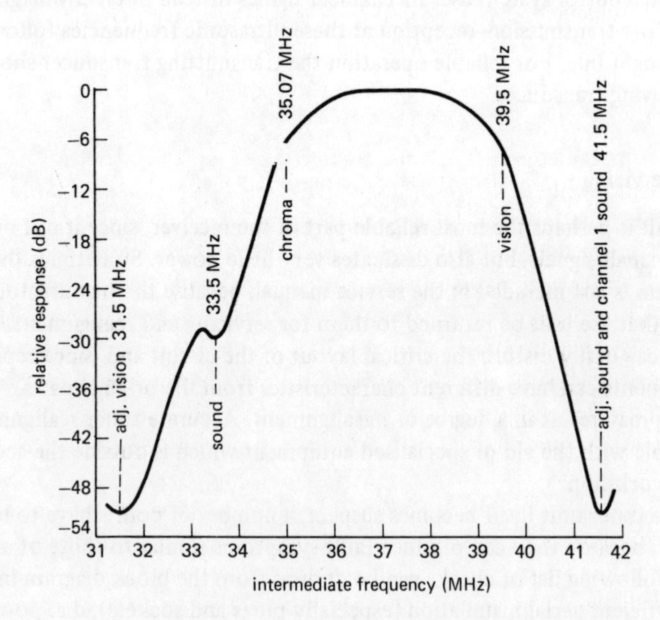

Figure 5.10 Typical i.f. response curve of a TV receiver

The recommended *standard vision i.f.* in the UK is 39.5 MHz; its choice is based on signal bandwidth, frequency allocation to other channels, etc. The oscillator in the tuner must operate at a frequency which is 39.5 MHz above the vision carrier of the required channel. For example, if the receiver is tuned to channel 33, whose allocated vision carrier is 567.25 MHz, then the *oscillator frequency* is 567.25 + 39.5 = 606.75 MHz. Channel 33 sound carrier is 6 MHz above the vision carrier at 573.25 MHz, and the chrominance information is centred 4.43 MHz above the vision carrier at 571.68 MHz.

The mixer in the tuner produces a number of signals whose frequencies correspond to the differences between the oscillator frequency and the respective carriers. The resultant intermediate frequencies are fed to the i.f. strip as follows:

vision i.f., 39.5 MHz; *sound i.f.*, 606.75 – 573.25 = 33.5 MHz; *chrominance i.f.*, 606.75 – 571.68 = 35.07 MHz. Although these figures are for channel 33, exactly the same values would be obtained for any other channel in the UK.

Signal frequencies corresponding to the carriers of the unwanted adjacent channel numbers can also reach the mixer; this is caused by the lack of perfect selectivity in the tuner r.f..amplifiers. Adjacent channel numbers are not used by the same transmitter in order to prevent interference, but they may be received in an area which is within the coverage of a number of transmitters.

In the above example adjacent channels are: channel 32 (vision 559.25 MHz, sound 565.25 MHz) and channel 34 (vision 575.25 MHz, sound 581.25 MHz). The mixer/oscillator tuned to channel 33 will produce the following additional i.f.s:

channel 32 vision: 606.75 – 559.25 = 47.5 MHz;
channel 32 sound: 606.75 – 565.25 = 41.5 MHz;
channel 34 vision: 606.75 – 575.25 = 31.5 MHz;
channel 34 sound: 606.75 – 581.25 = 25.5 MHz.

It can be seen from the above results that channel 32 sound and channel 34 vision i.f.s are separated by only 2 MHz from the required channel vision and sound, respectively. This situation is made worse by the presence of the sound carrier frequency of VHF channel 1 (405-line system), which is also 41.5 MHz. The unwanted signals can interfere with the reception of the desired channel either in the form of patterning on the screen or as 'noisy' sound. The extreme frequencies in the above list (47.5 MHz and 25.5 MHz) are considered to be too far away from the wanted spectrum to cause any problems. The same results for adjacent channel i.f.s would be obtained if other groupings were considered. This is, of course, due to the fixed spacings between the frequency allocations for the various channels and their carriers.

The frequency response of the i.f. strip must be arranged so as to suppress the unwanted signals of the *adjacent sound* (41.5 MHz) and *adjacent vision* (31.5 MHz). A typical i.f. response graph is given in figure 5.10–it will be noted that it has a 'flat' centre portion and very steep sides. The region of the relatively constant output is governed by the frequencies of the video bandwidth; the steep sides offer the necessary selectivity, or rejection of unwanted signals. Comparison of the i.f. response graph with that of the transmitted frequency spectrum (figure 2.8) indicates a reversal in the relative positions of the sound and chroma with respect to the vision carrier. This is, of course, due to the frequency subtraction in the mixer stage of the tuner.

Let us consider the i.f. response to the desired frequencies. The output at 39.5 MHz is approximately one-half of the maximum possible value (half voltage corresponds to –6 dB). The necessary reduction of the vision carrier amplitude is caused by the *vestigial sideband transmission* described in chapter 2. As indicated in figure 2.7, the low video frequencies, up to 1.25 MHz, appear in both the upper and the lower sidebands. Therefore the energy transmitted in the frequency range close to the vision carrier is twice the amount corresponding to the higher video frequencies. The imbalance must be corrected in the receiver–partly in the i.f. strip and partly in the video amplifiers (to be discussed in chapter 6).

The response to the chrominance information, centred around 35.07 MHz, is also 6 dB below the maximum. This arrangement helps to minimise the effects of the 4.43 MHz interference pattern on the screen. In colour receivers the chroma signal is amplified separately in the decoder at 4.43 MHz; therefore the reduction at 35.07 MHz can be compensated for.

The response at the sound i.f. of 33.5 MHz is considerably depressed (between 20 and 30 dB) to prevent pattern interference on the picture owing to the vision-sound and chrominance-sound beat frequencies. Again, the sound information is amplified separately at 6 MHz, and the present rejection, provided it is not excessive, does not affect the audio quality.

5.9 Practical I.F. Amplifier Circuits

The reader will recall that a resonant circuit can discriminate either in favour of or against signals of a certain frequency. Therefore, if a tuned circuit is placed in the collector of a transistor amplifier, a substantial gain will be obtained at the resonant frequency. Unfortunately, the response of the resultant simple *tuned circuit amplifier* cannot follow the required i.f. response of a TV receiver. It is not possible to achieve the necessary bandwidth *and* good selectivity in the simple amplifier arrangement.

A number of practical methods can be adopted to modify the circuit basic response to suit the requirements. One type of arrangement uses *coupled tuned circuits*, shown in figure 5.11(a). Inductors L_1 and L_2 form a transformer, but the primary and the secondary windings are independently tuned by capacitors C_1 and C_2. The amount of *magnetic coupling*, or interaction, between the windings is made variable by means of adjustable cores. The shape of the resultant frequency response curve depends on the amount of coupling between the two tuned circuits.

(1) A single sharp peak is produced–as for a normal tuned circuit. This arises with so-called *loosely coupled* circuits.
(2) Two peaks appear in the response curve with a dip between them (a 'double hump' response). This is due to the so-called *closely coupled* circuits.
(3) Level response occurs over a reasonably wide frequency range owing to *critically coupled* circuits.

The three possibilities are represented by the response curves in figure 5.11(b). Response (3) above is of most interest here. Unfortunately, practical alignment of tuned coupled circuits is difficult, because of the very critical interaction between the circuits.

Another method of increasing the bandwidth is by means of *staggered tuning.* This is represented in block diagram form in figure 5.12(a). Three i.f. amplifier stages are shown to be tuned to slightly different frequencies within the required passband. The resultant response curve, when expressed in dB, is the sum of the

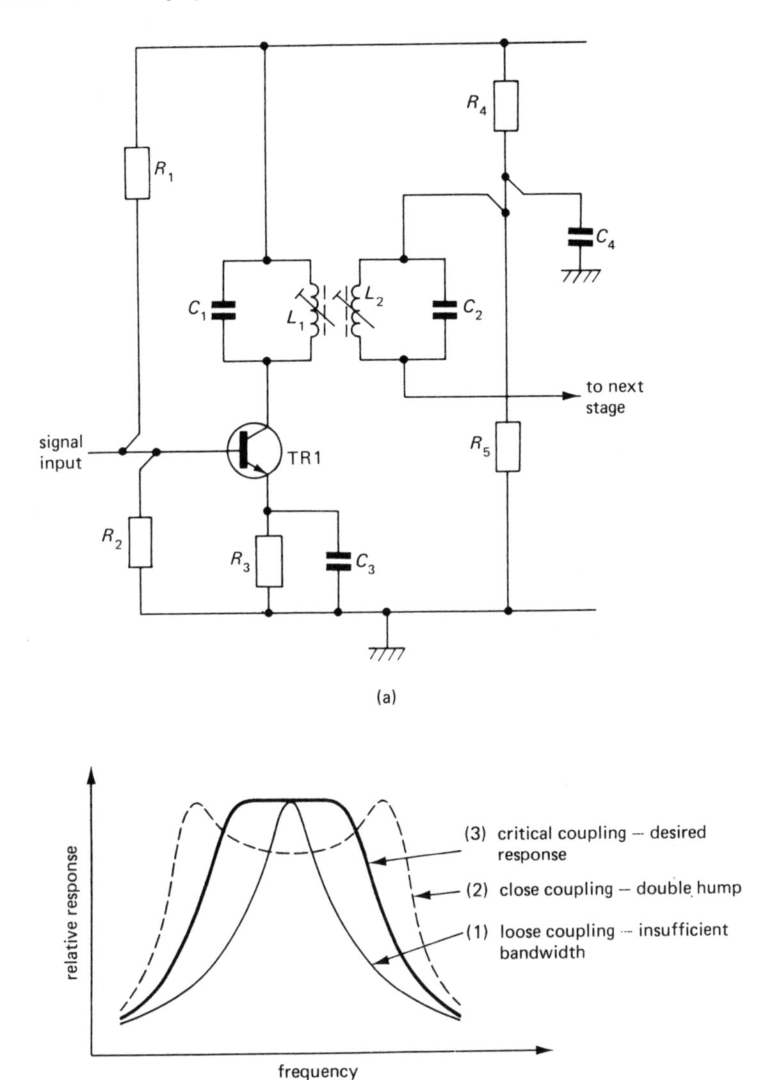

Figure 5.11 Coupled tuned circuit amplifier: (a) circuit diagram; (b) response curves produced by different tuning of L_1 and L_2

three individual curves. A relatively flat response with reasonable selectivity can be achieved, as given in figure 5.12(b).

The most popular method employs *broadband* or *wideband tuning* in conjunction with suitable response shaping *wave traps*. The i.f. amplifier in this arrangement also uses a resonant circuit in its collector, often without any provision for tuning adjustments. A *damping resistor* is connected across the circuit to damp, or to

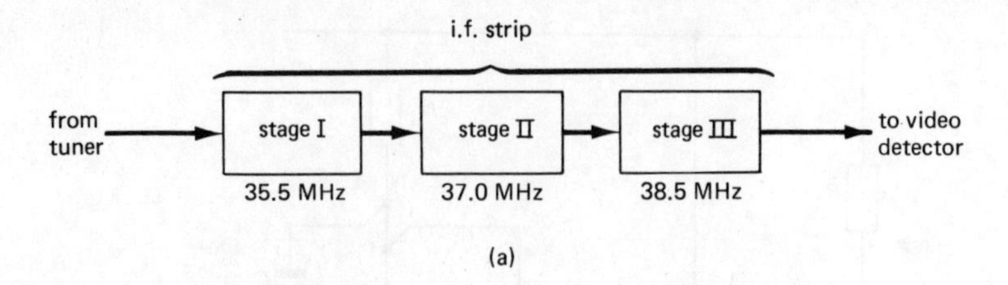

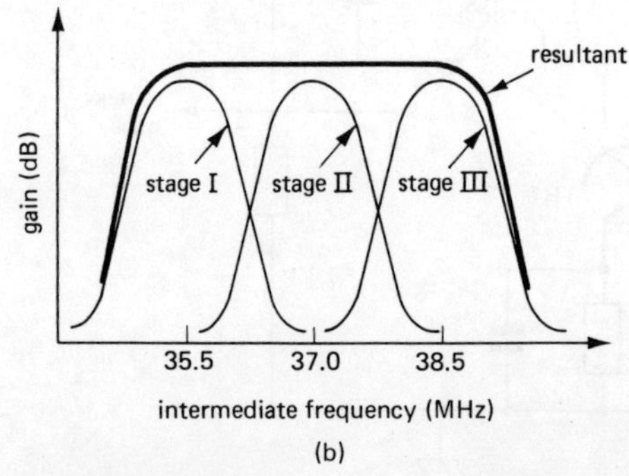

Figure 5.12 Principle of staggered tuning: (a) block diagram; (b) individual response curves of each stage added to produce the resultant of the i.f. strip

flatten, the response by 'spoiling' the Q factor. This arrangement satisfies the need for a considerable i.f. bandwidth, but its selectivity is very poor–especially where tuning adjustments are not provided.

The required shaping of the response curve of broadband amplifiers is by means of separate circuits which are made highly frequency selective. Traps, or filters, of this kind are often grouped at the input to the i.f. section. Basically they consist of suitably connected series and parallel resonant circuits. A series tuned circuit has a very low impedance at resonance, while a parallel circuit has a high impedance. Should a parallel circuit be placed *in series* with the signal path, it attenuates all information at its resonant frequency. A series tuned circuit placed in the same position would enhance the signal. On the other hand, if a series resonant circuit is connected *across* the signal path, its low resistance reduces the gain at that frequency; a parallel circuit in that position tends to boost the resultant output.

Another form of frequency selective network consists of a series-parallel combination which gives rise to two resonant frequencies–one of them enhances the response and the other one reduces it.

Finally, a popular rejector circuit is the so-called *bridged T* network formed by a combination of capacitors, resistors and inductors. Two types of these components (e.g. capacitors and resistors) are connected to form a letter T; the upper arm of the T is then bridged by the third type of component (in this case it would be an inductor). This filter design offers a very high degree of attenuation especially

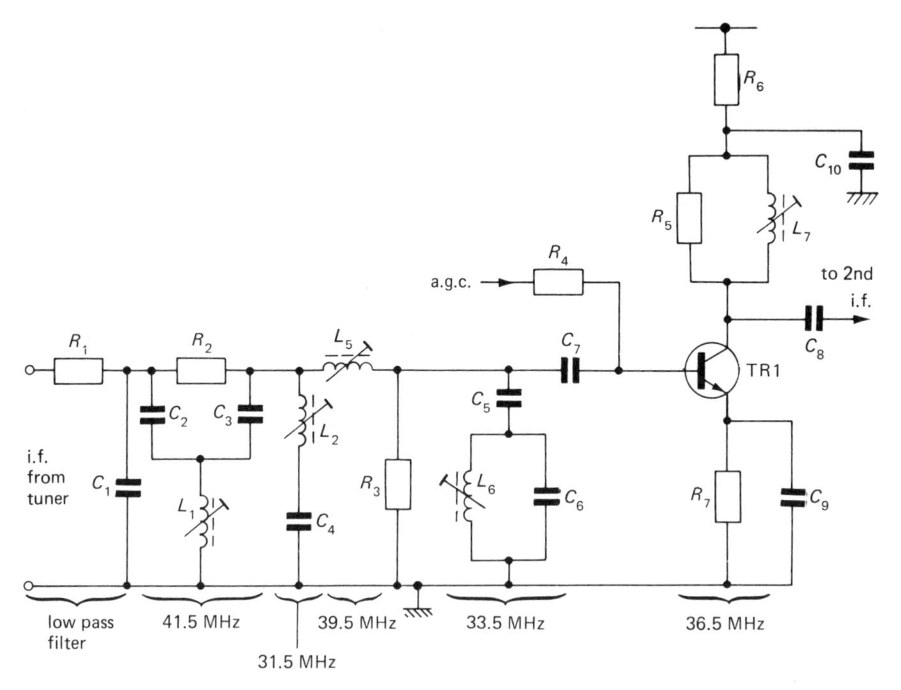

Figure 5.13 Tuned circuit i.f. amplifier with response shaping filters (Decca)

useful in suppressing adjacent channel frequencies.

An example of an i.f. amplifier with its input selective network is shown in figure 5.13; the resonant frequencies of each section are stated below their respective circuits. R_1, C_1 form a general low pass filter at the input; C_2, C_3, L_1, R_2 form a bridged T network (R_2 is the 'bridging' component); C_5, C_6, L_6 produce a series–parallel tuned circuit (series resonance occurs at 33.5 MHz as indicated, and parallel resonance would then automatically take place at a higher frequency to lift the response curve up). Inductor L_5 is part of a parallel resonant circuit which is tuned by the inevitable stray capacitance associated with the coil. A similar situation arises in the collector circuit of TR1, where the inductor L_7 is also tuned by the circuit stray capacitance in addition to the coupling capacitor C_8. The i.f. amplifier itself, TR1, is tuned to 36.5 MHz, which corresponds nearly to the centre of the 'flat' portion of the response curve (see figure 5.10).

Resistor R_3 in the input circuit and R_5 across the collector tuned circuit are damping resistors; these prevent the possibility of self-oscillations (instability), which may occur in high gain, frequency selective amplifiers. In some circuits separate damping components are not needed, since the relatively low input resistance of the following transistor stage provides a sufficient degree of damping.

Frequency selective circuits can also be connected in the *feedback path* of an amplifier, either to enhance or to suppress particular frequencies.

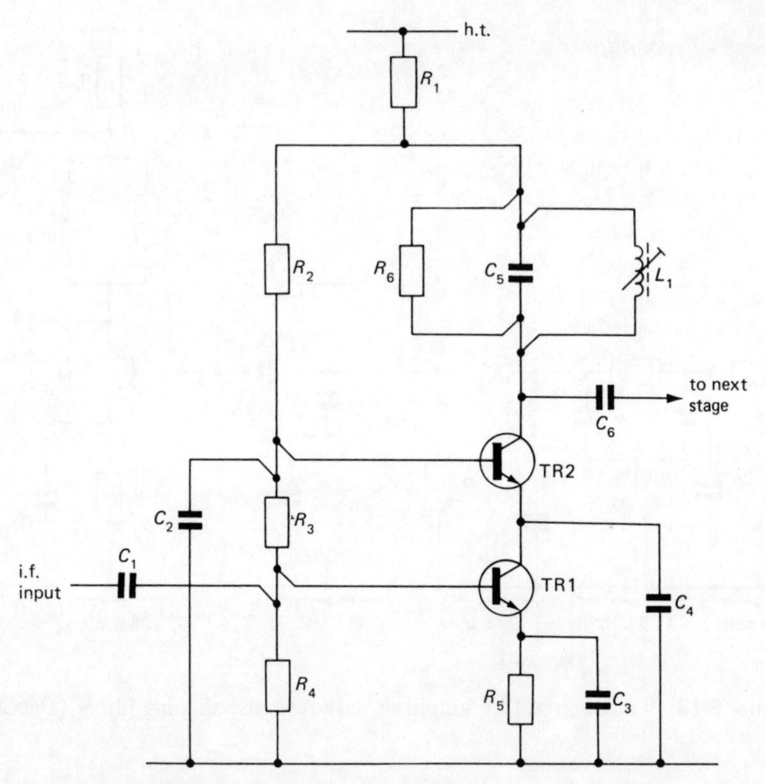

Figure 5.14 Cascode i.f. amplifier: TR1 (common emitter amplifier) connected directly to TR2 (common base amp.–base decoupled via C_2) (Rank)

Certain designs adopt values of trap frequencies slightly different from the standard values quoted so far. However, some interaction obviously exists between the filters, and it is the overall i.f. response which must be maintained.

The common emitter connection is usually used in i.f. stages (apart from integrated circuits), but occasionally a *cascode* amplifying circuit is included. An example of such an arrangement is shown in figure 5.14. Transistor TR1 is a common emitter amplifier whose collector feeds directly to the emitter of TR2. Since the base of TR2 is fully decoupled via C_2, this transistor operates in the common base mode; the output from the stage is at the collector of TR2. A cascode amplifier offers high gain with good stability, and it is particularly useful in high frequency applications.

A recent introduction which provides the necessary i.f. response shaping is the *surfacing acoustic wave filter*. The device consists of two *transducers*–one at the input, which converts the i.f. signal into mechanical (acoustic) vibrations, and another one at the output, where the incoming surface waves are reconverted into electrical signals. The term 'acoustic' denotes that the signal is propagated along the surface of the device in the form of mechanical waves. The construction of the filter resembles that of an I.C.; the transducers are a comb-like, interlocking structure deposited on the surface of a small chip or *substrate* made of a special material. Filtering action takes place because the amount of energy converted by the transducers depends on the frequency of the electrical signal. They can be designed to develop a maximum output at, say, 37 MHz, and a series of minimum outputs at 31.5 MHz, 33.5 MHz and 41.5 MHz. An amplifier is needed to compensate for the overall signal power losses in the device, but tuned circuits are no longer necessary and there are no alignment problems.

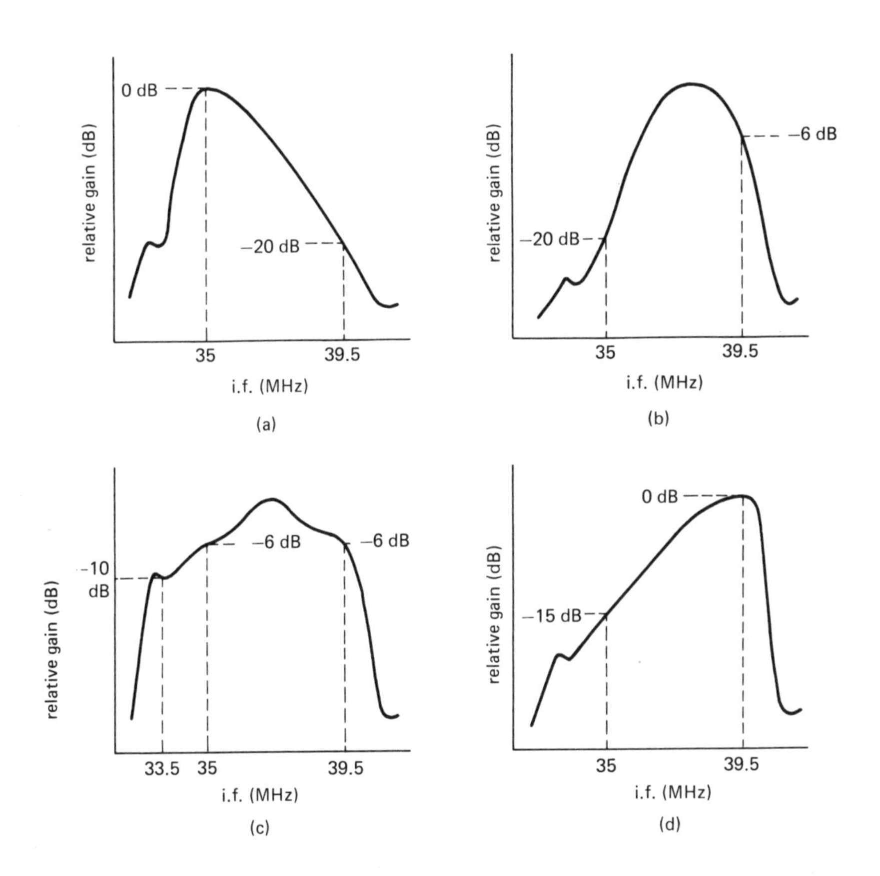

Figure 5.15 Examples of incorrect i.f. response: (a) insufficient response to the vision carrier (weak contrast, possibly poor sync.), excessive response to chroma signal (too much colour); (b) insufficient bandwidth (poor resolution of picture detail), insufficient chroma (no colour); (c) excessive peak in i.f. response (possible instability–ringing), excessive sound i.f. (beat patterns on the screen); (d) excessive response to vision i.f. (strong contrast, smears around coarse picture detail, vision buzz on sound), insufficient chroma (possibly loss of colour)

5.10 Alignment of I.F. Amplifiers

The purpose of alignment is to ensure that the frequency response of the i.f. strip conforms to the makers' design. Incorrect alignment can lead to defects in the picture (and in the sound!), as suggested by the examples of grossly misaligned response curves sketched in figure 5.15.

I.F. alignment, or more often realignment, may be necessary for any one of a number of reasons, including the replacement of components (usually those associated with the tuned circuits).

The alignment procedure involves the application of a number of signals at frequencies within the required range, say between 30 MHz and 42 MHz, to the input of the i.f. section. The resultant amplified output is then monitored at some convenient point in the receiver. Ideally, a complete response curve should be obtained; alternatively, the output only at certain specified frequencies can be checked against makers' recommendations. In any case, manufacturers' recommended procedures for alignment must be carefully considered.

The point at which the test signal from a signal generator is applied to the receiver i.f. section may vary between differing designs. Frequently it is injected at the beginning of the i.f. chain—that is, in the mixer circuit of the tuner (for example, TP1 in figure 5.4). In other cases the signal could be applied to the individual i.f. stages in turn. The signal generator must always be properly terminated, not only to preserve its correct calibration, but also to simulate the impedance levels found in such circuits. A diagram of a simple i.f. injection probe is shown in figure

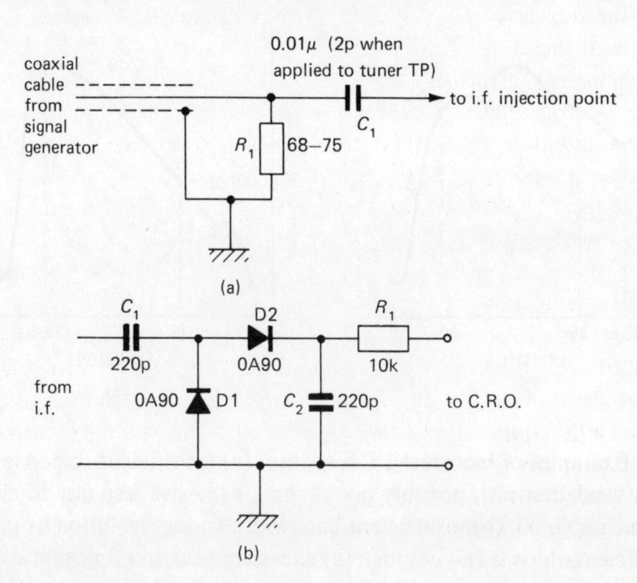

Figure 5.16 Examples of i.f. alignment probes: (a) i.f. injection probe; (b) i.f. demodulator probe (voltage doubler)—used when signal is to be taken directly from one of i.f. stages

5.16(a); the value of the d.c. blocking capacitor C_1 depends on the actual point of application into the circuit.

The output voltage from the i.f. stage can be obtained most conveniently from the vision detector load resistor or the video emitter follower (see chapter 6). The magnitude of the rectified signal can be measured using a d.c. voltmeter (preferably an electronic instrument or an oscilloscope). If necessary, a demodulator probe can be made up, as shown in figure 5.16(b), and connected to the output of the appropriate i.f. stage.

There are two basic methods of alignment: the wobbulator method and the spot frequency method.

Wobbulator method

A wobbulator, also known as a *sweep generator*, is a signal generator whose output frequency sweeps back and forth between certain limits. For vision i.f. alignment these limits are approximately ±5 MHz about a 'centre' frequency. If the wobbulator carrier frequency is set to 35 MHz, its output will swing from 30 MHz to 40 MHz. This signal is injected to the i.f. stages and the output from the i.f. strip is displayed on an oscilloscope (C.R.O.). The time base of the oscilloscope and the wobbulator action are synchronised–that is, the minimum frequency (say 30 MHz) of the generator coincides with the beginning of the C.R.O. scan. When the beam reaches the right-hand side of the screen, the wobbulator output corresponds to its maximum frequency (say 40 MHz). During the beam flyback period the output frequency is reduced back to its minimum value. The display on the screen will then be in the form of the i.f. response curve.

Some wobbulators utilise the C.R.O. time base waveform (available at the terminals marked–for example, X-OUT) to modulate the sweep. Alternatively, the wobbulator has a built-in time base oscillator which can be used to control both the generator frequency sweep *and* the oscilloscope horizontal deflection. In either case the sweep rate must be kept relatively low, which means that the C.R.O. time base can be set to a sweep speed of about 10 ms/cm. If the rate is too high, the results are likely to be incorrect since the receiver circuits may not be able to respond to such rapid changes in frequency.

A block diagram of the arrangement of instruments in the wobbulator method is shown in figure 5.17. It is important to be able to identify the position of the actual frequencies on the displayed response curve. This is done by injecting a separate signal at a required frequency together with the wobbulator signal. The two signals are mixed in the receiver and a marker 'pip' appears on the response curve in its appropriate position. The separate signal source is called a *marker generator*, which may either be built into the wobbulator or be connected externally via a low value capacitor, *C*.

The accuracy of the alignment depends upon the accuracy of the marker generator. The tuning of the i.f. coils is adjusted to produce a curve of the desired shape, as indicated by the positioning of the identifying 'pips'. To prevent overloading the

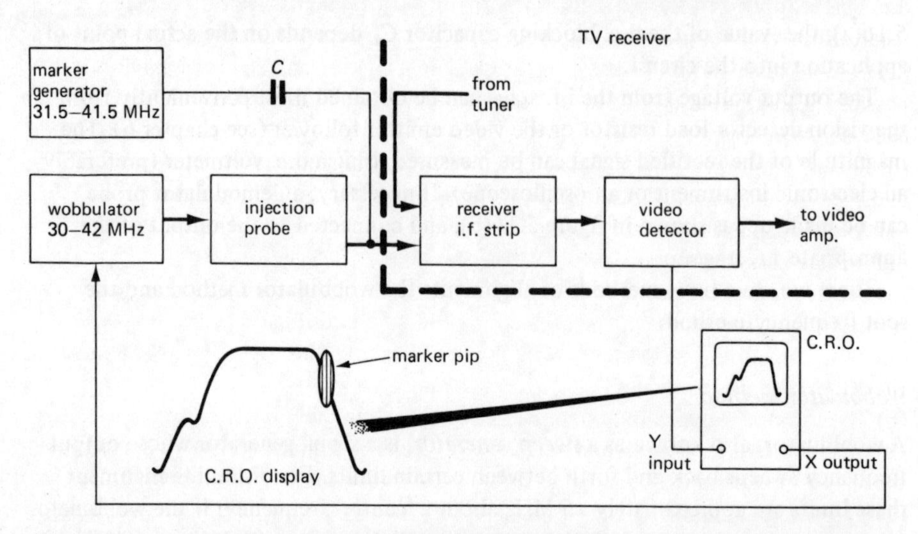

Figure 5.17 Block diagram of i.f. alignment setup. C.R.O. display can be either positive-going (as shown) or negative-going, depending upon polarity of the video detector

i.f. amplifiers, both the wobbulator and the marker generator outputs must be kept to a low amplitude. These should be just sufficient to produce a reasonable display on the C.R.O. screen. It is of interest to note that the oscilloscope display may appear 'upside down', depending upon the polarity of the vision detector diode.

The advantages of the wobbulator method are: speed of adjustment, good accuracy of alignment over the *complete* i.f. range and the ability to observe the interaction between the traps, coupled tuned circuits, etc.

Spot frequency method

This method requires an accurate a.m. signal generator and a suitable output meter connected in the manner explained in the general notes above. The applied signals should include the trap, vision, chroma and midband frequencies. Adjustments are then made until the correct response is achieved. Since an automatic frequency sweep is not provided by this method, it is necessary to sweep manually through the full frequency range. This provides a check that the response is symmetrical, without undue dips or peaks, that the bandwidth is sufficient, etc. Because of the uncertainty of the results obtained by this method, a number of manufacturers do not recommend its use.

General precautions

Irrespective of the method used, the following should be carried out to ensure correct alignment results.

(1) The automatic gain control (a.g.c.) must be rendered inoperative. Usually a fixed d.c. bias is applied to the a.g.c. line either from a suitable battery or from the receiver h.t. via a resistor or a test link according to makers' instructions.
(2) Coaxial cable should be used for the test leads, and it should be kept as short as possible. The input and output test leads must be kept well apart to prevent feedback between them.
(3) Special trimming tools should be used to adjust the cores in order to prevent damage. These tools are made of non-magnetic material–usually plastics.
(4) Allow both the signal generator and the receiver to 'warm up' for at least 15 min before carrying out any adjustments.
(5) Select the correct position of the tuning core, since there are often two different positions which appear to produce the required result. The accepted setting, unless one is advised to the contrary, is with the core closer to the top of the coil former; the alternative position could give incorrect bandwidth.
(6) The injected signal should be maintained at a reasonable level. Badly misaligned circuits may require a large input initially, but as they are brought into alignment, the signal strength must be reduced to prevent overloading.
(7) Ferrite cores should be finally secured with a core locking compound, any temporary components removed from the circuit and the operation restored to normal.

Warning The danger of *live* chassis must not be overlooked, since a number of instruments with earthed test leads will be used. The reader is advised to study the discussion on this aspect in chapter 14.

5.11 Automatic Frequency Control (A.F.C.)

A.F.C. maintains frequency stability of the oscillator in the tuner which, in turn, ensures constant vision i.f. This is a highly desirable feature, since in a properly aligned receiver the quality of reproduction will depend upon the stability of the intermediate frequency itself. The performance of a colour receiver can be especially affected by tuner oscillator drift. The chrominance signal is positioned on the slope of the i.f. response curve (figure 5.10) and it is also balanced with respect to the response to the vision and sound carriers. Should the actual value of i.f. shift, the respective carriers would effectively change in value. Consequently, this may result either in the loss of colour or in severe interference patterns.

Automatic frequency control is obtained by the application of a d.c. feedback signal to the tuner; the voltage fed back is proportional to drift in vision i.f. Clearly, the task of the a.f.c. circuit is similar to that of an f.m. demodulator, since, in either case, a voltage proportional to frequency deviation is produced. A block diagram of an a.f.c. arrangement is shown in figure 5.18, together with a typical circuit diagram. Part of the output from the final i.f. stage is applied to the *limiting amplifier* (TR1), which removes the amplitude variations from what is basically an a.m. signal. A limiting amplifier achieves its high gain usually with the aid of a

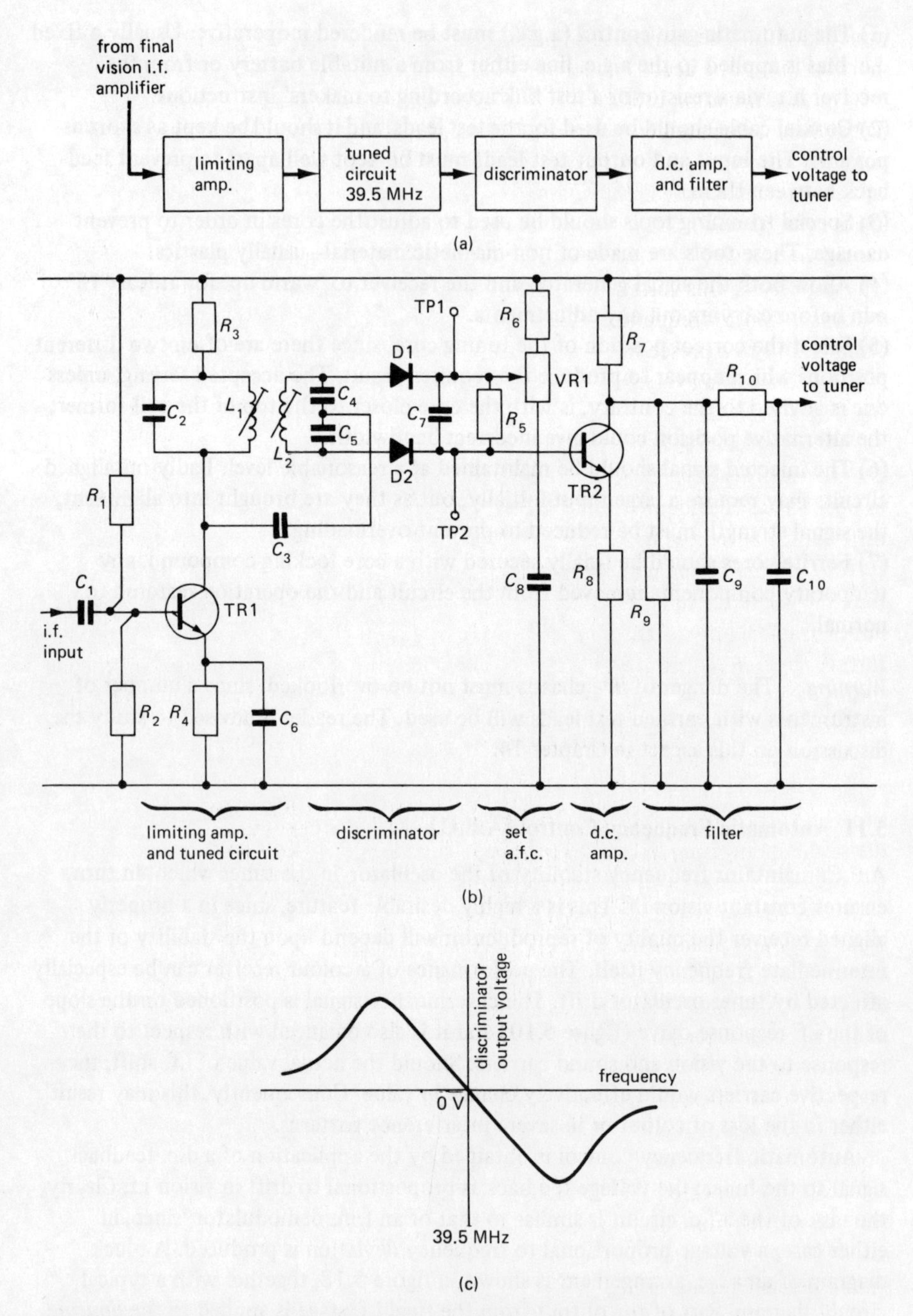

Figure 5.18 Automatic frequency control system: (a) a.f.c. block diagram; (b) a.f.c. circuit diagram; (c) discriminator response curve

tuned circuit. The carrier is now amplified to such an extent that the modulated portion cannot be accommodated by the amplifier.

The *discriminator* uses a Foster–Seeley type of f.m. detector formed by D1, D2 and the associated components. The collector circuit of TR1 is tuned to the required vision i.f. of 39.5 MHz. Consequently, when the incoming signal is exactly at that frequency, the drive to the detector is balanced and the net voltage across C_7 is zero. Should the incoming frequency deviate from 39.5 MHz, the diodes would conduct by differing amounts, resulting in a potential developing across C_7. The polarity of this voltage is dependent on whether the frequency drift is above or below the centre frequency. Therefore the discriminator output will either subtract from or add to a preset value which is controlled by the *set a.f.c.* potentiometer, VR1. The resulting voltage is amplified by the d.c. amplifier, TR2, and after filtering (R_9, C_9, R_{10}, C_{10}) the d.c. signal is fed to the tuner.

When a receiver is being tuned to a programme, the a.f.c. is usually disabled. This is done to ensure that the initial oscillator frequency is as close to the correct value as possible, as is evident from the quality of the reception. Any subsequent tuning drift will be corrected by the a.f.c., but it must be realised that the automatic feature can only cope with a limited frequency error. It is therefore possible to tune the oscillator to a frequency close to this limit, and the picture quality would not be immediately affected, owing to the corrective action of the a.f.c. If, however, further frequency drift occurred in the oscillator, a.f.c. would then lose control and tuning would be lost. This behaviour is also found in the other frequency control systems of a TV receiver (line time base oscillator and colour subcarrier oscillator).

An auxiliary contact can be provided on the channel selector buttons which automatically disables the a.f.c. [for example, by short circuiting terminals TP1 and TP2 together in figure 5.18(b)]. In other designs, channels are preselected by the servicing engineer (with the a.f.c. disabled!), and it is assumed that the viewer will not interfere with the settings.

It may be sometimes necessary to realign the a.f.c. detector circuit. A 39.5 MHz signal is then applied either to the base of TR1 in figure 5.18 or to the main i.f. injection point (see alignment procedure). A voltmeter is connected between the test points TP1 and TP2 and, with L_2 temporarily detuned, L_1 is adjusted for maximum instrument reading. Coil L_2 is then retuned so as to give zero voltmeter indication *across* C_7. The 'set a.f.c.' control (VR1) is adjusted to produce a specified value of d.c. as given in makers' instructions.

Ideally, the alignment of the a.f.c. tuned circuit should be carried out using a wobbulator. In this case L_2 would be adjusted to obtain a typical *S-curve* of an f.m. detector. The expected shape is shown in figure 5.18(c), from which it is evident that the output must be symmetrical about 39.5 MHz. If an ordinary signal generator is used and L_2 tuned as previously described, the symmetry of the response should be checked on either side of 39.5 MHz.

An a.f.c. detector can be incorporated into an I.C., usually, as part of the video *synchronous detector* circuit, to be discussed in chapter 6. A few receivers use I.C.s designed for normal f.m. sound detection, but the external circuitry has to be adapted to the requirements of TV a.f.c.

Fault-finding in a.f.c. circuits is relatively simple. The unit can be easily disabled to check whether normal tuning can be restored. Similarly, the a.f.c. voltage can be monitored to observe the effect of channel tuning on the control potential.

6 Vision Detectors and Video Amplifiers

6.1 Principles of Signal Demodulation

The i.f. signal in the receiver resembles the transmitted signal, except that the carrier frequency is much lower. The purpose of *demodulation*, or *detection*, is to separate the video information from the amplified i.f. signal. The waveform of the detected video signal represents an envelope of the amplitude modulated vision carrier. As a 'by-product' of the process of video detection, it is also possible to obtain both the intercarrier sound and the chrominance signals.

The video demodulator circuit is arranged so that the detected signal has either positive or negative polarity—that is, the video waveform is either positive- or negative-going with respect to the sync. pulse level. The required polarity is governed by the following factors.

(1) The method of applying the video signal to the picture tube—cathode- or grid-fed.
(2) The number and the type of video amplifiers—each common emitter (or common cathode) stage will invert the signal, while emitter (or cathode) followers are non-inverting.

In addition, the video detector has to be designed to provide the correct output polarity irrespective of the type of modulation—positive modulation (405-line system) or negative modulation (625-line system).

Figure 6.1 shows a block diagram of the path of the video signal in a 625-line receiver. In this case the signal is applied to the cathode of the picture tube via a

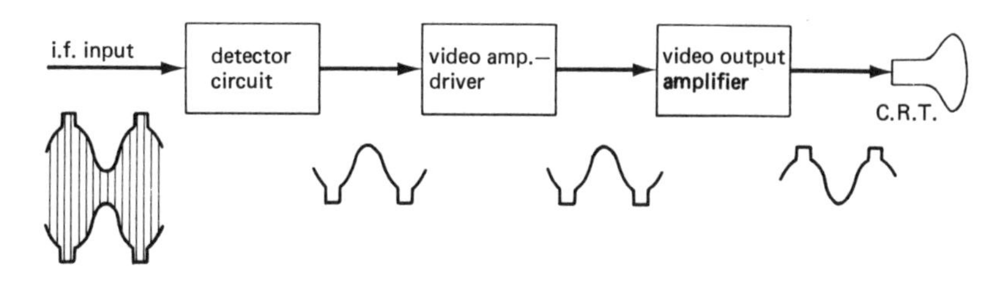

Figure 6.1 Block diagram of video path (see text for the waveforms)

driver (emitter follower) stage and the output stage (common emitter). Since the C.R.T. is cathode fed, it requires a negative-going video signal. As the video output stage is phase inverting, its input must be a positive-going signal. The latter is supplied by an emitter follower; consequently, the detector must also provide a positive-going video waveform. From reference to chapter 2 it can be seen that the positive-going video signal appears on the lower (negative) part of the modulated waveform.

The above comments apply to both valve and transistor receivers, except that in the case of valve equipment a single video stage is normally used (monochrome receivers). Similar reasoning governs the design of colour receivers or where I.C.s are employed.

There are two types of video detector circuits in use: (1) diode detector and (2) synchronous demodulator, both of which are described below.

6.2 Diode Detector Circuits

A diode detector is basically a half-wave rectifier whose input signal is the i.f. waveform. A closer study of the behaviour of the demodulator reveals, however, that the output produced consists of the following signals:

(1) The composite video signal (the modulating envelope).
(2) The d.c. level of the video signal.
(3) Remainder of the i.f. carrier (at 39.5 MHz).
(4) Signals at beat frequencies which are due to the non-linear diode characteristics —the detector behaves like a mixer so that the resultant outputs are
  (a) 39.5 MHz – 35.07 MHz = 4.43 MHz—this is the chroma signal.
  (b) 39.5 MHz – 33.5 MHz = 6 MHz—intercarrier sound signal.
  (c) 35.07 MHz – 33.5 MHz = 1.57 MHz—chroma/sound beat signal.

The composite video signal and often its d.c. level are fed to the video amplifiers. The remaining i.f. carrier is filtered out by a simple circuit, an example of which is shown in figure 6.2. Capacitors C_1 and C_2 together with the inductor L_1 form the i.f. filter. Since the value of their capacitance is very small, the capacitors may be replaced by the inevitable stray capacitance of the other components in the circuit. The design of the detector circuit in general, and of the filter in particular, is very critical, to ensure that the desired high video frequencies are not severely attenuated together with the unwanted i.f. In figure 6.2 the demodulated signal is developed across the *detector load resistor,* R_1, which is in series with a high video frequency peaking coil L_4. Since the total impedance of R_1 and L_4 rises with frequency, this arrangement produces an increased signal voltage at high video frequencies.

The 4.43 MHz chrominance signal can be steered to the decoder in a colour receiver, but it is unwanted in the video amplifiers of a monochrome set or in the luminance stages of a colour receiver. For that reason a suitable filter may be incorporated in a video amplifier circuit to prevent patterning on the screen.

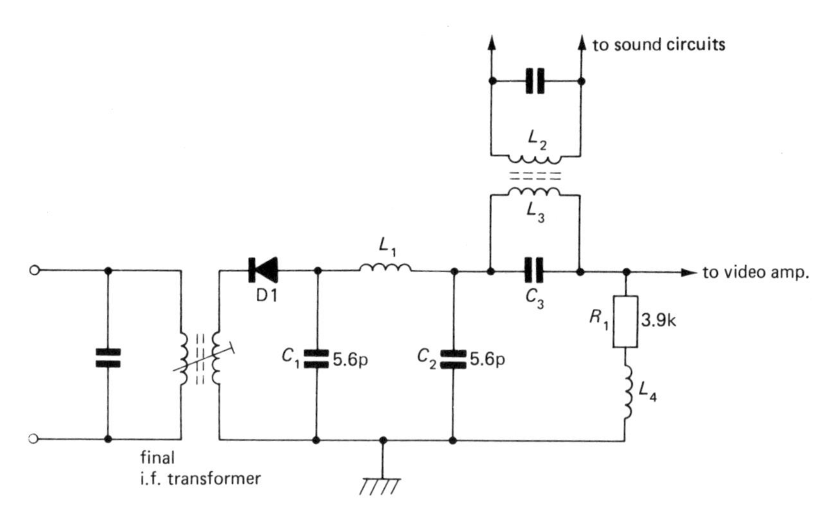

Figure 6.2 Diode video detector circuit

The 6 MHz intercarrier sound signal can be taken off at the detector and fed to the sound section of the receiver. However, a 6 MHz filter is placed after the take-off point, again to stop pattern interference appearing on the screen. In figure 6.2 the circuit L_3, C_3 forms a trap tuned to 6 MHz, and at the same time L_2, L_3 form a transformer which feeds the sound section.

Production of the 1.57 MHz beat signal is normally prevented by considerably depressing the i.f. response at 33.5 MHz, as explained in chapter 5. Otherwise, a rather objectionable and coarse interference pattern would appear on the screen (1.57 MHz is a relatively low video frequency). The entire video detector circuit is housed in a metal can to prevent radiation of beat frequency interference signals to other sections of the receiver.

6.3 Synchronous Video Demodulation

Synchronous demodulation means that the output is obtained by 'measuring' or sampling the amplitude of the video signal at least once every cycle of the i.f. In other words, the demodulator 'looks' at the amplitude of the incoming modulated i.f. in synchronism with the carrier. The sampling circuit behaves as though it were a switch which closes at the moment when the carrier reaches its maximum value. The output from the switch follows the modulating envelope, causing the original video signal to be recovered. The diagram in figure 6.3 shows the principle involved, the sampling switch being either a diode circuit or a transistor arrangement. The filter at the output smooths the changes between the sampling points to produce a continuous output waveform.

Because of the complexity of the circuitry, synchronous demodulation of the video signal is only economical in I.C. form. Figure 6.4 shows a block diagram of

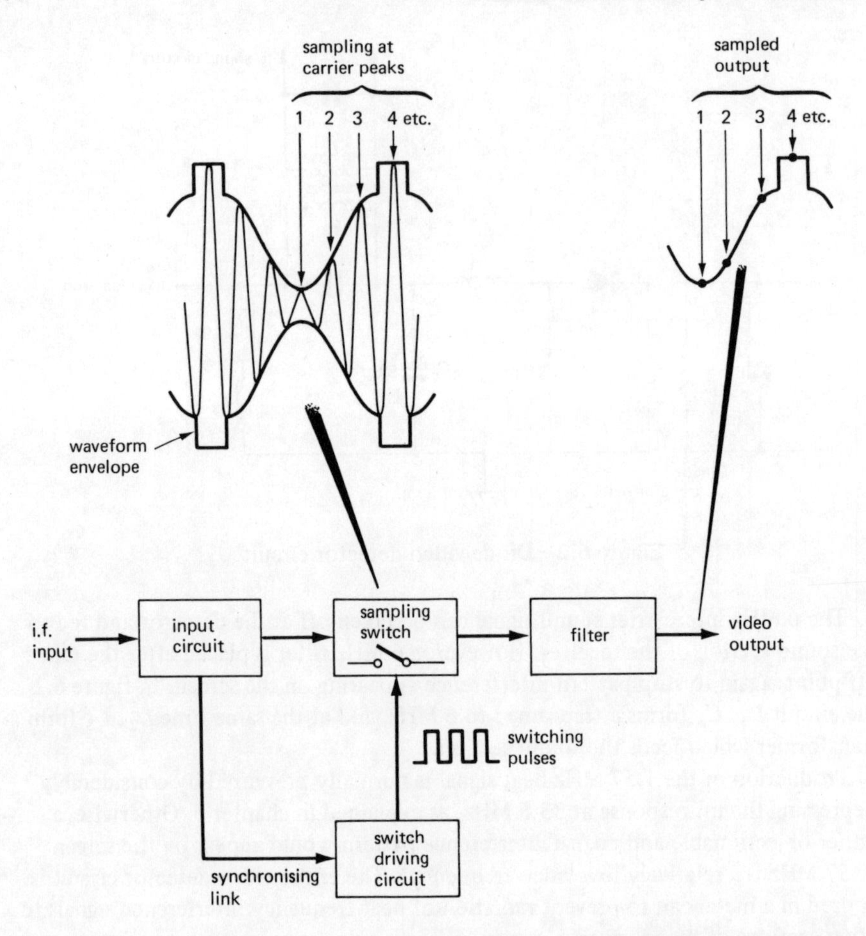

Figure 6.3 Principle of synchronous vision demodulator; switching pulses are derived from the vision carrier i.f. For practical reasons, the relationship between i.f. and picture line frequency is not correctly represented in the diagram

the Mullard TCA 270 I.C. together with some of the external components. The drive for the synchronous demodulator 'switch' is derived from the unmodulated part of the i.f. waveform (that is, below 18 per cent maximum amplitude). As described in the section on a.f.c., it is possible to obtain an unmodulated 39.5 MHz carrier by means of a limiting amplifier; in fact, this I.C. can also generate an a.f.c. signal (pin 11, figure 6.4).

The switching action of the synchronous demodulator is accomplished during both positive and negative half-cycles of the carrier; the negative voltage is then inverted and added to the positive half of the waveform. Examining the i.f. waveform in figure 6.3 shows that this process 'fills in' the gaps between the positive

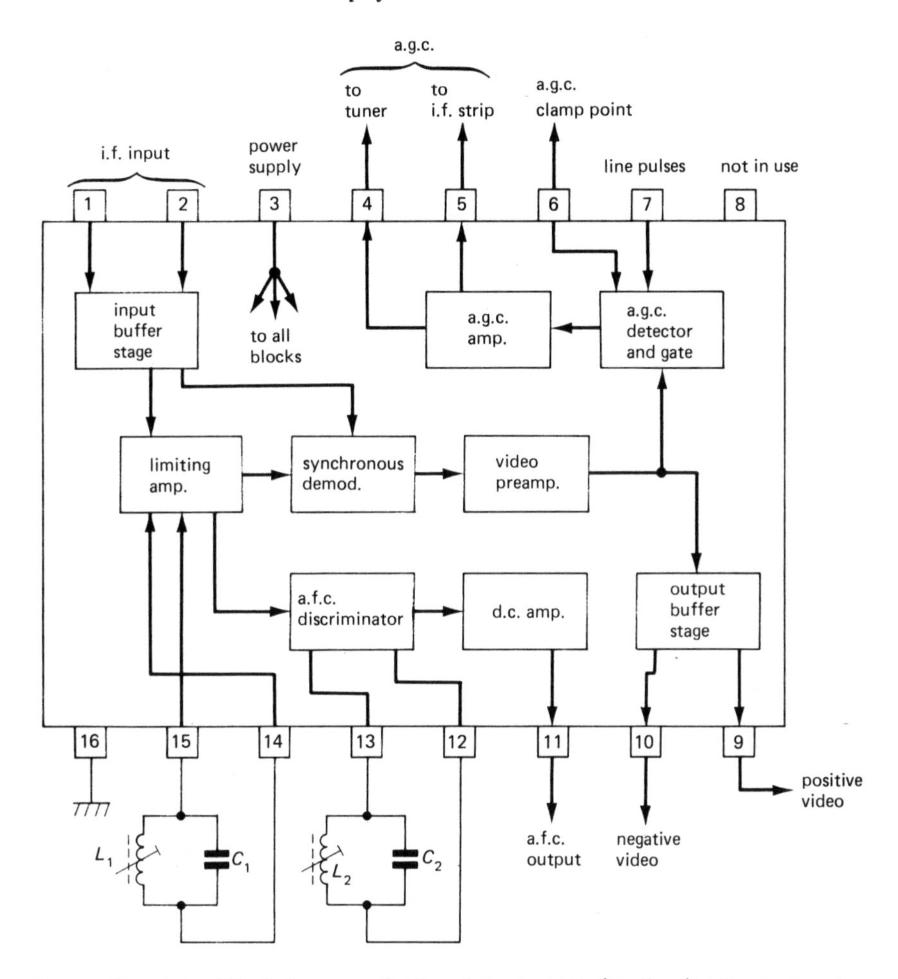

Figure 6.4 Simplified diagram of video detector I.C. (Mullard). The device also includes other functions as shown

sampling points, thus improving the detection of very fine detail. Owing to this double switching action, the video output now has a r.f. component at twice the original i.f. The filtering of the 79 MHz content is simpler and does not cause serious loss of high video frequencies, because the desired and the unwanted frequencies are well apart; the necessary filter circuit is external to the I.C.

This I.C. has a built-in video preamplifier offering video of either polarity, together with the intercarrier frequency and a.g.c. The external tuned circuit L_1, C_1 is associated with the limiting amplifier; L_2, C_2 is the *quadrature detector* coil for the a.f.c. (here the a.f.c. discriminator circuit employs quadrature detection, which will be described in chapter 8).

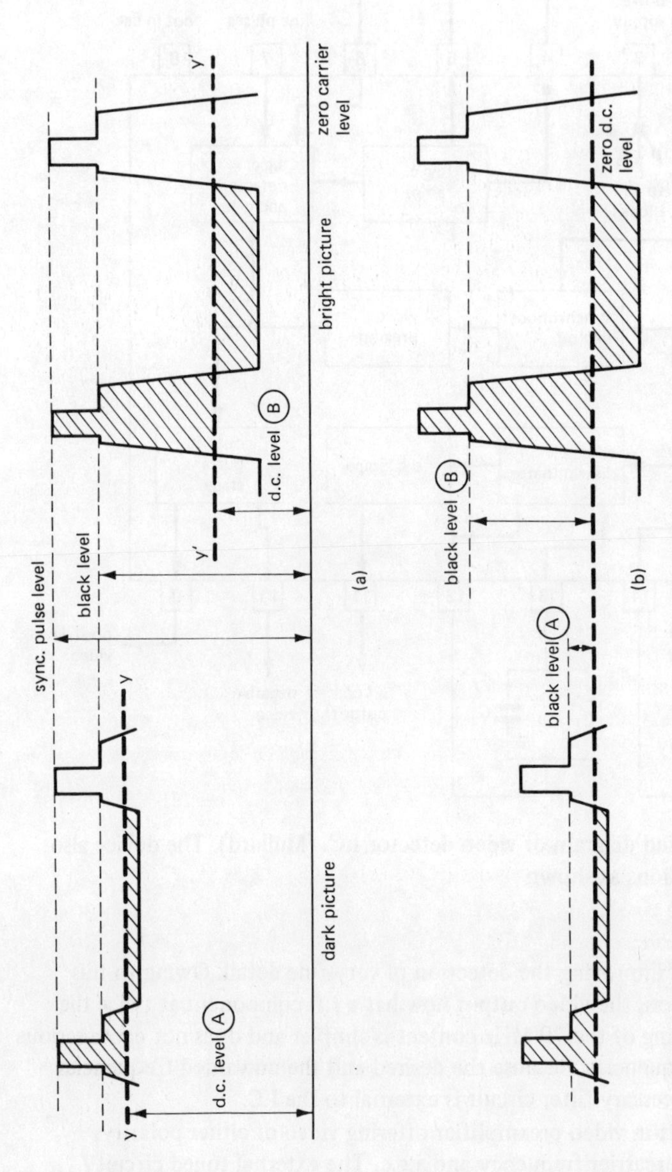

Figure 6.5 Effect of d.c. component in the video waveform (at tube cathode). *Note*: Shaded areas above and below the bold broken lines are of equal value. (a) Complete video waveforms—one line of dark picture (high d.c. level) and one line of bright picture (low d.c. level); the video black level remains constant. (b) Waveforms as for (a), but the d.c. level has been removed by video coupling capacitor; black level applied to the tube is no longer constant (see text for the effects on the picture)

Synchronous demodulation overcomes some of the disadvantages arising from the non-linear characteristics of the diode detector. For example, a diode detector can sometimes give an incorrect contrast ratio—that is, the output at low signal levels is not in direct proportion to the output at high levels of modulation (demodulating weak signals is especially prone to distortion). The previously mentioned beat frequencies generated by diodes can produce undesirable patterns on the screen, while the filtering of the residual i.f. after a diode detector is also a problem if a loss of high video frequencies is to be prevented.

Another form of synchronous demodulator, as used in the colour decoder, is discussed in chapter 7.

6.4 Requirements of Video Amplifiers

The video amplifying section of the receiver must have a relatively high gain and a wide frequency range. The amplitude of the detected signal is in the range 1-2 V peak-to-peak. The necessary signal swing to drive the tube from black to white varies from about 40 V (small screen) to 150 V (large screen).

Theoretically, the frequency response of video amplifiers should extend from d.c. to approximately 5.5 MHz. The requirement becomes evident if the effect of the various frequencies is considered on the appearance of the displayed picture.

(1) *D.C. level* is responsible for the overall illumination level of the picture. The same scene can convey the impression of either daylight or darkness, depending on the d.c. content. A similar effect is, of course, obtained by means of the ***brightness*** control, which alters the d.c. bias of the picture tube. The d.c. component of the video signal arises from the asymmetrical shape of the video waveform. As shown in figure 6.5(a), the d.c. level corresponds to the average value of the waveform—the shaded areas above and below the y-y and y′-y′ are equal. In the detected signal the d.c. level is high for a dark picture and low for a bright picture.
(2) *Low video frequencies* (up to 1 MHz) are responsible for coarse picture detail; most of the sync. pulse waveform is also contained in this range of frequencies.
(3) *Middle video frequencies* (from 1 MHz to 3.5 MHz) contribute to most of the average picture content.
(4) *High video frequencies* (from 3.5 MHz to 5.5 MHz) are responsible for fine picture detail and clear definition. These factors are of a special importance in large screen receivers; in a small portable TV set the size of the display is such that the eye cannot resolve the details from a normal viewing distance. Satisfactory display of Teletext also depends on high video frequencies.

6.5 Amplification of the D.C. Component

The d.c. component of the video waveform can be amplified only if the coupling between all the circuits from the detector to the tube is direct—that is, if coupling capacitors are not used.

Direct coupling in a multistage valve amplifier is not easy to design, since each succeeding valve requires an ever-increasing cathode bias to offset the d.c. anode voltage of the preceding stage. Direct coupling between the vision detector and the video amplifier can present problems if the resultant d.c. level is of negative polarity. A negative-going signal could drive the valve too far towards cut-off unless the no-signal bias was very small. In that case, however, the no-signal anode current could be excessive, and in a single-stage video amplifier it would also cause a low anode voltage of the output valve, resulting in a brightly lit screen.

Direct coupling between transistor amplifier stages is easier to design than is the case with valve circuits. The required forward bias can be derived from the preceding collector (or emitter) rather than from the h.t. supply. A disadvantage of this arrangement is that if one transistor shifts its operating point, the d.c. conditions in the remainder of the circuit will be affected. Various circuit arrangements are used to overcome that difficulty: for example, 'tying' the emitters to a potential divider will hold the voltages steady; applying negative feedback can stabilise the d.c. conditions; or using a black level clamp will maintain a predetermined d.c. level at the video output stage.

In view of these considerations the problem of the d.c. component can be approached in one of the following ways.

(1) A.C. coupling throughout (via capacitors)–the d.c. component is now lost, but the viewer can introduce it by means of the brightness control. This results in a simple circuit, but it can produce a picture of poor contrast whenever the illumination levels change in the transmitted programme.
(2) Part of the circuit is a.c. coupled, after which the video signal is *d.c. restored* and the remainder of the circuit is then directly coupled all the way to the tube.
(3) Direct coupling throughout.

The principle of *d.c. restoration*, or ***black level clamping***, depends upon charging a capacitor in one of the video amplifier stages (for example, a video coupling capacitor) to a predetermined voltage. The video waveform is superimposed upon that fixed d.c. level whose magnitude should correspond to the black level of the picture. The effect of a black level clamp is illustrated in figure 6.5(a)–if the clamp voltage forms the d.c. bias of a video amplifier, then the entire video waveform, including its own d.c. level, can be amplified.

When a black level clamp is not used in the receiver [see figure 6.5(b)], the coupling capacitor blocks the transmitted d.c. level of the waveform. Consequently, the correct black level corresponding to the cut-off voltage of the tube cannot be maintained. If the brightness control is adjusted correctly for the dark picture, then everything above the resultant black level (A) would indeed be black. Therefore, should the scene become bright, the tube would still remain cut off between the signal voltages corresponding to black levels (A) and (B). Conversely, if the receiver were adjusted for correct level (B), then a change to a dark scene would result in a picture which would be too bright.

Maintenance of the correct black level and the sync. pulse level is also important to the reliable functioning of many a.g.c. and sync. separator circuits, and it is vital in colour receivers.

The circuit in figure 6.6 shows a simple diode *d.c. restoring circuit* particularly popular in valve amplifiers. Without d.c. restoration, the video waveform has positive and negative excursions about the zero d.c. level (X′–X′). When applied to the clamp circuit, negative excursions of the video waveform cause the restoring diode D to conduct, charging the coupling capacitor *C*. The capacitor charges to a voltage which is equal to the peak of the sync. pulse because the discharge time constant $R \times C$ is relatively long and no significant loss of charge occurs between the pulses. The video waveform 'sits' upon this reference potential so that the d.c. level is now included, or restored, in the signal.

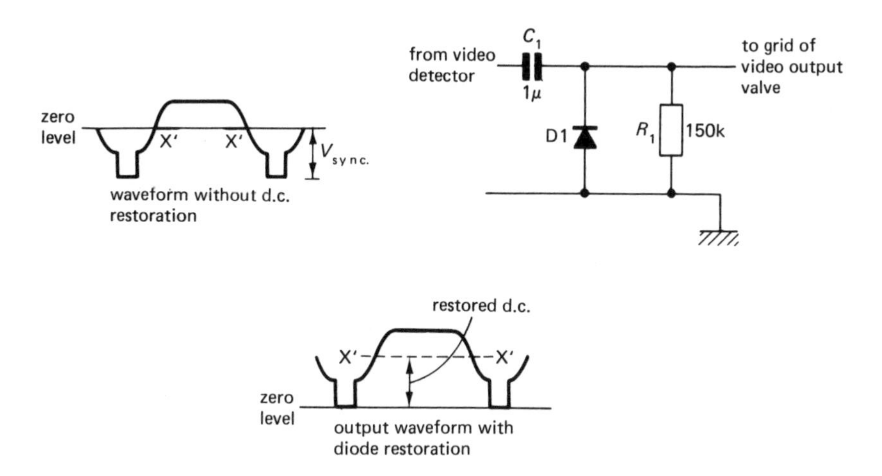

Figure 6.6 Simple diode restorer circuit. D1 clamps the tips of sync. pulses to zero level (chassis), which automatically gives constant black level, and the video d.c. level is restored

Since the d.c. restorer clamps the peaks of the sync. pulses, it follows that the black level of the video waveform would also remain constant. Unfortunately, the simple circuit described above has to adjust itself to any changes in the amplitude of the negative-going part of the waveform. The required adjustment, however, is rather slow because of the long time constant of the diode circuit. Consequently, the actual black level can be affected by variations in signal strength (e.g. when changing channels) or even by the picture content.

In order to ensure a stable reference point for the black level, *driven clamp circuits* are used. In these the video (or chroma) coupling capacitor is charged to a d.c. potential via a special circuit which is switched on by the line flyback pulses. The magnitude of this potential is often governed by the setting of the brightness control. Such a relatively complex arrangement is universally used in colour receivers. It is absolutely essential that the three video drives to the colour tube have stable black level potentials. Any drift in the d.c. bias of one video output amplifier with respect to that of the other two would alter the tube current associated with the particular gun and lead to unwanted colour tinting on the screen. A driven black level clamp effectively 'resets' the d.c. bias of each gun to a fixed level at the beginning of every picture line.

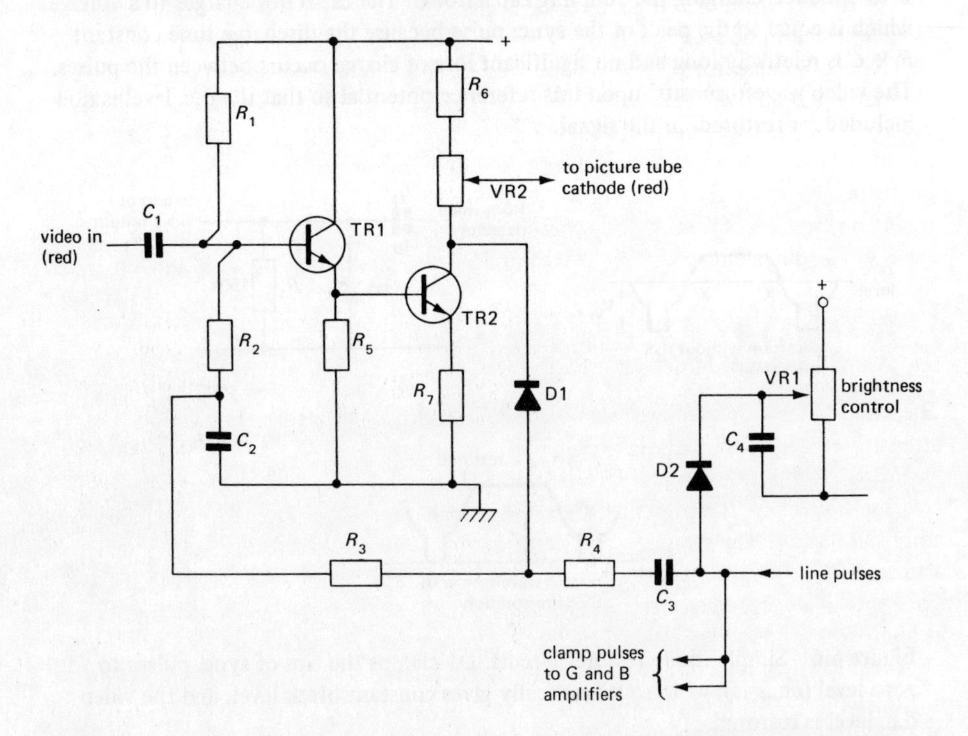

Figure 6.7 Example of a driven black level clamp circuit in the 'red' output stage of a colour receiver (a similar arrangement is repeated in the 'green' and in the 'blue' amplifiers). In this circuit brightness control effectively alters the black level of video waveform. (Decca)

An example of a driven black level clamp is shown in figure 6.7. Transistors TR1 and TR2 form the video output stage driving the red cathode of the tube. The base bias circuit for TR1 is similar to a conventional potential divider formed by R_1 and R_2. The difference is, however, that the lower end of R_2 is not connected directly to the chassis, but to the capacitor C_2. Positive-going line flyback pulses are fed to the circuit via C_3 and R_4; when the magnitude of each pulse is equal to the collector voltage of TR2, the clamp diode D1 conducts and the resulting current flow charges capacitor C_3. Therefore the voltage across C_3 is dependent upon the difference between the constant amplitude line pulse and the actual collector voltage, just as the video waveform is at its black level. During the line scan period the charge on C_3 maintains a relatively fixed potential at the junction of R_2, R_3 and C_2. This provides a reference voltage upon which the picture information is superimposed. The circuit is repeated for each of the remaining two colours, all three arrangements receiving the same clamping pulses.

The effectiveness of the clamp circuit in stabilising the black level at the video output stage can be seen by considering, for example, a fall in the collector voltage of TR2. Now capacitor C_3 in figure 6.7 charges to a higher voltage, owing to the increased potential difference, as outlined above. Consequently, the higher negative voltage on C_3 reduces the collector current of both TR1 and TR2. The resultant rise in the collector voltage of TR2 compensates for the original reduction in that potential.

In some receivers black level clamping is carried out within an I.C. Suitable capacitors, equivalent to C_2 or C_3 in figure 6.7, are then connected externally to the device.

6.6 Amplification of Middle and High Video Frequencies

The middle video frequencies do not present serious problems, and we devote our attention to the amplification of high video frequencies.

Readers will recall that the high frequency gain of an amplifier reduces with increasing frequency, the principal cause being stray capacitance, which shunts the amplifier output. The stray capacitance is unavoidably present in the circuit (see also section 5.1). This problem is worse in valve stages than in transistor amplifiers, since the high resistance of the anode load means that the stray capacitance offers, at high frequencies, a comparatively low reactance path to the chassis.

A technique known as frequency compensation is used to extend the bandwidth of the amplifier at the high frequency end of the spectrum. Two popular methods of applying compensation are: (1) the use of so-called *peaking coils*, which are connected in the output of the video amplifier, and (2) the application of *frequency dependent negative feedback* to the amplifier. Both techniques are used in the amplifier in figure 6.8, and are described below.

Inductor L_1 in figure 6.8 is a peaking coil. At middle and low frequencies the reactance of L_1 is small compared with the resistive load, R_4. At high frequencies the reactance of L_1 increases in value, so that the effective impedance of the anode load is increased; this results in an increase in gain with frequency. At some frequency the inductance of L_1 could be made to resonate with the stray capacitance and give maximum amplifier gain; damping resistor R_3 is used to prevent unwanted oscillations. When connected in the position shown, L_1 is known as a *shunt peaking coil*, since it effectively appears as though it were connected across the signal path. An alternative arrangement is to connect the peaking coil, together with its damping resistor, in series with the signal path. Such a *series peaking coil* could, for example, replace the parallel combination C_2, R_6 in figure 6.8. The action of the series inductor breaks up the stray capacitance of the circuit into two parts, C_{STRAY1} (associated with the valve) and C_{STRAY2} (associated with the picture tube). This effectively reduces the capacitance connected directly to the video output valve and the gain can be increased.

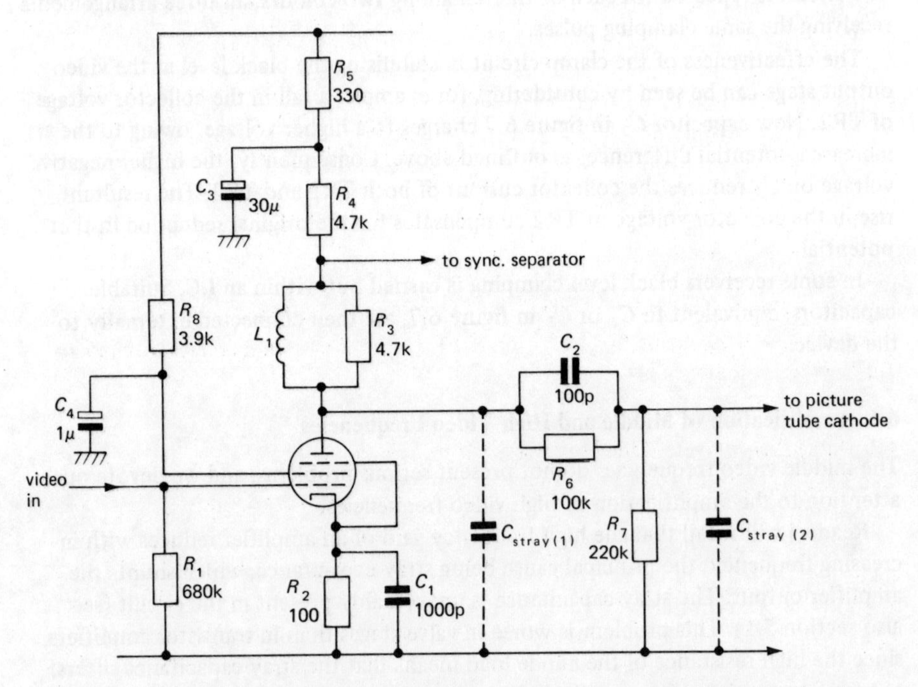

Figure 6.8 Video output stage with high frequency compensation given by the action of L_1, C_1, C_2 and the choice of value for R_4

Frequency dependent negative feedback is applied by means of the cathode resistor R_2 and its bypass capacitor C_1. At middle and low video frequencies the reactance of C_1 is high, so that R_2 is not effectively bypassed. The resultant negative feedback reduces the voltage gain of the amplifier at those frequencies. At high video frequencies the capacitive reactance of C_1 reduces, which gives a reduction in negative feedback, leading to an increase in amplifier gain.

The parallel combination R_6, C_2 also acts as a frequency compensating network. At low frequencies the reactance of C_2 is high, and the video signal is attenuated by the potential divider action of R_6 and R_7. At middle and high frequencies the reactance of C_2 reduces, so that an increasing proportion of the signal voltage is applied to the tube cathode.

It is of interest to point out here why the video output stage has a relatively high power rating. In order to maintain a good frequency response, it is necessary to provide a low circuit resistance—this minimises the effect of stray shunt capacitance. A low value of circuit resistance implies high peak values of current and, when taken with the high voltages involved, means that the power dissipation in the output stage must be rather high. On the other hand, the power taken by the picture tube itself is supplied by the line output stage and not by the video amplifier.

While one attempts to boost the high frequency gain, the low frequency response is sometimes suppressed. To redress the balance some form of low frequency compensation could be used. This is achieved in the circuit in figure 6.8 by means of C_3 and R_5. At middle and high frequencies R_5 is fully decoupled by C_3, so that it does not affect the amplifier gain. At low frequencies the decoupling action of C_3 is ineffective, which places R_5 in series with R_4 to give a high gain.

Practical methods of compensation used in other circuits depend on a number of factors, including the type of picture tube, the characteristics and the layout of the components used, the i.f. response, etc.

6.7 Luminance Stages in Colour Receivers

The luminance signal in colour receivers is the equivalent of the video signal in monochrome receivers. Therefore similar requirements regarding the bandwidth and gain have to be fulfilled in the luminance stages. (A block diagram of a typical luminance stage is shown in figure 6.18.)

As shown in previous chapters, the processing of the signal in a colour receiver takes two separate paths. Finally, however, both the luminance and chrominance are combined to produce the three colour signals. The bandwidth of the chroma signal is initially 2 MHz, reducing to 1 MHz after the detection; but the bandwidth of the luminance path is 5.5 MHz. This difference in the frequency response causes the signal to pass faster through the luminance amplifiers than through the decoder. Although the theoretical proof is beyond the scope of this book, the problem could be considered to be similar to the results of square wave testing of electronic amplifiers. When a square wave is applied to the amplifier input, the resultant output should also be a square wave. Owing to the deficiencies in the amplifier performance, however, this is not always the case. If the response to high frequencies is limited, the leading edge of the output waveform slopes, and the maximum amplitude is reached after a time delay, as shown in figure 6.9. (A similar effect also occurs at the trailing edge of the waveform.) The net result of this type of response in the TV receiver would be that the luminance signal would be applied to the tube a fraction of a second before the corresponding chrominance signal. The visual effect depends upon the size of the screen and on the actual time delay between the signals; it ranges from an impression of poor focus to one of double image as if suffering from 'ghost'. The time difference between the two signals depends on the amplifier design, and typically lies in the range 0.5–0.9 μs (500–900 ns). If the width of the picture is, say, 52 cm (approximately 20 in), and since the duration of one picture line is 52 μs, then the beam moves across the screen by 1 cm in 1 μs. Should the signal delay be 0.8 μs, the display would consist of two 'pictures' which would be 8 mm (approximately $\frac{1}{3}$ in.) apart.

To ensure that the two signals are correctly superimposed, the luminance path contains a *delay line.* The device is in the form of a long coil with copper strips placed between the winding and the coil former. The strips are connected to the chassis to form a distributed capacitance over the whole length of the coil. Such an arrangement is called a *transmission line*; one of its properties is the ability to reduce the speed of signal propagation. Figure 6.9 shows how a delay line, L_1, is connected in circuit; components L_2, L_3, R_1 and R_2 provide the correct terminations to the line. Faults in the terminating components could lead to multiple 'ghosting'.

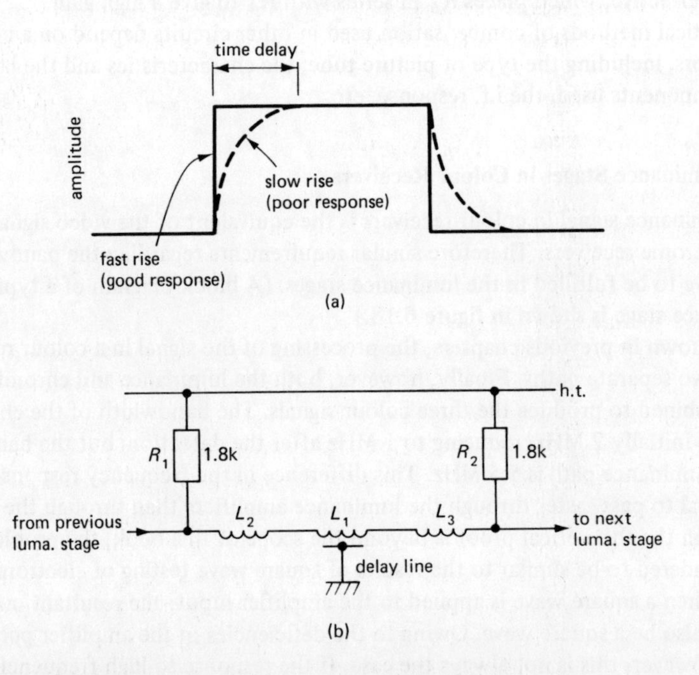

Figure 6.9 Principle of luminance delay line: (a) amplifier response to square waveform (see text); (b) practical circuit

An open circuit in the luminance path results in an extremely poor definition picture–due to the chrominance signal only (assuming that time base synchronisation has not been lost).

The luminance amplifier in a colour receiver also contains a 4.43 MHz *notch filter.* This is a tuned rejector circuit whose purpose is to remove the modulated chrominance information centred around 4.43 MHz. The presence of such a signal in the drive waveform to the picture tube could cause a very fine dot pattern on the screen. The 4.43 MHz signal could also produce desaturation of certain colours in the display. This would occur as a result of the action of the tube which rectifies the unwanted signal and adds it to the existing luminance information. The reader will recall that the only acceptable addition of signals is that of the correct value luminance matrixed with the demodulated colour difference information.

It will be noted that there are a few monochrome receivers equipped with a sub-carrier notch filter—again to prevent the dot interference pattern on the screen. On the other hand, the rectifying and signal adding properties of the tube could be beneficial to a monochrome display by improving the apparent contrast of the picture content.

The presence of the notch filter affects the high video frequency response of the luminance amplifier. It can be seen from the graph in figure 6.10 that the response curve has a notch around 4.43 MHz and a certain loss of fine picture detail is thus unavoidable. Some designs make provisions for disabling the notch filter during a monochrome transmission (when the *colour killer* operates) so as to maintain good picture quality.

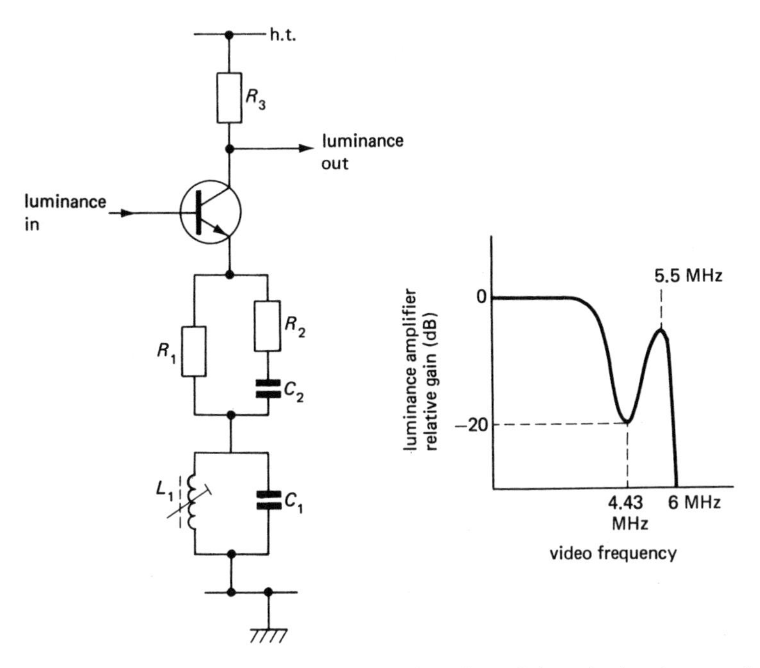

Figure 6.10 Example of 4.43 MHz notch filter (L_1, C_1) in the luminance channel; the graph shows the effect of the filter on frequency response

6.8 R, G, B Drive Circuits

It was mentioned in section 6.1 that the picture tube can be either cathode- or grid-fed. A colour tube has three electron guns, one for each of the primary colours, red (R), green (G) and blue (B). The light output associated with a particular colour is the result of the addition of two signals: the luminance signal and one of the colour difference signals. This addition can take place in an electronic circuit known as the *R, G, B matrix*, or it may take place within the tube itself.

Receivers which use an R, G, B matrix external to the tube are called *R, G, B drive receivers*. In these the three primary colour signals are individually applied to the cathodes of the tube. This method is most suited to transistorised output stages, and it is widely used for two principal reasons. Firstly, the cathode drive voltage needed to control a given beam current is lower than that necessary on the grid of the tube; consequently, the required voltage rating of the output transistors is reduced. Secondly, the low output impedance of the transistor amplifiers, compared with their valve counterparts, allows the full 5.5 MHz bandwidth to be easily obtained.

The alternative method of driving the picture tube is known as the *colour difference drive.* In this arrangement, as discussed in section 6.9, the matrixing of the luminance with the colour difference signals takes place within the tube.

An example of an R, G, B matrix circuit is illustrated in figure 6.11. The three colour difference signals are amplified by TR1, TR2 and TR3, respectively; R_2, R_{11} and R_{13} are the amplifier load resistors. The luminance input signal to the matrix is

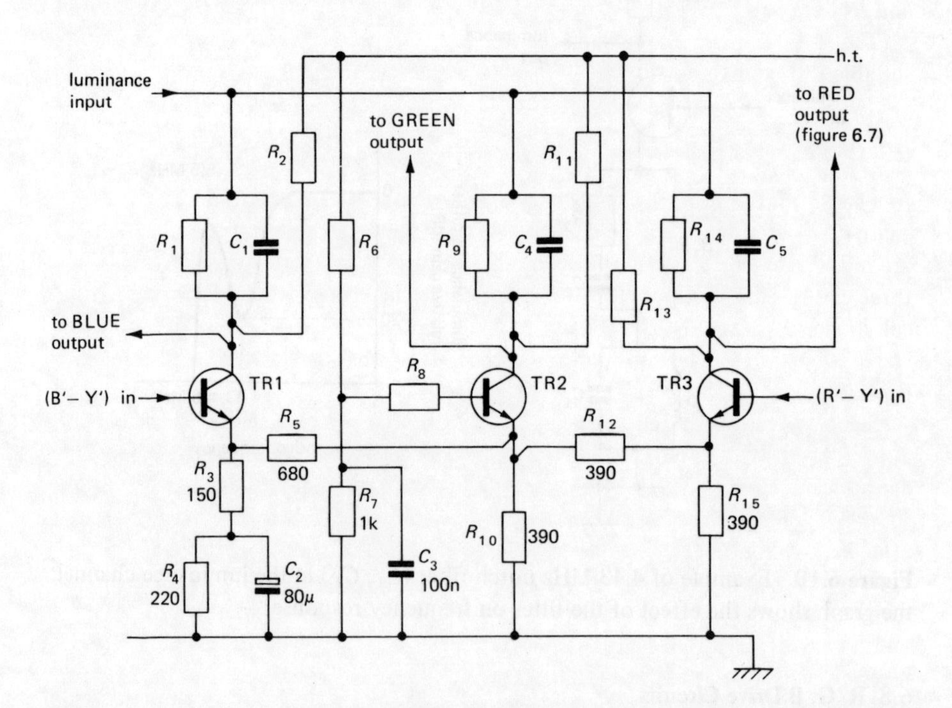

Figure 6.11 (G′ – Y′) and R′, G′, B′ matrix (Decca). (G′ – Y′) is formed at the emitter of TR2 from (B′ – Y′) and (R′ – Y′) fed from TR1 and TR3, respectively; R′, G′, B′ signals appear at the collectors

fed via frequency compensating networks (R_1C_1, R_9C_4 and $R_{14}C_5$) so that the signal addition takes place at the collectors of the transistors. The resultant three primary colour signals, after further amplification (see figure 6.7), are applied to the cathodes of the tube.

The circuit shown in figure 6.11 includes the *(G – Y) matrix*. The reader will recall that the transmitted chroma signal contains only the (B – Y) and (R – Y) components. The third colour difference signal, (G – Y), is recovered in the receiver by the addition of differing proportions of the other two signals. This addition, or matrixing, is done via R_5 and R_{12}–note the unequal values of resistance which give the required proportions. The (G – Y) signal is then amplified by the common base amplifier containing TR2.

The gain of the colour difference amplifier is also adjusted to provide *deweighting* of the demodulated signals. In order to bring the signal outputs to a common level, the (B – Y) amplifier needs a higher gain compared with the (R – Y) amplifier. This compensates for the unequal *weighting factors* applied at the transmitter. In the circuit in figure 6.11 the deweighting is brought about by including a higher value of unbypassed resistance in the emitter of TR3 than is the case with TR1. The d.c. biasing remains the same for each amplifier; therefore the values of ($R_3 + R_4$) and R_{15} are nearly equal to each other.

The R, G, B matrix, together with the (G – Y) matrix, is often provided in a suitable I.C., an example of which is discussed in chapter 7.

6.9 Colour Difference Drive

In a colour difference drive receiver the picture tube is used to combine, or to matrix, the luminance and colour difference signals. The colour difference signals are fed to the three grids, and the luminance signal is simultaneously applied to all three cathodes. The resultant beam currents are then proportional to the primary colour signal content. Although this technique is widely used with valve output stages, there are also a few designs which employ transistorised colour difference drive circuitry.

Signal voltages required on the grid of the picture tube are about 50 per cent above those found in circuits where cathode drive is used. Consequently, the voltage and the power ratings of transistors needed in a receiver with a colour difference drive would have to be relatively high. The reliability of valves is not seriously affected by such elevated voltages; on the other hand, the design of transistor circuits tends to favour the lower operating potentials. A comparison of the two methods of driving the tube shows that the R, G, B drive requires three full bandwidth (5.5 MHz) output stages at lower voltages, while the colour difference receiver needs one 5.5 MHz output stage and three relatively high voltage amplifiers of only 1 MHz bandwidth.

The colour difference output amplifiers logically belong to the colour decoder, therefore their arrangement is considered in chapter 7. The associated luminance output stage is represented by the diagram in figure 6.12. Basically, it is a video output amplifier, the signal being fed to the tube cathodes via drive controls VR2 and VR3. These adjustments are part of *grey scale tracking*, which is described in chapter 12. Other features of the circuit that are worthy of note are

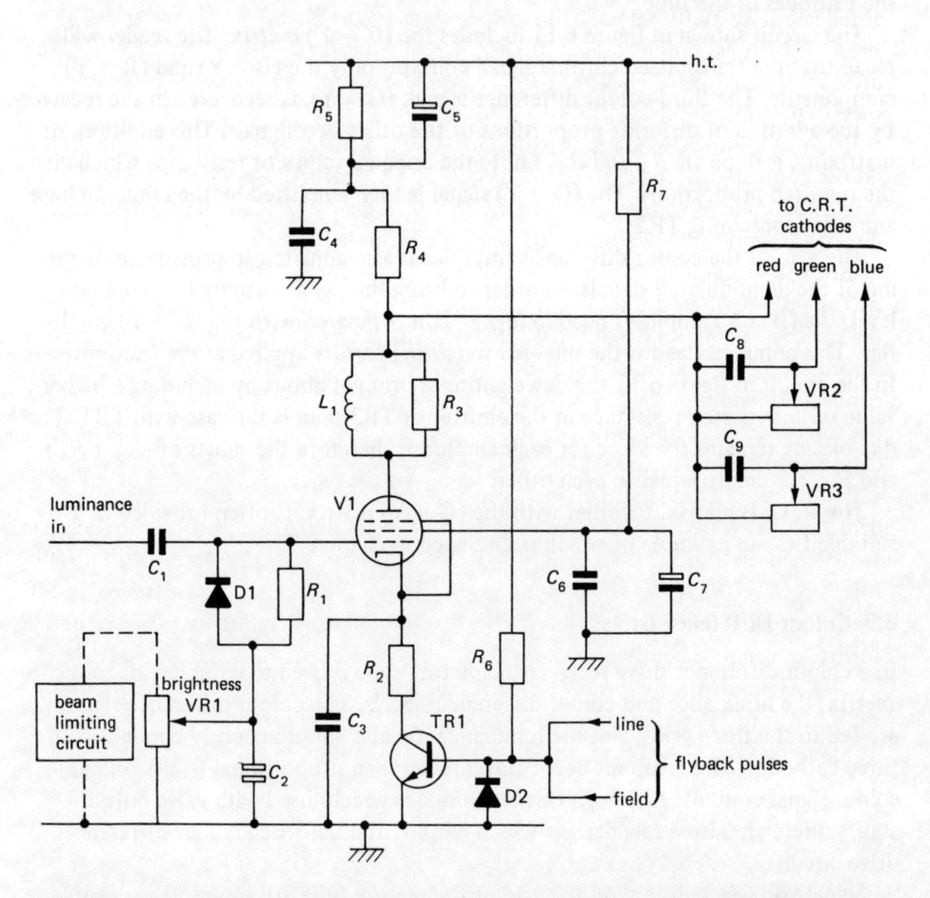

Figure 6.12 Luminance output stage in a colour difference drive receiver. Circuit includes flyback blanking, brightness control, beam limiting, d.c. restoration, high frequency compensation and tube drive controls (part of grey scale tracking arrangement)

(1) the d.c. restoring circuit (C_1, D_1, R_1);
(2) series peaking coil circuit (L_1, R_3);
(3) middle and low frequency compensation (C_4);
(4) very low frequency compensation (C_5);
(5) high frequency compensation (C_3, C_8, C_9).

The action of TR1 and the beam limiting circuit will be discussed later in this chapter.

6.10 Brightness Control

Brightness control effectively changes the d.c. level or d.c. bias applied to the picture tube. The setting of the control is normally such that the screen is dark in the absence of any picture detail (this corresponds to the black level!). Occasionally a preset brightness potentiometer is provided in addition to the viewer's control; the preset is then adjusted according to the makers' instructions, since it often aims at specified voltage levels.

In most monochrome receivers the brightness control alters the d.c. potential on the control grid, the cathode of the tube being left free for video signals. The brightness control voltage can also be applied to the cathode of the tube—either directly, or indirectly via the video amplifying stages.

Sometimes the control circuit features a facility for *switch-off spot suppression*. At the instant when the receiver is switched off, deflection of the beam ceases, but the cathode of the tube is still hot and the e.h.t. is also present owing to the charge stored on the final anode capacitor. These conditions allow the beam current to continue flowing for a while, resulting in a stationary bright spot in the centre of the screen. In time, the phosphor coating of the tube would become damaged and create a dark patch in the middle of the picture.

One form of spot suppression circuit, shown in figure 6.13, has a voltage dependent resistor (VDR) in the 'earthy' end of the brightness control. When the receiver is switched on, normal h.t. is applied to the circuit, and the resistance of the VDR is low, effectively connecting the lower end of R_2 to the chassis. Upon switching off the h.t. decays, the resistance of the VDR rises, and the result is that the brightness control appears to be disconnected from the chassis. That way the remaining positive h.t. is momentarily fed to the grid to increase the beam current and discharge the e.h.t. capacitance before the scanning has finally collapsed.

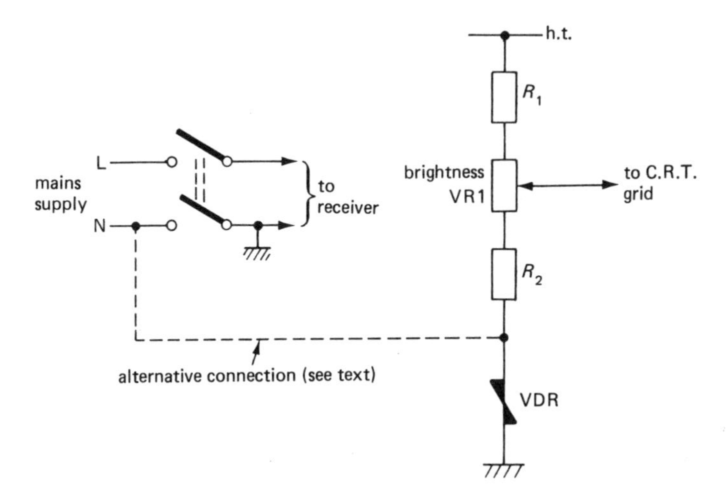

Figure 6.13 Brightness control incorporating switch-off spot suppression

The alternative method of connection, shown by a broken line in figure 6.13, also provides switch-off spot suppression. Instead of using a VDR, the 'earthy' end of R_2 is connected to the neutral on the mains side of the ON/OFF switch. When the receiver is switched on, the brightness control is connected to the chassis via the switch and operates normally. Opening the mains switch breaks the chassis connection and the h.t. is fed to the control grid. This circuit, however, presents a potential hazard; should the mains supply connections be reversed, the inside of the receiver would be live despite a double pole switch!

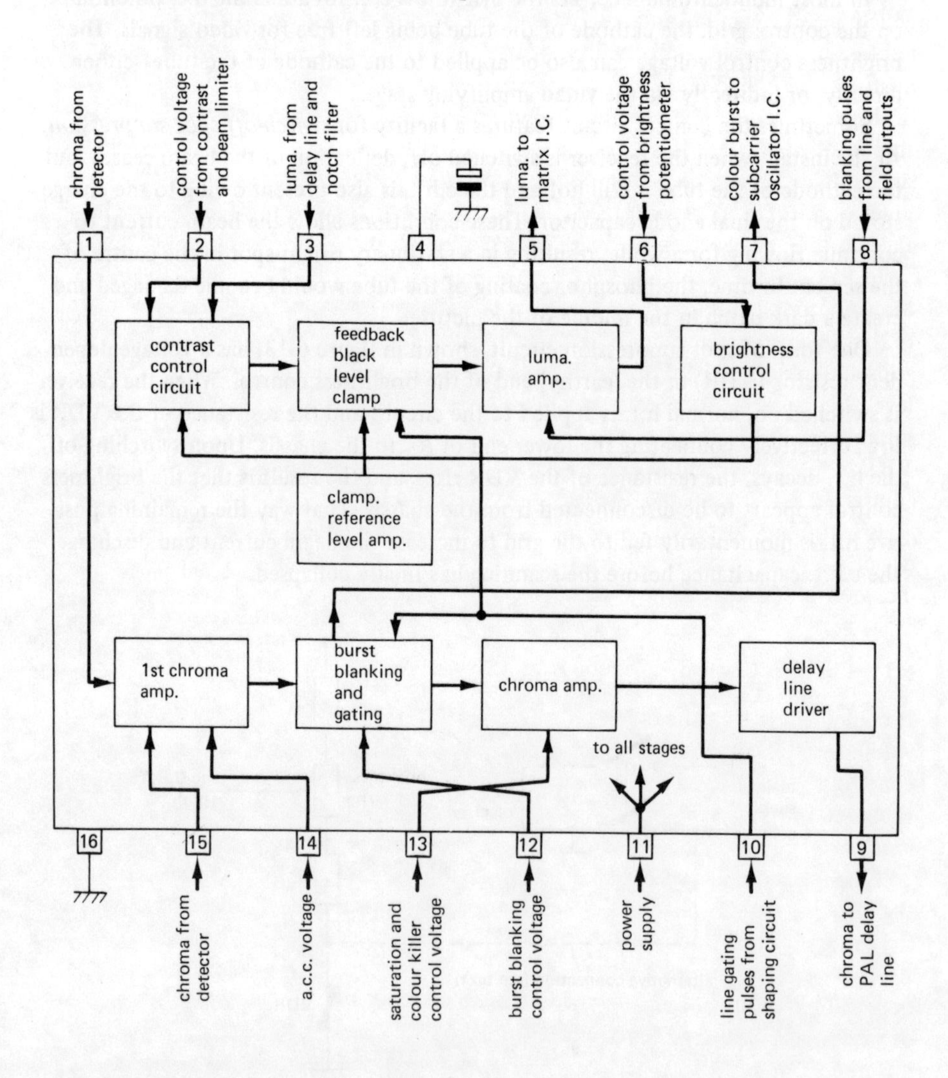

Figure 6.14 Luminance and chrominance I.C., TBA 560 C (Mullard). Functions associated with the chroma path are described in chapter 7

Another method of spot suppression is based on the use of a large value capacitor which normally decouples the slider of the brightness control; on switching off, the capacitor maintains a positive voltage on the grid of the tube.

In some receivers the supply to the brightness control is between the positive h.t. rail and a negative supply. The negative supply can be obtained by rectifying (and smoothing) the negative-going flyback pulses—a VDR can be used for this purpose.

Brightness control in colour receivers can present some problems. The colour tube has three guns whose outputs must be altered equally to prevent mistracking, which, in turn, would cause colour changes. The brightness control can be connected to a black level clamp circuit as shown in figure 6.7. The action of this arrangement was described in section 6.5, where it was explained how the overall d.c. bias of the output stage was controlled by constant amplitude line pulses. In fact, the amplitude of those pulses is governed by the setting of the brightness control VR1, since diode D2 clips their magnitude to the value of the potential appearing at the slider of VR1. (The clipping diode conducts as soon as the amplitude of the pulse exceeds the voltage from the brightness control.)

In R, G, B drive receivers the brightness control can also be connected to the three grids strapped together; the arrangement then functions in a manner similar to that of a monochrome tube circuit. Where colour difference drive is used, the grids are fed with separate signals; therefore brightness could be varied via a black level clamp. Alternatively, the d.c. level of the luminance output amplifier can be suitably altered by the brightness control, as shown in figure 6.12. Potentiometer VR1 adds a d.c. potential at its slider to the d.c. restoring circuits which finally adjusts the d.c. bias of the output valve.

Brightness control can also be carried out in an I.C.; a block diagram of a suitable device is shown in figure 6.14. Again, the picture brightness is varied by altering the d.c. level of the luminance signal.

6.11 Contrast Control

It could be said that the effect of contrast control is the video equivalent of a volume control, its purpose being to vary the amplitude of the video signal. Picture contrast can be altered by controlling the amount of video fed from one stage to the next, or by varying the gain of a video amplifier. Alternatively, contrast control is achieved by changing the gain of the i.f. strip, either directly or via the a.g.c.

Examples of contrast control techniques are shown in figure 6.15. In circuit (a) the video output is collected at the slider of VR1 and applied to the tube. In circuit (b) the amount of video signal fed from TR1 to TR2 is adjustable by means of VR1. An important feature of this arrangement is that the contrast control is connected between two points of equal potential marked x and y in the diagram. The choice of values of R_2, R_3 and the operating point of TR1 ensure that there is no voltage difference along the track of VR1; this prevents any changes in the d.c. biasing of TR2 whenever the contrast control is altered. The reader will recall that variations of this bias would affect the picture brightness.

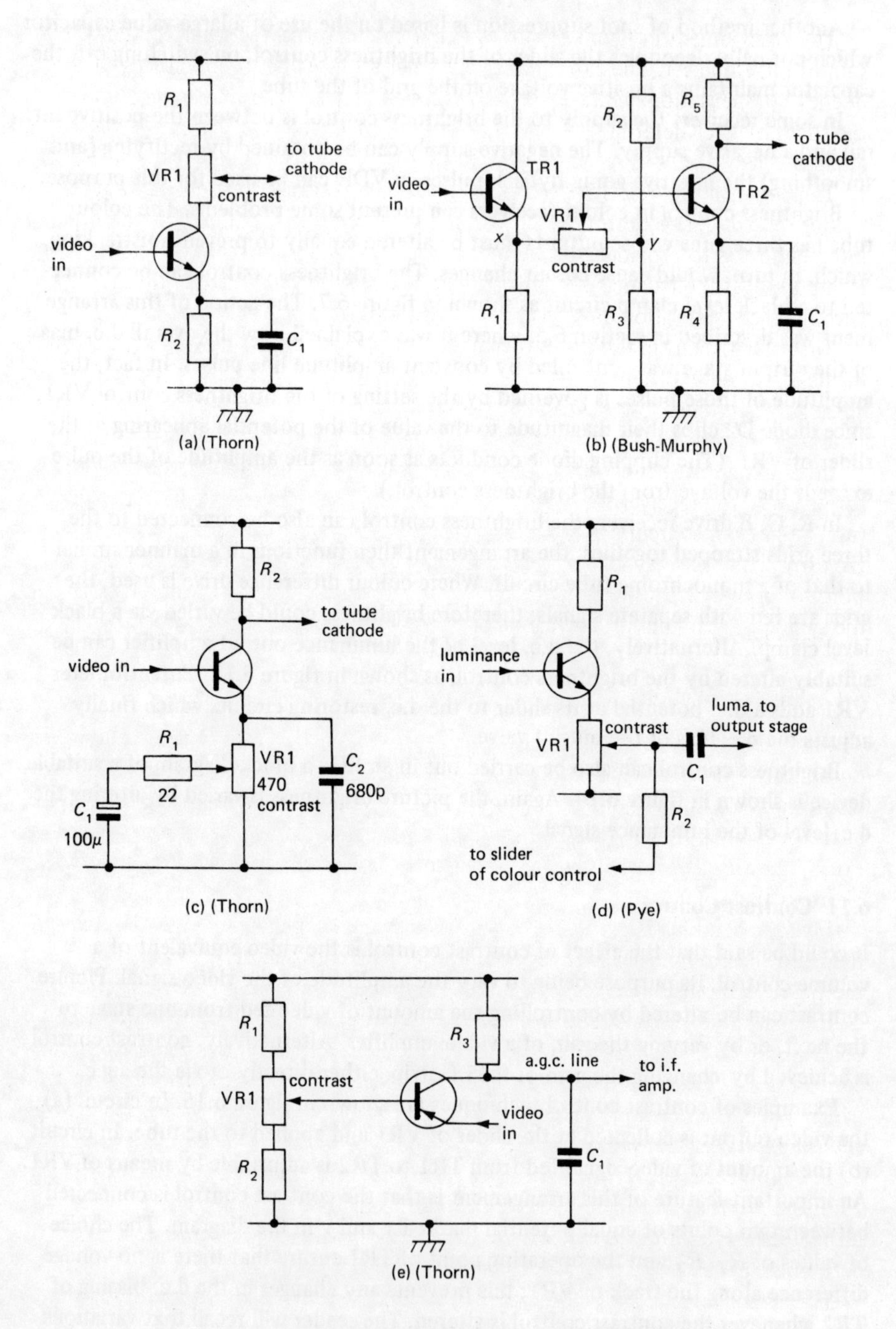

Figure 6.15 Examples of contrast control arrangements

The arrangement in figure 6.15(c) uses variable negative feedback to alter the gain of the amplifier. The capacitance of C_2 has a low value to provide high frequency compensation, while C_1 is a large value bypass capacitor. The position of the slider of VR1 determines the amount of negative feedback within the stage and alters the gain.

The circuit in figure 6.15(d) shows a method of contrast control found in a colour receiver. Here both the contrast and the colour (saturation) controls are interconnected to provide a degree of tracking between the two adjustments. This is important if the balance between the luminance (contrast control) and the chrominance (saturation control) is to be maintained.

The arrangement in figure 6.15(e) achieves contrast control by altering the bias of the a.g.c. detector/amplifier transistor, which, in turn, varies the voltage on the a.g.c. line, thus changing the gain of the i.f. strip. Contrast control can also be incorporated in an I.C., as shown in figure 6.14.

Incorrect adjustment of the contrast control can have serious effects on the performance of the receiver. Insufficient contrast can produce a weak and noisy picture, and in some circuits it can cause no colour as well as poor synchronisation. Excessive contrast may lead to overloading of the video stages, again with a possibility of poor synchronisation. The latter would arise if the signal were so large that the sync. pulses could not be accommodated within the amplifier output.

6.12 Flyback Blanking

The purpose of flyback blanking is to cut off the electron beam of the picture tube during each line and field flyback interval. If this was not done, bright flyback lines would spoil the display on the screen. In fact, the presence of *Teletext* information at the beginning of each field can affect the brightness level at the top of the screen. For that reason the field flyback blanking period is often extended to include the Teletext time.

Flyback blanking is performed by flyback pulses derived from both the line and field time bases. In order that they may be applied either directly or indirectly to the tube, such pulses must be of a correct amplitude, suitable polarity and duration.

In most monochrome receivers the suppression pulses are communicated to the control grid of the C.R.T. so as to override momentarily the setting of the brightness control. Apart from the pulses being passed through a resistor–capacitor network which offers the necessary shaping and timing, a blanking transistor may be used to act as a switch and/or pulse amplifier.

In colour receivers flyback blanking pulses are frequenctly fed to the brightness control circuit; again they overcome the setting of the control and the tube is ultimately driven towards cut-off. Similar problems to those discussed in section 6.10 on brightness control apply to the methods of flyback blanking.

The arrangement shown in figure 6.12 causes the luminance output valve to be switched off by the blanking pulses. Transistor TR1 in the cathode circuit of the valve is normally biased into saturation via R_6, effectively connecting the lower end of R_2 to the chassis. Negative-going flyback pulses are applied to the base of TR1 so that the cathode circuit of V1 is briefly interrupted. This causes the anode voltage of the valve to rise to h.t., and the resultant high positive potential communicated to the cathode of the tube cuts the beam off. Diode D2 protects the transistor base–emitter junction against negative-going pulses.

Flyback blanking can be also carried out inside an I.C., as shown in figure 6.14.

6.13 Beam Limiting

Beam limiting is used in TV receivers to restrict the magnitude of the picture tube current to a specified value. Some method of restricting the beam current is used in most, if not in all, colour receivers and it may be occasionally employed in monochrome receivers. An excessive tube current can overload the line output stage and cause a reduction in e.h.t. which, especially in colour receivers, could spoil the quality of the picture (poor convergence, purity and focus). A considerable increase in beam current might cause overheating of the shadow mask which would lead to serious and permanent colour purity errors. Since the tube current has to be supplied by the line output stage via the e.h.t., the overload condition may also damage the e.h.t. rectifiers.

An excessive beam current could be caused by a fault in the circuits supplying the tube; alternatively it may occur on the peak whites, especially if the brightness, contrast or black level clamps were incorrectly adjusted. The limiting circuit is designed to sense the magnitude of the beam current and then develop a control signal which reduces either the brightness or the contrast, or lowers the power supply voltage.

Examples of beam limiting circuits are given in figure 6.16. In circuit (a) the beam current is sensed by transistor TR1, whose base potential is governed by the voltage drop across a low value resistor, R_4. This resistor is in the cathode circuit of the line output valve, V1, as the line output stage is also responsible for the current taken by the picture tube. Transistor TR1 is normally cut off by the positive bias on its emitter from VR2; therefore the brightness control circuit is not affected. An excessive beam current causes the cathode voltage of V1 to rise above the emitter potential of TR1 so that the transistor turns on. The resultant conduction of TR1 lowers the voltage at the brightness control (point X at VR1), effectively reducing the picture brilliance. Capacitor C_1 bypasses the high frequency line scan current.

An alternative method of beam current limiting is shown in figure 6.16(b) and it is used in a fully transistorised, R, G, B drive receiver. The beam current is sensed in the 'earthy' lead from the e.h.t. overwind, since the path of this current is from the e.h.t. winding, then through the e.h.t. rectifier and the picture tube to the chassis. The return path from the chassis to the winding is via diodes D1 and D2; it would appear from the above direction of the current flow that D1 conducts in the reverse direction. In fact, D1 acts as a switch being turned on by the forward bias via R_1 from the positive h.t. rail. It can now be seen that two opposing currents flow

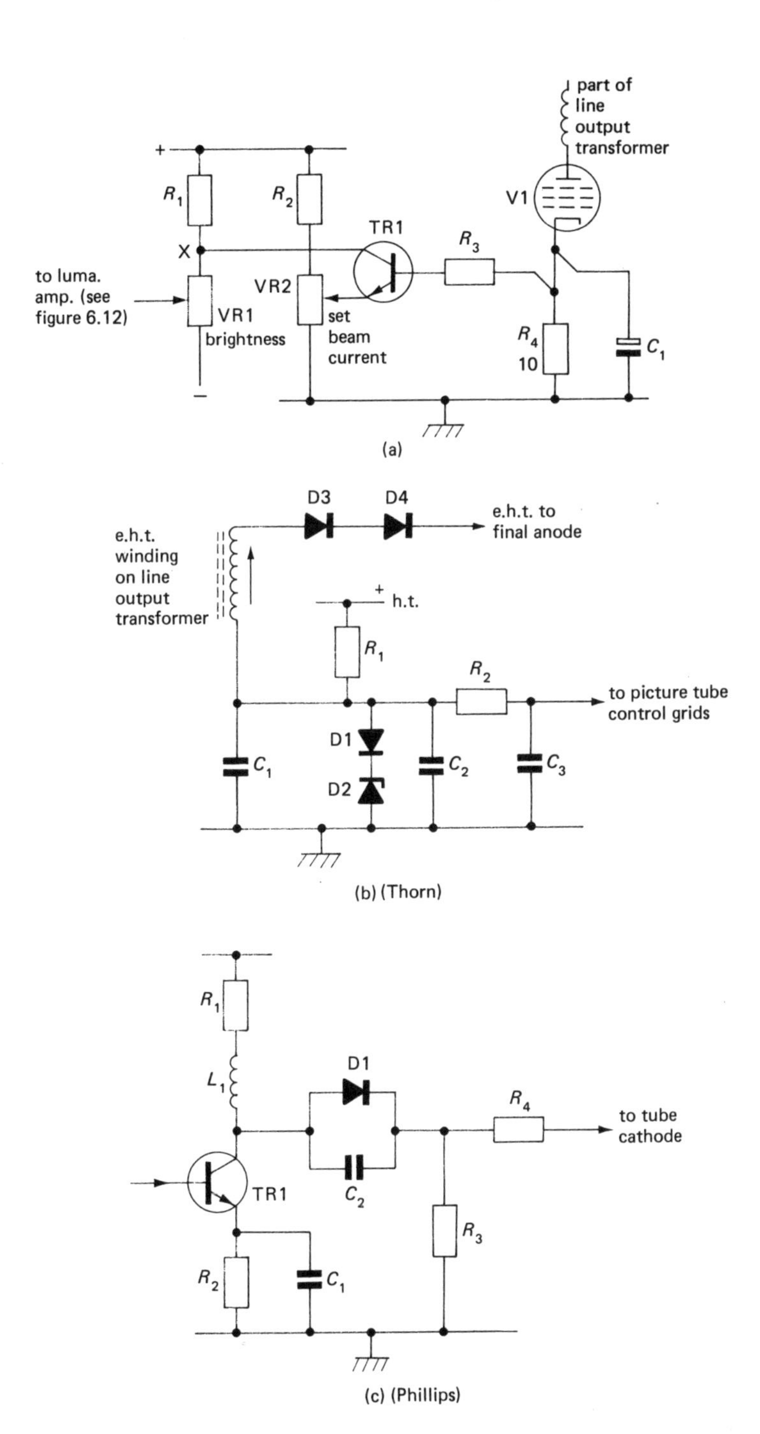

Figure 6.16 Examples of beam limiting arrangements

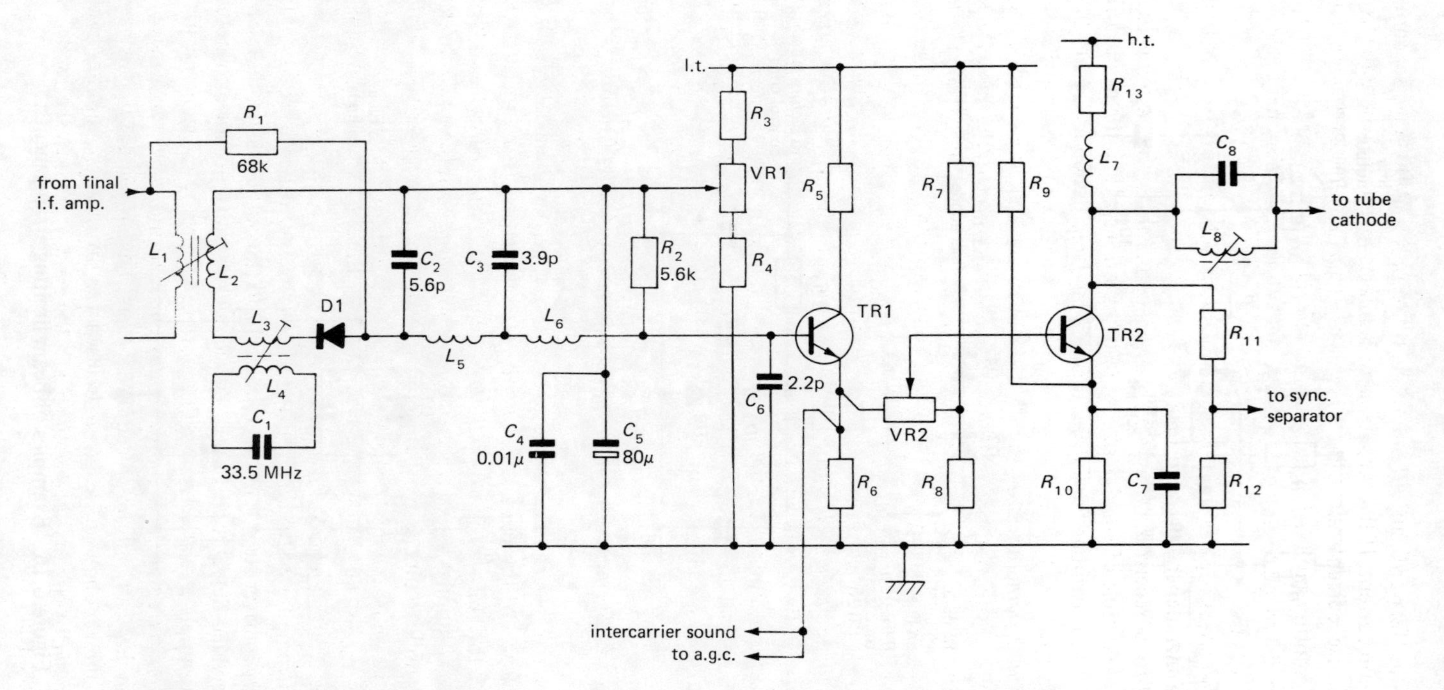

Figure 6.17 A complete monochrome video detector/amplifier circuit (Rank)

through D1—the 'forward' current due to the h.t. supply and the 'reverse' current due to the beam current. If the tube demand increases to a point where the 'forward' and 'reverse' currents through D1 are nearly equal in value, the diode resistance rises, preventing any further increase in the beam current. The circuit responds to a change in the average value of the beam current, its a.c. component is bypassed by C_2 and C_3.

In colour receivers the beam limiting circuit restricts the tube current to between 1 mA and 1.5 mA, depending upon the size of the screen.

In monochrome receivers a simple form of beam limiting circuit is sometimes used, a typical example of which is shown in figure 6.16(c). Here the beam current flows from the tube cathode via R_4 and the parallel combination of R_3 and the video output stage (TR1). When the voltage at the collector of TR1 is higher than the potential at the junction of R_3 and R_4, the beam limiting diode D1 is biased ON and the circuit functions normally. An increase in beam current raises the voltage at the cathode of D1 and causes the diode to turn off. The tube is then left with its own cathode bias developed across R_3 and R_4, which should restrict the current; the video waveform can still be fed to the tube via C_2. This circuit can also limit the beam current under conditions of excessive tube drive caused by the video output stage. In particular, a fault in the amplifier which results in a very low voltage at the collector of TR1 would otherwise lead to a considerable beam current. However, under these circumstances the anode voltage of the limiting diode falls below the cathode potential and D1 turns off. The beam current is again limited by the tube self-bias developed by R_3 and R_4.

6.14 Typical Arrangements of Video and Luminance Amplifier Circuits

In monochrome receivers which use a valve output stage the diode detector is a.c. coupled to the grid of the valve, usually with d.c. restoration at this point. A pentode is used as the video output amplifier, which, apart from feeding the cathode of the tube, also supplies the sync. separator. Frequently both the a.g.c. and the intercarrier sound circuits derive their input from the video output stage. Another popular alternative is to feed the signal to the sound section from the video detector. In either case a 6 MHz intercarrier sound trap is provided in the video path to prevent pattern interference on the picture.

In monochrome transistor receivers the video section consists of two amplifying stages—namely a driver stage (emitter follower) and the output stage (common emitter). The video detector is directly coupled to the emitter follower, whose function is to supply a relatively large base current needed by the video output transistor. The driver stage offers a high input impedance to prevent excessive loading of the detector and the preceding i.f. stage. The low output impedance of the emitter follower allows this stage to be used as a distribution amplifier in order to feed a number of circuits such as the a.g.c., the sync. separator, the sound stages and the video output itself.

A practical version of a monochrome video detector and amplifier circuit is shown in figure 6.17. From earlier description the reader will recognise the functions of many components in the arrangement, but some additional features need the following explanation. Turning to the video detector first: diode D1 produces a negative polarity output voltage, although its video signal content is positive-going (see section 6.1). Transformer L_1, L_2 is the final i.f. transformer; L_3, L_4 is a trap to attenuate the sound carrier as part of the i.f. response shaping system. The i.f. filter after the detector is formed by L_5, L_6, C_2, C_3 and C_6. The video signal is developed across the load resistor R_2, whose chassis return connection cannot be made directly, as it would upset the d.c. biasing of TR1. Instead the upper end of R_2 is decoupled via C_4 and C_5; two capacitors are used–a high value electrolytic capacitor to bypass the low and middle frequencies and a relatively low value C_4 to decouple at high frequencies. (Electrolytic capacitors, by virtue of their construction, are not very effective at high frequencies.)

The emitter follower in figure 6.17 is biased via R_3, VR1, R_4; the preset control VR1 must be carefully adjusted, since it affects the operating point of the entire video section (d.c. coupling throughout!). An incorrect setting would clip either the peak whites ('*white crushing*') or the sync. pulses. As the intercarrier sound is taken off at the emitter follower stage (TR1), the 6 MHz signal is filtered out by C_8, L_8 in the feed circuit to the tube.

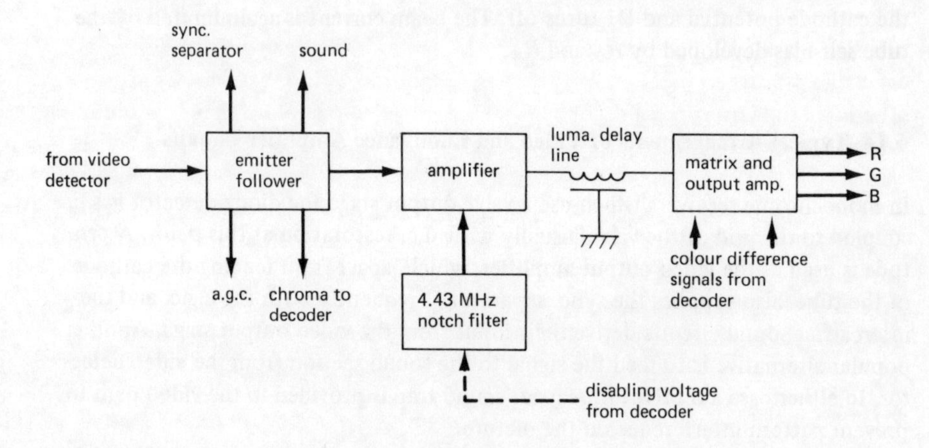

Figure 6.18 Block diagram of the luminance stages of a transistorised receiver (R, G, B drive)

Where integrated circuits are used, a considerable amount of signal processing is done within the I.C. itself, depending, of course, on the type of device. It is usual, however, to find the video output stage as a separate transistor amplifier because of the relatively high voltages required for driving the tube.

In colour receivers the luminance channel follows the pattern outlined for monochrome receivers unless various specialised I.C.s are used, which are described in chapter 7. For quick reference a block diagram of a luminance channel is shown in figure 6.18.

6.15 Automatic Gain Control (A.G.C.)–Basic Principles

The chief purpose of a.g.c. is to maintain a relatively constant picture contrast irrespective of changes in the signal propagating conditions, programme selection, etc. (Many receivers have a contrast control in the form of a preset adjustment not normally accessible to the viewer.) An efficient a.g.c. is also essential in order to maintain a constant level of synchronising pulses, which ensures reliable triggering of the receiver time bases.

The a.g.c. system is based on a feedback arrangement which includes within its loop the r.f., the i.f. and the video amplifier stages of the receiver. The system develops a d.c. control voltage which is dependent upon the amplitude of the detected video waveform. The resultant control signal is applied to the i.f. amplifiers (and often to the tuner r.f. amplifier), causing the gain to be adjusted accordingly.

To ensure satisfactory operation, the a.g.c. must respond to the strength of the received signal. This is relatively easy to do in radio receivers, where the d.c. level of the detected signal is simply a measure of its strength. However, in the case of television, the d.c. level (average value) of the detected video depends not only on the signal strength, but also on the picture content. Despite this limitation, *mean level a.g.c.* has been used, especially in receivers designed for 405-line operation. The more modern type of a.g.c. derives its control voltage from the peak amplitude of the detected video; for this reason there can either be a *peak level a.g.c.* or its variant, known as the *gated (keyed) a.g.c.* system.

The design of the a.g.c. controlled i.f. (or r.f.) stage is such that its gain varies with the applied base bias. In most applications the gain decreases with an increased forward bias, and the system is known as the *forward a.g.c.* A few circuits employ *reverse a.g.c.*, where the gain also decreases, but the base bias must then be reduced. Both effects are discussed later in this chapter.

6.16 Mean Level A.G.C.

The principle of mean level a.g.c. is shown in figure 6.19. The video input signal to the circuit may be derived either from the sync. separator or the emitter follower or from the vision detector. As the input signal varies about its d.c. level, capacitor C charges to the mean level of the video waveform. The time constant of the circuit, $R \times C$, is relatively long, so that the voltage on the capacitor is the average of many picture fields. (The value of C is in the order of several μF, while that of R is a few kΩ).

This type of circuit has several disadvantages. Firstly, it is affected by long term variations in the transmitted d.c. level, as could occur, for example, when the scene

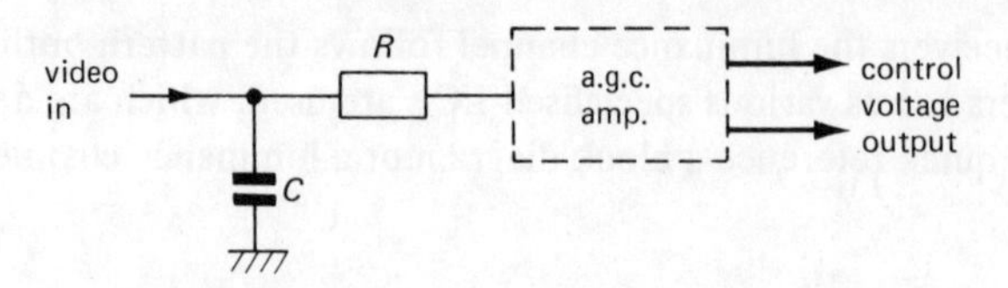

Figure 6.19 Principle of mean level a.g.c. Capacitor charges to the average (mean) value of video waveform; resultant voltage may be amplified before being fed to the i.f. strip

illumination changes from dark to bright. Secondly, the system is slow-acting, since the control voltage is averaged over a number of fields in order to reduce the effect of fluctuations in picture content. Thirdly, the mean level a.g.c. is prone to incorrect action known as 'blocking', or 'lock-out', which could occur if the receiver tuning is switched from a weak signal to a strong one. Because of its slow response, the a.g.c. allows a very powerful signal to be fed to the video amplifier, driving it into saturation. The outcome of a 'lock-out' is a poor quality picture with a possible loss of synchronisation. Finally, mean level a.g.c. has proved to be unsatisfactory with negative picture modulation and intercarrier sound. This is because the a.g.c. 'sees' bright picture content as though it were brought about by insufficient signal strength owing to the reduced d.c. level (see figure 6.5). Consequently, the gain increases, with the possibility of the video signal breaking into the sound signal.

6.17 Peak Level A.G.C.

In peak level a.g.c. a capacitor is charged to the peak value of the video signal. This voltage corresponds to the tip of the synchronising pulse, which, in turn, coincides with the maximum amplitude of the vision carrier. As the above reasoning implies, this type of a.g.c. responds to the signal strength and is not affected by the picture content. The time constant of the capacitor charging circuit can be made relatively short, leading to a fast response to changes in signal strength–for example, aircraft 'flutter' can be prevented.

Two examples of peak a.g.c. circuit are shown in figure 6.20. In circuit (a) diode D1 is the *a.g.c. detector*, which is biased so that it conducts during the most negative part of the video waveform–namely during the sync. pulse period. At the same time the base bias resistor, R_2, of the transistor provides sufficient current to drive TR1 into saturation. Owing to the connections of diode D1, polarity of the 'detected' signal stored on C_1 causes the transistor to come out of saturation. Consequently, the collector voltage of TR1 is low in the absence of any signal, but it increases with the rise in the amplitude of the received signal. The output from TR1 is smoothed by the filter C_3, R_5, C_4 and is used as the bias voltage for one of the i.f. amplifiers. The point of conduction of D1 and TR1, and hence, the gain of the i.f. strip, can be varied by means of a separate d.c. voltage derived from the contrast control.

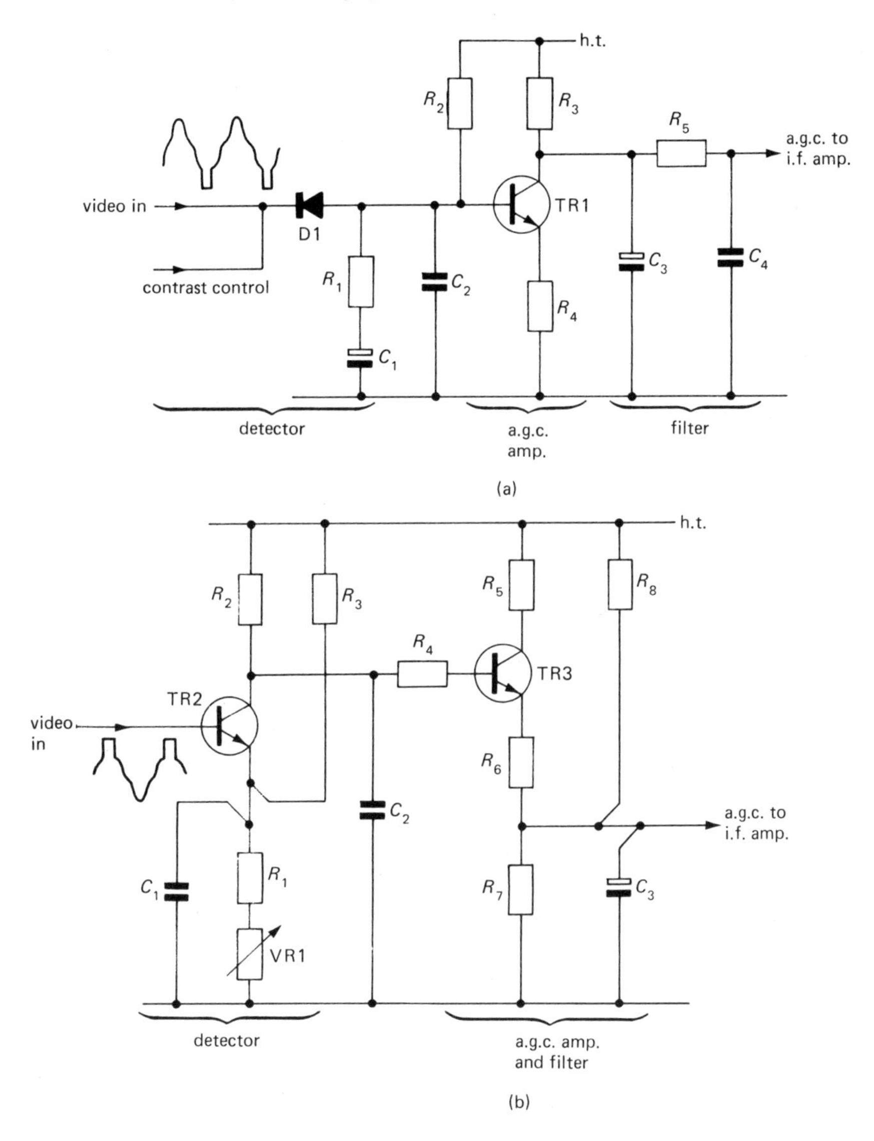

Figure 6.20 Peak level a.g.c. circuits: (a) a.g.c. diode detector and amplifier circuit (GEC); (b) a.g.c. transistor detector and amplifier circuit (Decca)

The circuit in figure 6.20(b) uses transistor TR2 as the a.g.c. detector. Unlike the preceding arrangement, in the absence of any signal TR2 would be biased off by means of R_3 connected between the emitter and the positive supply rail. When the video waveform is applied to the base of TR2, the positive-going sync. pulses overcome the initial reverse bias and the transistor conducts. The output from the detector is fed to the a.g.c amplifier circuit, TR3. This is a p–n–p transistor which conducts whenever TR2 is made conductive by the sync. pulses. Hence, an increase in the value of the video input signal causes both transistors to pass more current, and the a.g.c. voltage increases in value. The control voltage is smoothed by C_3 and is 'clamped' to a minimum level (when TR3 is not conducting) by the potential divider R_8, R_7. Similarly, the maximum voltage is also restricted by the potential divider of R_6, R_7. Contrast control is provided in this circuit by means of VR1, which, in conjunction with R_1 and R_3, determines the point of conduction of TR2.

6.18 Gated A.G.C.

Gated, or keyed, a.g.c. is a variant of the peak level a.g.c. system. The chief disadvantage of peak level a.g.c. is that it can be affected by the field synchronising

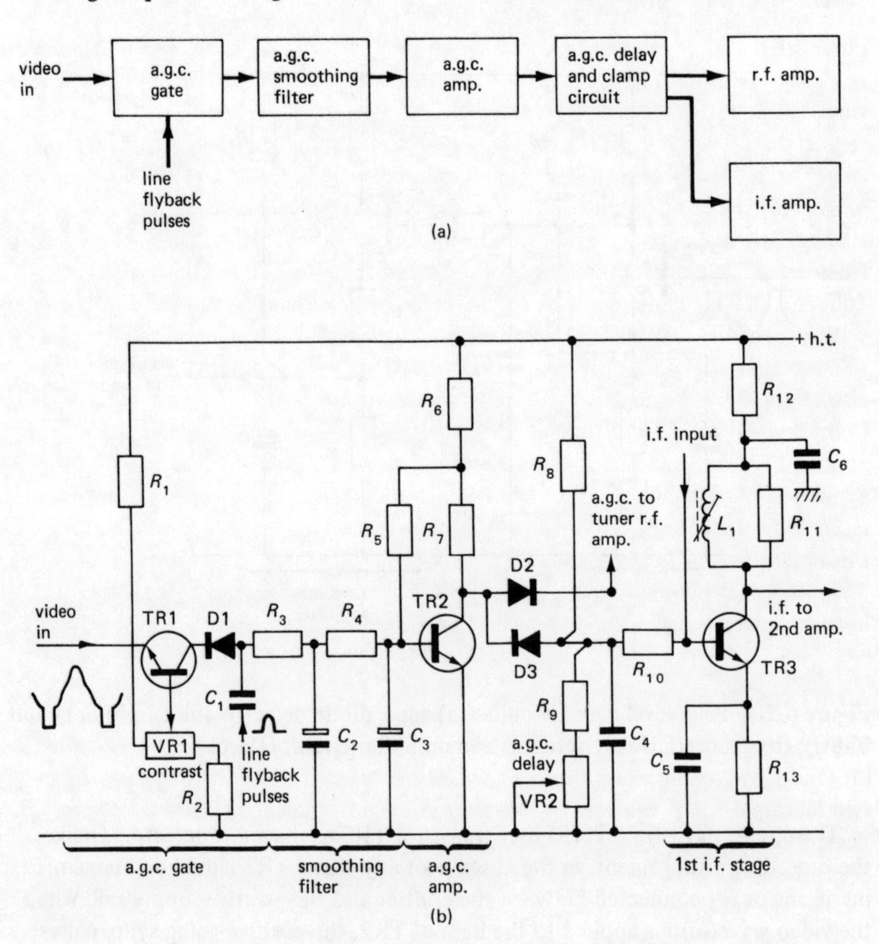

Figure 6.21 Gated a.g.c. system: (a) block diagram; (b) circuit diagram (Bush)

sequence and by interference pulses. Such pulses extend into the sync. pulse level (negative picture modulation) and the control system might respond to their presence as if a strong signal were being received. The long time constant of the a.g.c. detector and filter helps in reducing the effect of unwanted signal; alternatively, low value filter capacitors can be used to remove interference from the a.g.c. detector-amplifier circuit.

Gated a.g.c. is used to overcome the above difficulties while maintaining fast speed of response to any changes in signal strength. The system is designed to sample the amplitude of the video waveform at specified intervals. With negative modulation the sampling points occur in the sync. pulse period, which, as noted earlier, corresponds to the peak of the vision carrier and, consequently, reflects the signal strength.

A block diagram of gated a.g.c. is shown in figure 6.21(a). The a.g.c. gate is biased ON by the line flyback pulses, so that it conducts only during the line sync. pulse interval. The output from the gate is therefore proportional to the magnitude of the synchronising pulses. Since the gate is open for only 15 per cent of the total time, the possibility of interference breakthrough is also reduced. A filter is still required, but, as the ripple is now at the line frequency, smaller values of capacitance can be used, thus increasing the speed of system response.

A gated a.g.c. circuit is given in figure 6.21(b), in which TR1 is the gated transistor. This device will not conduct until a positive-going flyback pulse is applied from a winding on the line output transformer via C_1 and D1. The pulse at the collector of TR1 coincides with the negative-going, sync. pulse portion of the video waveform at the emitter. Therefore the transistor conducts, allowing C_2 to charge to a voltage dependent upon the magnitude of the sync. pulses; this potential is also affected by the amount of base bias given by VR1—a preset contrast control. Components R_4 and C_3 form a smoothing filter which, in conjunction with R_5, applies d.c. bias to the a.g.c. amplifier transistor TR2. The a.g.c. voltage appears at the collector of TR2 and is fed to the tuner r.f. amplifier via the *delay diode* D2, and to the first i.f. stage (TR3) via the *clamp diode* D3. The operation of this circuit is considered further in section 6.20.

Because of the relative complexity of the gated a.g.c. system, it is more popular in the receivers employing integrated circuits (see figure 6.4).

6.19 Forward and Reverse A.G.C.

The a.g.c. potential, well smoothed and fully decoupled, is applied to the controlled i.f. (and r.f.) amplifier to alter its gain. The change in amplification is caused by two factors. Firstly, the gain of transistors depends upon the collector current; devices are selected for a.g.c. control stages in which this effect is very pronounced. Secondly, the amplifier gain varies with the collector-to-emitter voltage—the lower the voltage the lower the gain.

In practical circuits both effects are employed together, as shown in figure 6.21(b). In the i.f. stage, resistor R_{12} is fully decoupled to chassis via C_6; when the a.g.c. signal is applied to the base of TR3, the resultant change in collector current alters the p.d. across R_{12} and changes the collector-emitter voltage of TR3. Readers will, of course, note that the bias applied to TR3 consists of a d.c. potential supplied by the resistor chain R_8, R_9, VR2, modified by the a.g.c. signal.

The effect of the a.g.c. voltage on the gain of the controlled stage can be seen from the general curve relating the amplifier gain to the base bias current. The curve is shown in figure 6.22, and it can be noted that there is a region of maximum gain–if the bias is either reduced or increased, the result is a reduction in amplifier gain. When the a.g.c. is applied so that an *increase in bias* causes the gain to fall, it is known as *forward a.g.c. Reverse a.g.c.* exists when a *reduction in bias* reduces the amplifier gain. The reader will observe that the two modes of a.g.c. operation correspond to either side of the maximum gain shown on the curve.

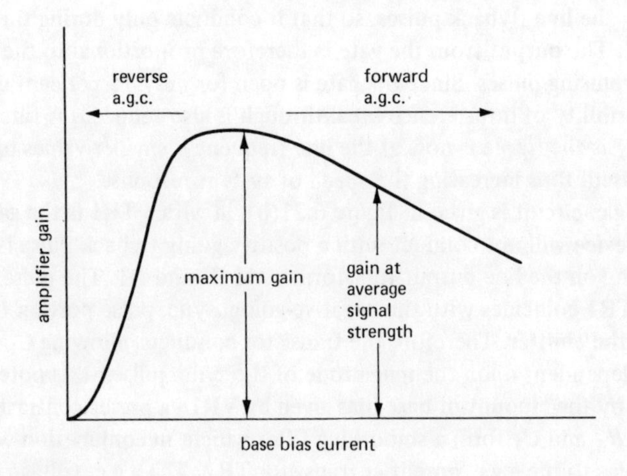

Figure 6.22 Effect of amplifier bias on the stage gain

The most common form is forward a.g.c., as may be anticipated from the shape of the curve in figure 6.22. The gradual and linear reduction in gain to the right of the maximum suggests that the response is proportional to the applied control signal over a wide range.

With reverse a.g.c. the change in gain is very rapid and is also non-linear. This form of control is sometimes used in tuner circuits, as it offers a slight advantage over forward a.g.c. in u.h.f. amplifiers. The curve in figure 6.22 implies that reverse a.g.c. does not produce a wide change in transistor currents and, consequently, the voltages remain relatively constant. In particular, the base-collector potential difference is important, since this junction is reverse biased with the result that there is a base-to-collector capacitance effect (generally similar to the action of a varicap diode). It is, therefore, of some advantage at very high frequencies to maintain this capacitance at a constant value.

To decide which type of a.g.c. is used in a particular circuit, one must consider the effect of a change in the signal strength upon the conduction of the controlled stage. Returning, for example, to the circuit in figure 6.21(b), the reader will observe that an increased signal strength makes the sync. pulses more negative-going at the input to TR1. This causes C_2 to charge to a more negative potential, which, in turn, reduces the conduction of the n–p–n transistor, TR2. As a result, the collector voltage of TR2 rises and so does the forward bias on the i.f. amplifier transistor TR3 (disregarding for the moment the effect of D3 and the associated circuit). Since the object was to *reduce* the gain, but as the control signal *increases* the bias of the i.f. stage, the arrangement represents forward a.g.c. Using a similar approach in the case of circuits in figure 6.20, it can be seen that both examples also provide forward a.g.c.

6.20 Direct and Delayed A.G.C.

It was mentioned previously that the a.g.c. signal is always applied to the i.f. strip, and, in some receivers, it is also fed to the r.f. amplifier in the tuner. In most circuits the i.f. amplifier is connected directly to the a.g.c. line, so that changes in the control voltage are communicated to the given stage. For this reason the a.g.c. to the i.f. strip is referred to as being *direct.*

In practice the range of change of the control voltage must be restricted. The i.f. transistor must not be driven either into saturation on very strong signals or towards the unstable region near the maximum gain (see the curve in figure 6.22). To avoid this possibility, there are provisions for clamping the a.g.c. voltage to a fixed level. Looking at the circuit in figure 6.21(b), it can be seen that a *clamp diode*, D3, is used. Should the signal strength increase to such an extent that the voltage at the collector of TR2 exceeds that at the junction of R_8 and R_9 (see section 6.19), the diode becomes reverse biased. This leaves the bias of TR3 under the control of the potential divider R_8, R_9, VR2, and any further a.g.c. action cannot take place.

A.G.C. applied to the tuner, on the other hand, should not become effective until the full range of i.f. gain control has been exhausted. For this reason tuner a.g.c. is known as *delayed a.g.c.*; the delay in this case is in terms of signal strength. Such an arrangement is desirable, since the tuner mixer/oscillator ought to be presented with the largest signal possible in order to swamp the unwanted noise generated in this stage. Any reduction in the tuner amplification may take place only when the i.f. gain is at its minimum and the signal would still cause overloading (the symptoms being excessive contrast, poor synchronisation, interference patterns on the screen and possibly a buzz on the sound).

In the circuit in figure 6.21(b) the delay is performed by diode D2; normally it is reverse biased because the collector of TR2 is at a lower voltage than that of the base of the r.f. amplifier. (The r.f. amplifier, not shown here, is fed from its own potential divider within the tuner unit.) A very strong signal causes TR2 collector voltage to increase to such an extent that D2 is biased ON. This voltage change is then applied to the base of the relevant transistor in the tuner, reducing its gain accordingly. The delayed a.g.c. ought to operate at the point when the clamp diode, D3, switches off–a suitable adjustment to the a.g.c. crossover is provided by VR2.

In other circuits the delay can be performed by a transistor which is normally biased OFF. When a suitable a.g.c. voltage has been reached, it overcomes the reverse bias and introduces an appropriate degree of tuner control.

Tuner a.g.c. is not always used; instead some form of manual control can be adopted which performs a similar function by altering the base bias of the controlled transistor.

6.21 Fault-Finding in Video Amplifiers and A.G.C. Circuits

The appearance of the display on the screen serves as a good indication of the possible fault area, or the nature of the defect. Therefore, the picture may show such symptoms as smearing (due to instability or to poor low frequency response), lack of fine detail (poor high frequency response), incorrect contrast or brightness, and so on.

A video amplifier fault can also result in a blank, but controllable, raster, suggesting a defect in the path of the video signal (assuming other sections of the receiver are in order). This can sometimes be localised by checking for the presence of sound and sync. pulses with respect to their take-off points in the video stage.

In d.c. coupled video amplifiers (or directly coupled sections of this part of the receiver) a fault can often affect the display by creating either a very bright, uncontrollable raster, or no raster at all. In either case the d.c. voltages of the tube are upset; if the tube is cathode-driven, then a fault which cuts off the video output stage raises the cathode voltage to h.t. and the beam current is also cut off. Conversely, when in the same arrangement the video output voltage falls to a very low value, the beam current is a maximum, and frequently the operation of the brightness control has no effect. The situation is reversed if the video output stage feeds the grid of the tube–a high grid voltage results in a bright screen, while a low voltage cuts the tube off.

Similar remarks apply to colour receivers, except that in many cases it is only one gun which is affected; colour casts are then produced of either a primary or a complementary hue (see chapter 7).

A.G.C. fault-finding can present difficulties because the system is based on a feedback loop. Consequently, a defect anywhere within the loop can produce very similar symptoms, and it is necessary to eliminate those sections which are unlikely to be responsible.

The a.g.c. line voltage can be measured while either tuning to different channels or unplugging the aerial. If a range of readings is obtained, and provided they are within the makers' limits, the basic action of the a.g.c. circuit would appear to be correct.

The a.g.c. voltage can also be clamped to a fixed d.c. level; the correct procedure would be given in the i.f. alignment notes in the service manual. If this clears the fault symptoms, then the a.g.c. detector-amplifier circuit could be responsible.

An oscilloscope is useful, especially when servicing a gated a.g.c. system. Since the waveforms involved are not symmetrical, they have a d.c. level, and a multimeter switched to a d.c. range is useful in measuring the resulting potentials.

Faulty smoothing and decoupling capacitors in the a.g.c. circuit may not necessarily affect the d.c. readings, but they can produce various defects in the display. For example, a rapidly fluctuating picture, or a display broken into wide horizontal bars of varying intensity, could be traced to faulty capacitors.

7 Chrominance Signal Stages

7.1 Processing the Chroma Signal

The chrominance signal forms part of the composite video, from which it is separated in the receiver and amplified before demodulation. It will be recalled (see chapter 2) that the modulated chroma is without its carrier. The relative complexity of the colour decoder is due to the necessity to reinsert the 4.43 MHz subcarrier.

The regenerated subcarrier must be of the correct frequency, and also in the exact phase relationship with the transmitted signal. Owing to the quadrature modulation of the U and V signals, there are two subcarrier feeds which are 90° out of phase with each other, and in the PAL system there is a 180° phase change from one line of the picture to the next. A suitably amplified chroma signal is then detected using the subcarrier to produce the outputs proportional to the original colour difference signals (R′ – Y′) and (B′ – Y′). The third colour difference (G′ – Y′) is formed from the proportions of the other two. Finally, the luminance information, Y′, must be added to the three colour difference signals. The resulting signals produce light outputs from the screen of the picture tube related to the red, green and blue content of the original televised scene.

The reader can refresh his memory by referring to chapter 4 (figure 4.1; also section 4.5), where the block diagram of the decoder was included as part of a complete receiver.

In modern receivers integrated circuits are used to perform most of the decoding, so that it is even more important to appreciate the functions of the individual blocks and the associated waveforms.

The block diagram in figure 7.1 expands the original arrangement shown in figure 4.1; the waveforms are based on a colour bar signal (chapter 3), the amplitudes being expressed in volts, peak-to-peak, which may be applicable to an average receiver.

Figure 7.1 Block diagram of a PAL decoder of a receiver which uses R, G, B drive. Amplitudes of typical colour bar waveforms are measured peak-to-peak; repetition times, or frequencies, also shown as appropriate

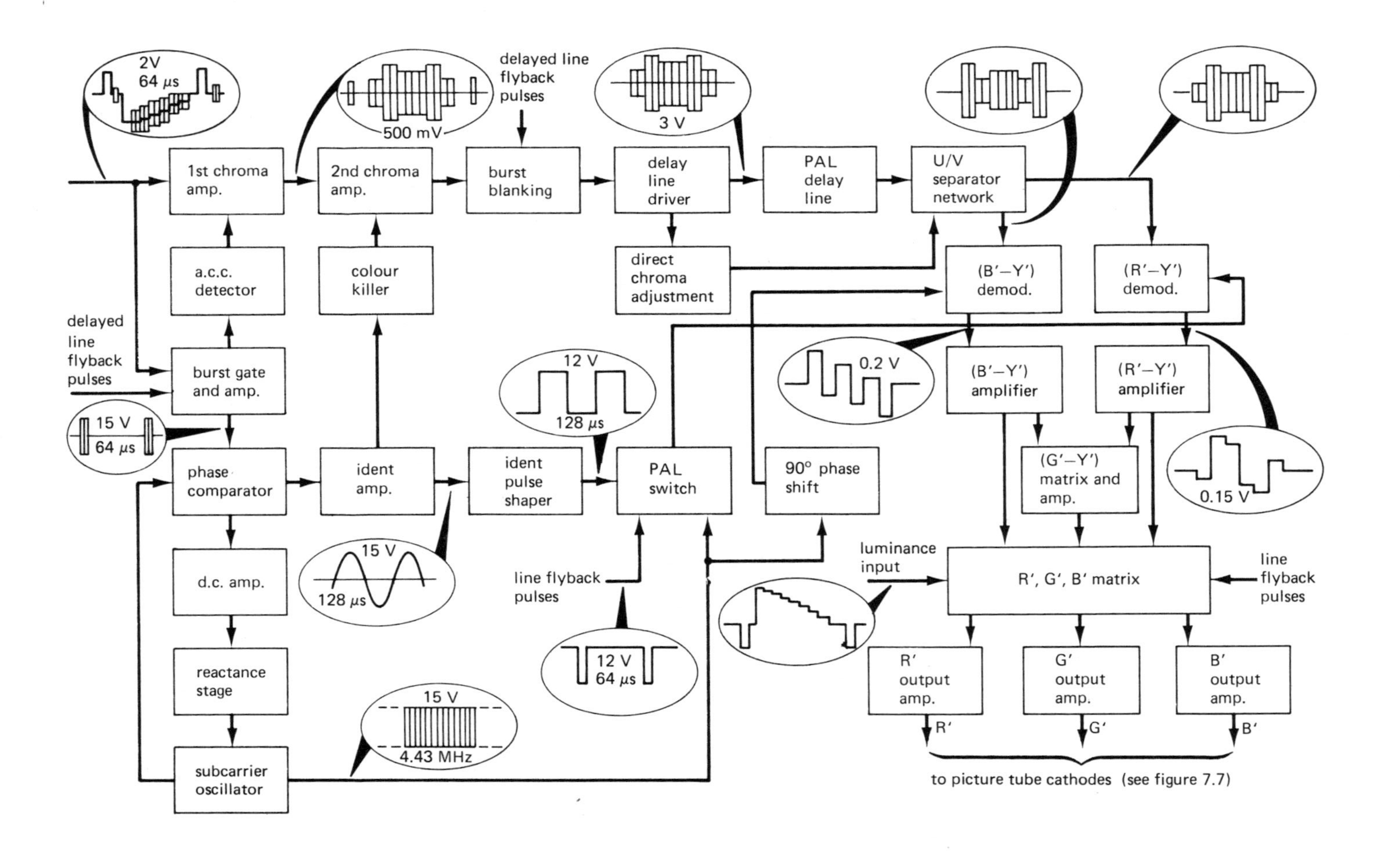
2V
64 μs
1st chroma amp.
500 mV
2nd chroma amp.
delayed line flyback pulses
burst blanking
3 V
delay line driver
PAL delay line
U/V separator network
a.c.c. detector
colour killer
direct chroma adjustment
(B′−Y′) demod.
(R′−Y′) demod.
delayed line flyback pulses
burst gate and amp.
0.2 V
(B′−Y′) amplifier
(R′−Y′) amplifier
12 V
128 μs
15 V
64 μs
phase comparator
ident amp.
ident pulse shaper
PAL switch
90° phase shift
(G′−Y′) matrix and amp.
0.15 V
d.c. amp.
15 V
128 μs
line flyback pulses
luminance input
R′, G′, B′ matrix
line flyback pulses
12 V
64 μs
reactance stage
R′ output amp.
G′ output amp.
B′ output amp.
15 V
4.43 MHz
subcarrier oscillator
R′
G′
B′
to picture tube cathodes (see figure 7.7)

7.2 Chrominance Take-Off Arrangements

The chrominance signal can be separated from the composite video by means of a simple diode, whose non-linear characteristics make it act as a mixer. The output from such a circuit consists of signals whose frequencies result from the subtraction of the signal frequencies at the input. The feed to the colour decoder is normally from the vision i.f. strip—for example, from the last i.f. amplifier or the vision detector circuit. The vision i.f. at 39.5 MHz, when mixed with the chroma sidebands corresponding to 35.07 MHz, generates the required signal centred around 4.43 MHz. The reader will recall (see chapter 5) that there is also one other i.f.—33.5 MHz—from the sound carrier. Therefore this will mix with both the vision i.f. and the chroma i.f. to give rise to frequencies of 6 MHz and 1.57 MHz, respectively. The former is the intercarrier sound signal, and the latter is the chroma/sound beat; both are unwanted in the decoder. In order to select only the 4.43 MHz signal, tuned circuits are used to act as bandpass filters (often in conjunction with rejector circuits resonant at 6 MHz or 33.5 MHz, depending on their position in the signal path).

The chroma take-off point can be in the form of a separate diode circuit which resembles that of the vision detector, its function being to produce the 4.43 MHz signal. It is sometimes referred to as the *chroma detector*, but one must not confuse this with the actual demodulators which will be found at a later point in the decoder.

Since an ordinary vision detector can also produce the chroma beat frequency, it is sometimes possible to find the take-off point early in the luminance channel—e.g. immediately after the detector or at the emitter follower stage.

7.3 Chroma Amplifiers

Before any further processing of the chrominance signal can take place, it is essential to raise its level from, say, 200 mV to about 2 V. This is done in the chrominance amplifiers, which may either be within an integrated circuit or in perhaps two or three cascaded transistor amplifier stages. Each stage normally has a tuned circuit resonant at 4.43 MHz and giving the necessarv overall bandwidth of 2 MHz. To secure the optimum response, correct alignment of this section is important. The exact procedure can only be obtained from the makers' manual, since the response curve here may also be dependent upon the alignment of the vision i.f. strip. The principle, however, follows the outline given in the wobbulator method in chapter 5; the details, however, must be adapted to suit the decoder.

The first chrominance amplifier usually offers variable gain to compensate for the strength of the signal at this point. The d.c. biasing can be altered by means of the *automatic chroma control* (a.c.c.) potential. Its action is similar to the a.g.c. in vision i.f. amplifiers, and the reader is referred to chapter 6 for full discussion of this topic. The derivation of the a.c.c. voltage will be described later, in section 7.7.

In order to prevent coloured 'noise' appearing on the screen, it is desirable to

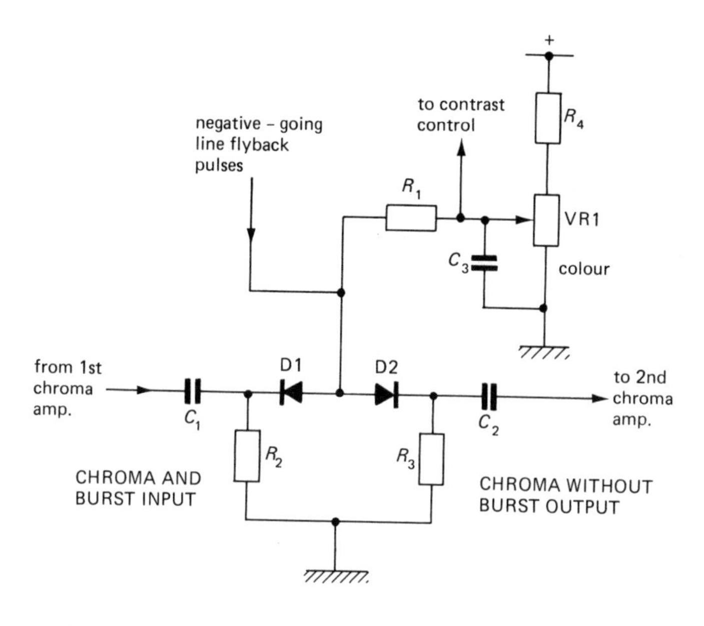

Figure 7.2 Colour control and burst blanking circuit

shut down the chroma path during a monochrome transmission. By removing the d.c. bias from one of the chroma amplifiers, the signal is blocked at this stage. A separate circuit, called the *colour killer*, is responsible for developing a suitable bias only during colour reception. The action of the colour killer also removes the effects of a number of decoder faults from the display, so that the viewer is left with a good monochrome picture.

As mentioned above, the gain of one of the amplifiers is controlled automatically, but the final balancing between the chrominance and the luminance is carried out by means of a viewer's control called COLOUR or SATURATION. In effect, its setting governs the amplitude of the composite chrominance signal, which, in turn, affects the demodulated colour difference signals and, hence, the saturation of the displayed colours. The adjustment is normally subjective: in an otherwise correctly set-up receiver the aim is to achieve a pleasing reproduction of 'skin tones'. The possible circuit arrangements are similar to those of contrast control in video (luminance) amplifiers (see chapter 6). Examples of saturation control methods are given below.

(1) Varying the gain of one of the chroma amplifiers—for example, by altering the a.c.c. bias.
(2) Varying the amount of chroma fed from one stage to the next; a simple variable resistor can be used or, alternatively, a diode attenuator of the form shown in figure 7.2. Diodes D1 and D2 are connected 'back-to-back' with a variable supply

fed to their junction via R_1. Now a diode offers a resistance to the signal, but its value depends upon the actual amount of applied d.c. bias–low bias current means high resistance, and vice versa. This effect is due to the non-linear characteristics of the diode. Unfortunately, the non-linearity could also distort the signal; that is why there are two diodes connected in opposition to cancel out the distortion. This circuit avoids problems that may arise with the routing of the high frequency signal from the decoder panel to the remote control.

It is desirable to maintain the relative amplitudes of the chroma and the luminance signals constant to preserve their transmitted ratio; therefore the colour and the contrast control may be coupled electrically or mechanically. Changing one of them will have a similar effect on the other.

The diode attenuator circuit in figure 7.2 can also be used for the removal of the colour burst from the composite chroma. At this stage the burst has no further use, and its continued presence may have disadvantages. The burst, if demodulated, could cause a spurious colour patterning on the left-hand side of the screen. The demodulated burst could, in some cases, affect the performance of the black level clamp. In figure 7.2 negative line flyback pulses derived from the line output transformer are fed to the junction of D1 and D2. Their amplitude is sufficient to cancel out the forward bias from VR1, causing the diodes to be reverse biased. The chroma path is shut off for the duration of the pulse, which, in turn, is arranged to coincide with the timing of the burst.

7.4 PAL Delay Line

It has already been mentioned that the PAL system, although based on the American NTSC, improved the quality of the received colour picture. This is due to the alternation of the phase of the V signal from line to line.

We already found in chapter 3 how different colours contain their specific values of (R′ – Y′) and (B′ – Y′); as shown in the phasor diagrams in figure 7.3, these two components give a resultant chroma of a specific phase angle with respect to the horizontal reference. If this angle changes, the (R′ – Y′) and (B′ – Y′) components also alter, giving rise to a different hue. Should such a change be spontaneous in the transmitter–receiver chain, then the resultant colour is incorrect because it differs from that seen by the camera.

Let us assume that the phase angle is increased slightly owing to the various imperfections in the whole system. As shown in figure 7.3, during one line when the V signal is unswitched (diagram a), the result would be an increase in the (R′ – Y′) component and a decrease in the (B′ – Y′). During the next transmitted line, when the V signal is switched through 180° (diagram b), the additional phase angle reduces the inclination of the chroma phasor, so that the (R′ – Y′) is now reduced but (B′ – Y′) is increased. It must be pointed out that after the demodulator the (R′ – Y′) signal of the switched line is reversed so as to coincide with the (R′ – Y′) of the previous line. Instead of the desired chroma, we could get one line

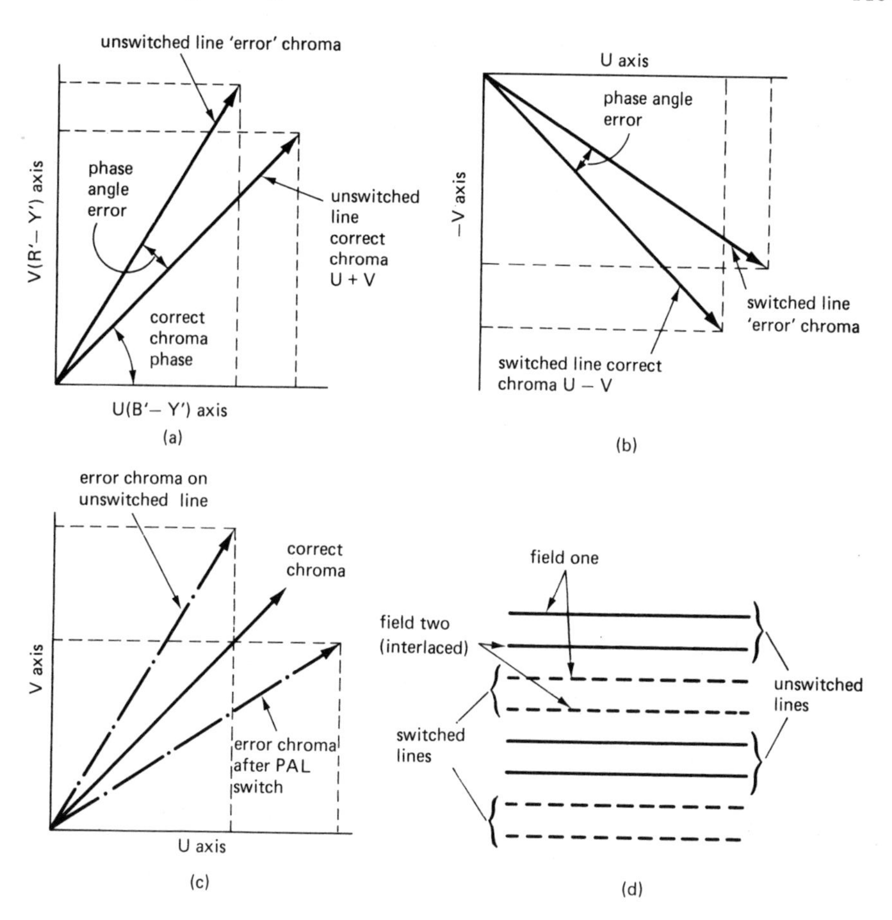

Figure 7.3 Chrominance phase errors. (a) Effect of phase error on the phasor diagram of unswitched line (U + V); note that the U and V components are now different, suggesting a change in the reproduced hue. (b) As above, but for the switched line (U – V). (c) After PAL switching, the V component from diagram (b) is reversed, and the two consecutive lines correspond to different colours. (d) Hanover blinds produced on the screen are in pairs of lines, owing to scan interlacing. The continuous lines correspond to the chroma detected during unswitched lines; the broken lines are produced by the chroma on alternate lines

of one colour followed by a line of another colour. Because of the interlaced scanning the lines of the second field, similarly affected, will fall in between those of the previous field. The effect on the screen will be a display of coarse horizontal lines of two different colours. These lines are known as *'Hanover blinds'* or *'Hanover bars*' (the PAL system, which is prone to this defect, was invented at Hanover in Germany).

Seemingly, we have not yet improved the quality of the display at all. In fact, it is possible to cancel out the effects of the errors in the phase relationship between the signals comprising the composite chroma. The process of cancellation can take one of two forms: optical and electronic.

Optical cancellation. The Hanover blinds, as described above, are allowed to be displayed on the screen. However, the interesting point is that, from a normal viewing distance, the different colour lights produced by these bars tend to merge to create the impression of the original hue, but somewhat desaturated (pale). This is a rare method of operation, except for test purposes or under certain fault conditions. We can visualise that if such phase errors were relatively large, the two sets of bars would be of so vastly different colours that the optical averaging would not be effective, giving rise to most objectionable line structure. When the receiver does rely on this system, it is called *simple PAL*, or *PAL-S.*

Electronic cancellation. The chrominance signals corresponding to two consecutive lines are matrixed to produce two outputs. These outputs are equal to the sum of and the difference between the chroma signals during the switched and the unswitched lines, respectively. To do this we have to store, or delay for 64 μs, the signal received during the previous line, so that it can then be processed together with the information arriving during the current line period. The more precise value of the delay is 63.943 μs, which is slightly less than one line interval because it must correspond to a whole number of the colour subcarrier half-cycles (567 in this case). On the other hand, the slight difference from 64 μs does not affect the quality of the picture. A block diagram of a circuit used in this method is shown in figure 7.4(a). This system is known as PAL-D (de-luxe, or delay line).

The delayed signal is the preceding line. Let us assume that it is unswitched; then the composite chroma = (U + V). The current line must be with its V component reversed, so that chroma = (U – V). The addition of the two yields

$$(U + V) + (U - V) = 2U$$

the corresponding subtraction gives

$$(U + V) - (U - V) = 2V$$

At this stage we can see that the method separates the two quadrature signals (the increased amplitude does not change the principle). Indeed, this is a beneficial by-product of the process, although correctly adjusted demodulators are also capable of separating the respective colour difference waveforms.

The cancellation of phase errors can be proved by the phasor diagram in figure 7.4(b). Taking the two chroma signals which occur if phase errors were present, as shown in figure 7.3(c), we find that after the delay line and the ADD/SUBTRACT matrix, the U and the V components will add separately in the manner shown in the above equations to give the output signals. The resultant chroma formed from the two pairs of components lies along the correct chroma phasor; therefore its

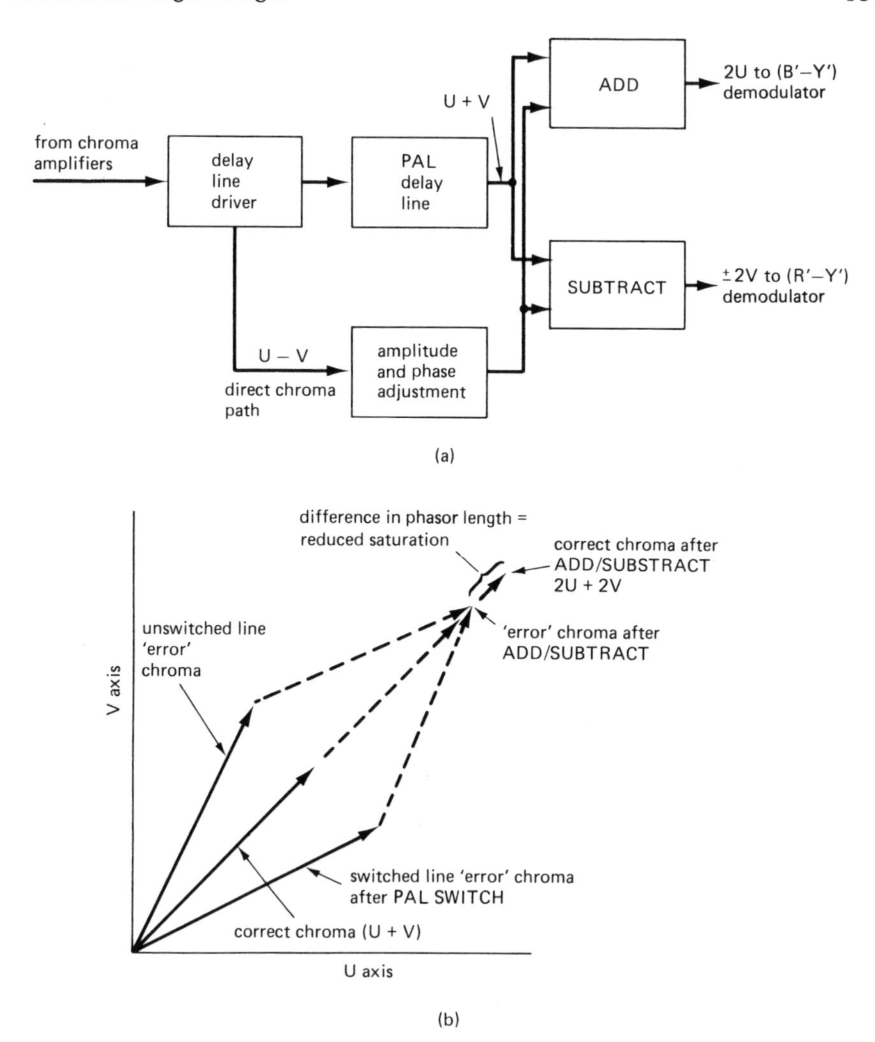

Figure 7.4 Effect of PAL delay line on phase errors: (a) block diagram of PAL delay line circuit; (b) phasor diagram showing the result of electronic averaging; the U and the V components of chroma add in each case (figure 7.3 shows how the actual chroma phasors were obtained)

phase is unchanged, but the amplitude is slightly less than it would otherwise be. This means that the ratio of U to V is constant, giving the same hue of slightly reduced saturation (see chapter 1 for the definitions).

The chrominance delay line is a separate circuit component made of special glass to very critical dimensions, with two piezoelectric transducers attached to it. The

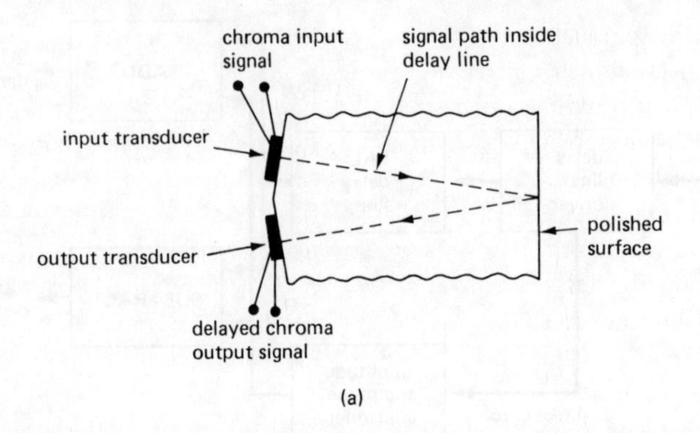

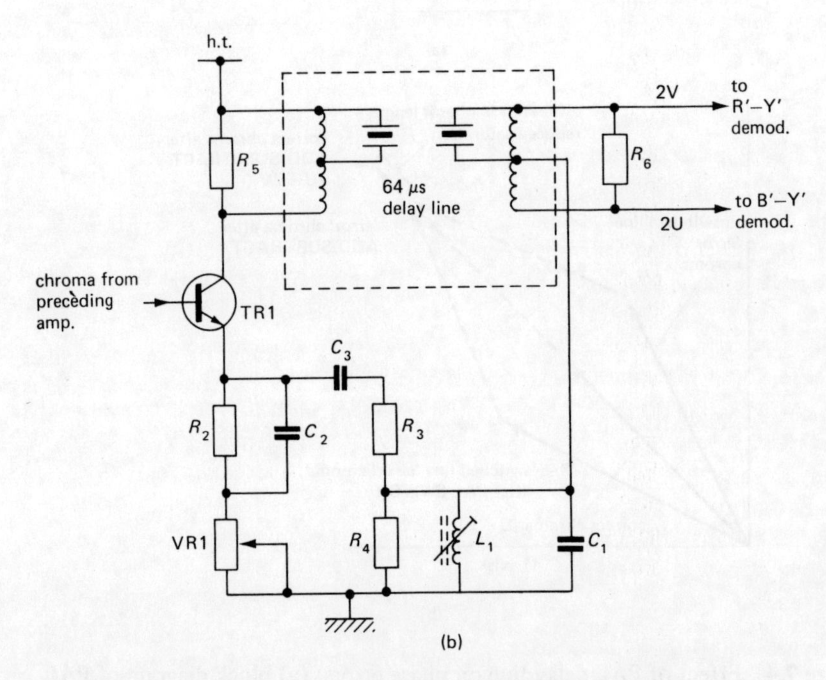

Figure 7.5 PAL delay line: (a) principle of construction; (b) example of the associated circuit

simplified diagram in figure 7.5(a) illustrates the principle involved. One of the transducers converts the incoming electrical signals into mechanical vibrations, which spread through the glass body to be picked up by the second transducer. The latter reconverts them back into an electrical signal (the idea is somewhat similar to that of a record player cartridge!).

The speed of propagation of these vibrations in the glass is relatively slow, and therefore the required time delay in the signal can be achieved. The type of delay line used in the luminance section of the receiver (see chapter 6) would be impractical, as the delay now is almost a hundred times longer.

There is a loss of signal power in the delay line, which must be made good by a separate amplifier stage—a *delay line driver* (occasionally an amplifying stage may follow the delay line).

A typical circuit diagram of the arrangement shown in block form in figure 7.4(a) is given in figure 7.5(b). The delay line is supplied with an amplified signal from the collector of the driver transistor TR1, while the direct chroma path is taken from the emitter. The transducers are connected to the remainder of the circuit via autotransformers which are usually an integral part of the delay line unit. They provide impedance matching, and at the same time the output autotransformer is used as an adding and subtracting circuit. Being a centre tap coil, the winding produces two outputs of equal amplitude but in phase opposition. The direct chroma is connected to the centre tap, so that it adds to the delayed signal in one half of the winding and subtracts in the other half. There are now two outputs from the outermost terminals of the transformer, which correspond to the separated U and V signals in the manner shown above. The reader will appreciate that the correct addition of the delayed and the direct chroma only takes place if the amplitude and phase of each signal at the point of matrix are equal. Since the signals take two different paths, adjustments are provided; L_1 is primarily used to equalise the phase and VR1 is chiefly used for amplitude compensation. Sometimes the coupling coils to the delay line itself have adjustable cores to achieve the necessary equalisation. If the settings are incorrect, instead of cancelling the errors a misadjusted circuit would introduce them, and the display will produce Hanover blinds.

If the delay line develops a fault, the receiver continues to function, since the direct chroma path is still available, but Hanover bars are again likely, because the decoder reverts to the PAL-S mode. Sometimes it may even be desirable to check the adjustments of the demodulator feeds or the subcarrier oscillator with the receiver PAL delay line being inoperative (the process of electronic averaging can mask the defects in the demodulating action).

7.5 Synchronous Demodulators

The reader would have already come across the idea of synchronous demodulation described in chapter 6 in connection with video detectors. There the process was desirable, but not strictly necessary, since a simple diode detector will operate reasonably well. The chrominance signal presents an entirely different problem, because of the suppressed subcarrier. As outlined in chapter 3, a simple diode detector would produce an output which bears no resemblance to the original information; hence, synchronous demodulation is absolutely necessary. This process requires a sampling circuit which 'looks' at the signal at the time when the transmitter subcarrier passes through its maximum amplitude. The demodulator behaves as though it were a switch which closes at that instant, and transfers the sampled amplitude to the output. The switching action must be synchronised with the transmitter colour subcarrier in order to ensure that the output waveform is of correct shape. The sampling process is steered by the subcarrier regenerated in the receiver, and is synchronised with the transmitted signal by means of the colour burst.

The composite chrominance signal consists of the U and V components. These were originally modulated by means of what effectively appeared to be two carriers of the same frequency, but 90° out of phase with each other (quadrature modulation). The delay line circuit can separate the two components, which doubtless helps in improving the action of the demodulators and simplifying somewhat their design. But it has to be emphasised that a correctly adjusted synchronous demodulator can also separate the required component from the composite signal, as would be the case in the PAL-S receiver. Quadrature modulation means that one subcarrier feed is reaching its maximum amplitude while the other one is passing through zero. It follows that the demodulator will 'see' only the maximum value when it is being switched by its own, appropriately synchronised, subcarrier, resulting in only one colour difference signal output. When the subcarrier amplitudes change a quarter of a cycle later, it will be the other demodulator which produces an output corresponding to its required signal while the previous demodulator is switched off.

Practical circuits are relatively simple because the switching subcarrier has already been obtained in another part of the decoder. (In a video synchronous detector, the switching carrier had to be produced from the modulated signal itself, so that it was economically feasible only by means of an I.C.)

Two arrangements are shown in figure 7.6. In diagram (a) diodes D1 and D2 are simultaneously switched by the subcarrier, whose phasing has been correctly adjusted to ensure synchronisation with the relevant feed in the coding circuit at the transmitter. When the diodes conduct, the chroma coupling capacitor C_2 will charge to the amplitude of the chrominance signal at the time. During the next half-cycle of the switching subcarrier the diodes are reverse biased and are non-conducting. The voltage acquired by C_2 appears at the output across R_3. The parallel L_1/C_1 is a filter circuit, which removes the subcarrier, so that the output waveform is in the shape of the modulating envelope. Resistors R_1 and R_2 are in series with the diodes to limit the current during the switching operation.

In diagram 7.6(b) a four-diode circuit is used to act as a switch. All four diodes conduct simultaneously when the junction of D1 and D2 is positive with respect to the junction of D3 and D4. On inspection, the reader will notice that the bridge circuit is placed in series with the signal path, which means that when the diodes conduct, the bridge circuit looks as though a switch were closed to allow chroma to pass through it to charge capacitor C_2. During the next half of the subcarrier cycle the diodes are reverse biased, and the voltage from across C_2 is transferred via the filter to the output.

The diagrams shown refer to (B′ – Y′) demodulators, the (R′ – Y′) counterpart being identical; the only difference is the arrangement of the subcarrier feed between the oscillator and the respective demodulators to accommodate the 90° phase shift as well as PAL switching. This aspect of the circuit will be discussed in later sections.

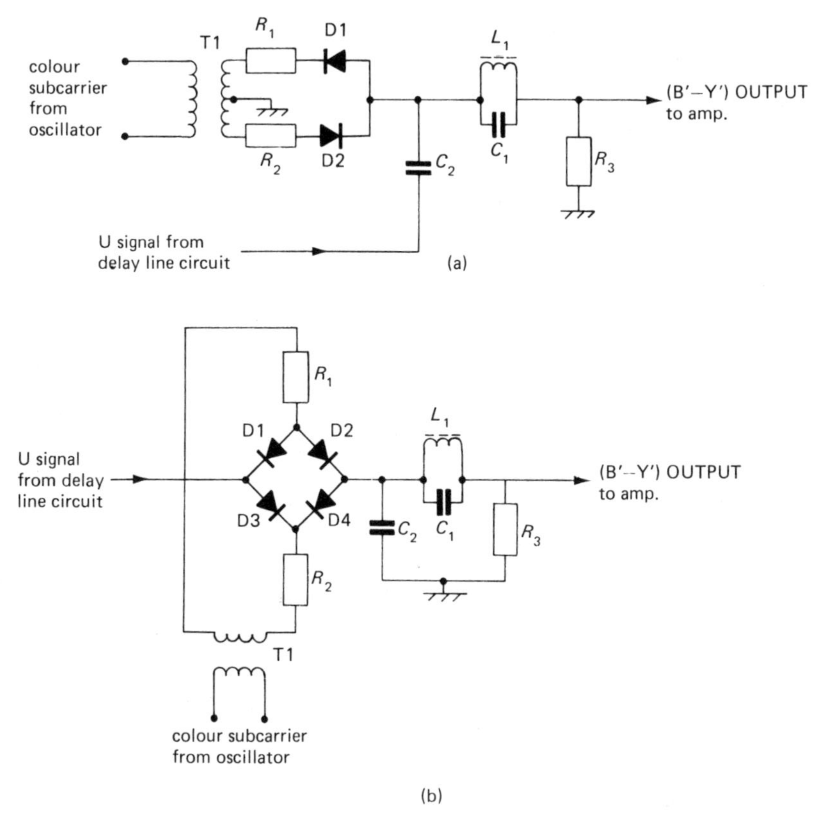

Figure 7.6 Examples of synchronous demodulator circuits: (a) two-diode circuit; (b) four-diode circuit

7.6 (G′ – Y′) and R, G, B Matrix

As mentioned in earlier chapters, the third colour difference signal, (G′ – Y′), is not transmitted, since it can be 'reconstructed' in the receiver from the specified portions of the demodulated (R′ – Y′) and (B′ – Y′). This is done in the circuit called the *(G′ – Y′) matrix*. Basically it consists of an amplifier which receives signals from the other two colour difference amplifiers. The (R′ – Y′) and (B′ – Y′) amplifiers are fed with signals from their respective demodulators, their gains being unequal, to compensate for the weighting factors.

In practice, the majority of receivers use R, G, B drive; therefore the luminance and chrominance paths are combined before being fed to the tube. This is done in the stage called the *R, G, B matrix*. One part of the circuit acts as the (G′ – Y′) matrix, while the other part is the luminance/colour difference matrix; here the three primary colour signals–red, green and blue–are produced. The reader should now refer to chapter 6 and look at the circuits in figures 6.11 and 6.7. In figure

6.11 the unequal value resistors R_5 and R_{12} determine the specified proportions of the (R′ – Y′) and (B′ – Y′) needed to form (G′ – Y′). In this way the (G′ – Y′) input is applied to the emitter of TR2, and the amplified output becomes available at its collector.

If we look at the emitter circuits of TR1 and TR3 (figure 6.11), the (B′ – Y′) and (R′ – Y′) amplifier transistors, we note that the total resistance in each case is 390 Ω or nearly so ($R_3 + R_4$ and R_{15}); these values are required to maintain correct d.c. conditions. However, the gain of the (B′ – Y′) amplifier is made greater, since the 220 Ω (R_4) of the emitter resistance is bypassed by the capacitor C_2. In the emitter of the (R′ – Y′) amplifier R_{15} is unbypassed and the resultant negative feedback reduces the voltage gain. These differences compensate for the unequal weighting factors in the U and V signals.

The luminance signal is introduced in figure 6.11 at the three collectors via R_1/C_1, R_9/C_4, R_{14}/C_5; this signal is added to the amplified colour difference signals at these points to produce the blue, green and red waveforms. The signals at these points are of an insufficient magnitude to drive the tube, and must be amplified. A simple output circuit was shown in figure 6.7, together with the associated black level clamps; a full description was given in the accompanying text.

It is interesting at this point to look at the R, G, B drive waveforms based on a 100 per cent saturated colour bar signal, whose detailed description was given in chapter 3. The reader will recall that the colour sequence is white, yellow, cyan, green, magenta, red, blue and black. To produce the corresponding light output, the cathodes of the picture tube must be driven by the waveforms in figure 7.7. As discussed in chapter 6, the video signal is negative-going; in other words, the sync. pulses must take cathodes towards h.t., leaving the control grids at a fixed lower potential. A falling cathode voltage during normal picture information reduces the net bias between the tube electrodes and the beam current increases in proportion.

In view of our earlier discussion in chapter 6, the reader will accept the concept of *peak white* when all three guns are driven to their maximum output, as shown in the white bar.

To produce a yellow colour, only the red and the green (see chapter 1) electron guns are driven; the blue is switched off at its black level.

Cyan requires blue and green drive, with the red cathode at its black level.

The red bar is the result of the red gun alone being driven, the other two being at their black level. The reader is invited to verify the appearance of the three waveforms for the remaining colours: green, magenta, blue and black.

There are small ‘notches’ shown between the adjacent ‘ON’ parts of the waveforms; these may be seen on the screen of an oscilloscope when one views the drive waveforms. They are the resultant of the addition of signals in the R, G, B matrix, but the circuits do not always respond sufficiently fast, and the transition is delayed. On the screen it often shows up as a darker band between the adjacent bars.

Waveforms of similar shape may be found at different points between the R, G, B matrix and the tube cathodes; their direction may be inverted, depending on

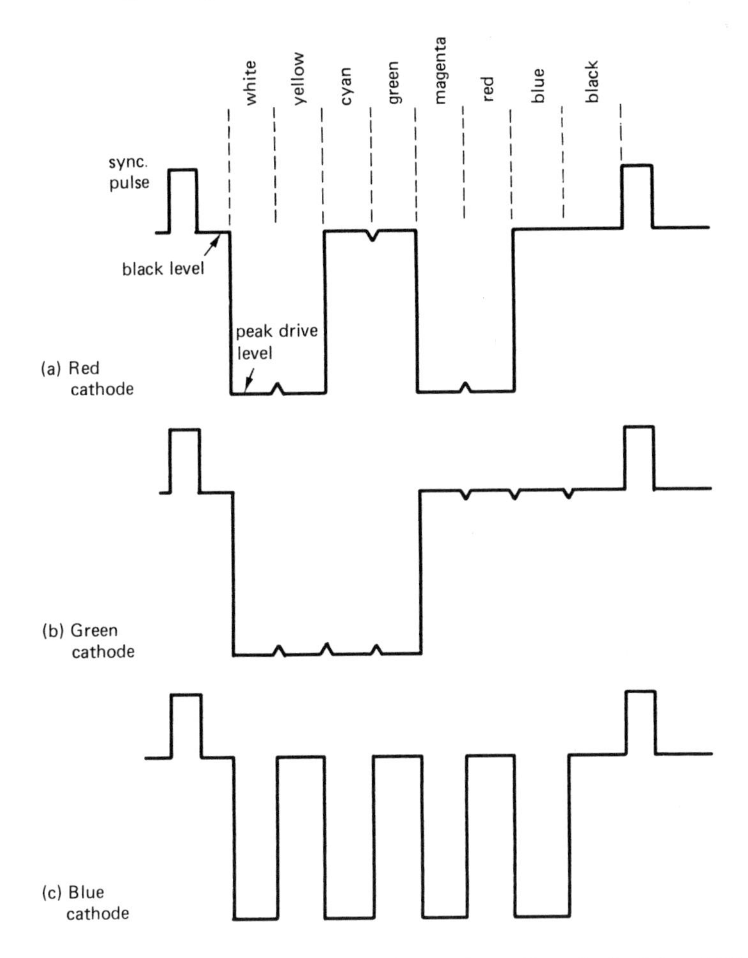

Figure 7.7 Cathode ray tube drive waveforms; R, G, B cathode drive, 100 per cent saturated colour bars

whether the amplifying stages reverse the polarity (common emitter) or not (emitter follower).

The waveforms in figure 7.7 are based on theoretical considerations, but in practice various overshoots in the vicinity of the sync. pulse may be displayed on the oscilloscope (depending on the behaviour of the actual receiver circuitry). For practical purposes these may be ignored, because the flyback blanking (chapter 6) suppresses any effects of spurious pulses on the actual picture. It has been mentioned earlier that, for simplicity, all the waveforms illustrated in this book refer to 100 per cent saturated colours. Some colour bar test signals offer waveforms corresponding to reduced saturation; if this is the case, then, whenever the gun is shown as switched off, there is, in fact, a drive signal present (except for black, of course), its amplitude being well below the peak level. The reason is that a desaturated colour contains some white light in addition to its own hue (chapter 1), but to produce white we must have *all three* guns at least partly switched on. Many service manuals may show such desaturated waveforms instead of the examples given here.

The illustrations in this book suggest that the amplitudes of the signals are equal in each case. As will be shown in chapter 11, the actual drive required is different for each gun and is adjustable as part of so-called *grey scale tracking* (see VR2 in figure 6.7). However, the peak-to-peak amplitude of the signal, although dependent upon the size of the tube screen and other constructional details, is typically between 100 and 150 V.

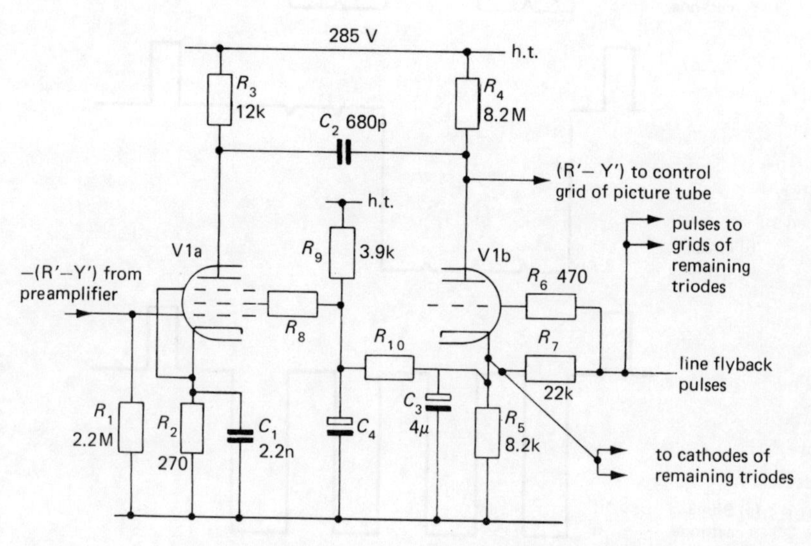

Figure 7.8 Colour difference output stage (Mullard)

A few receivers of earlier design feature a valve colour difference drive circuit, of the type shown in figure 7.8.

Preceding the valve is a transistorised preamplifier stage, which also includes the (G′ – Y′) matrix. In that respect the circuit follows the basic principle outlined above. Valve V1a, the pentode section of a double valve, acts as an amplifier to raise the signal voltage level from about 7 V to, say, 200 V. The output signal at the anode of V1a is coupled via C_2 to the anode of V1b, and from there to the control grid of the picture tube (the red gun in this case). The triode section, V1b, is a black level clamp whose control grid is driven by line flyback pulses supplied from a suitable winding of the line output transformer.

Owing to the complexity of the preceding decoder, by this point in the circuit the colour difference waveforms would have lost their true zero reference level, and this must now be restored. It is essential not only to preserve the black level and, hence, the average d.c. level of the luminance (or video) signal (described in chapter 6), but also to maintain a correct reference point for each colour difference signal. Otherwise a situation arises in which the drive magnitude to the respective grids could be either excessive or insufficient; this results in either primary or complementary colour casts appearing over the entire picture (even to be present on a monochrome programme). To ensure a correct 'black level', a large positive pulse (typically 50 V) is applied to the grid of the triode which coincides with the beginning of each picture line (during back porch when the colour difference signal is zero). The valve is driven into saturation, and its anode voltage drops to that of the cathode, which is fixed by the potential divider R_9, R_{10}, R_5. This allows the capacitor C_2 to charge to a predetermined voltage from the h.t. supply. During the picture information period V1b is cut off, and the relevant grid of the C.R.T. is held at a direct potential given by the charge stored in C_2; the colour difference signal being superimposed on this level. Capacitor C_2 discharges only very slowly via the high value resistor R_4, so that the reference level is substantially constant until the arrival of the next clamping pulse. In some circuits the cathode voltage of V1b is made variable to form part of the brightness control arrangement.

The circuit shown in figure 7.8 is for the (R′ – Y′) output stage; the remaining two colour difference output amplifiers are identical, the cathodes of the clamp triodes being connected together and the line pulse feed is also shared.

The luminance part of the receiver using colour difference drive was discussed in chapter 6, and the corresponding luminance output stage was illustrated in figure 6.12. The matrixing of the luminance and colour difference signals takes place within the picture tube, whose light output is proportional to the voltage difference between the control grid and the cathode. If the grid is driven by (R′ – Y′) and the cathode is driven by (–Y′) (minus sign denotes negative-going video), the net signal applied to the tube is (R′ – Y′) – (–Y′) = R′; similarly for the other two guns.

The waveforms applied to the grids of the tube are of the now familiar shape (see chapter 3) of (R′ – Y′) (typically 180 V peak-to-peak), (B′ – Y′) (220 V) and (G′ – Y′) (100 V). The associated d.c. voltages at the grids are then in the order of 120 V, and originate from the triode clamps described earlier.

R, G, B matrixing and other functions described here can also be performed within an integrated circuit, which often includes other forms of signal processing. We shall look at this type of decoder in a later section.

7.7 Regeneration of the Colour Subcarrier

The subcarrier associated with the chrominance signals was suppressed at the transmitter but, as we saw in section 7.5, the subcarrier is necessary to operate the synchronous demodulators and it must be regenerated in the receiver. This is done using the colour burst, which is a transmitted sample of the original subcarrier

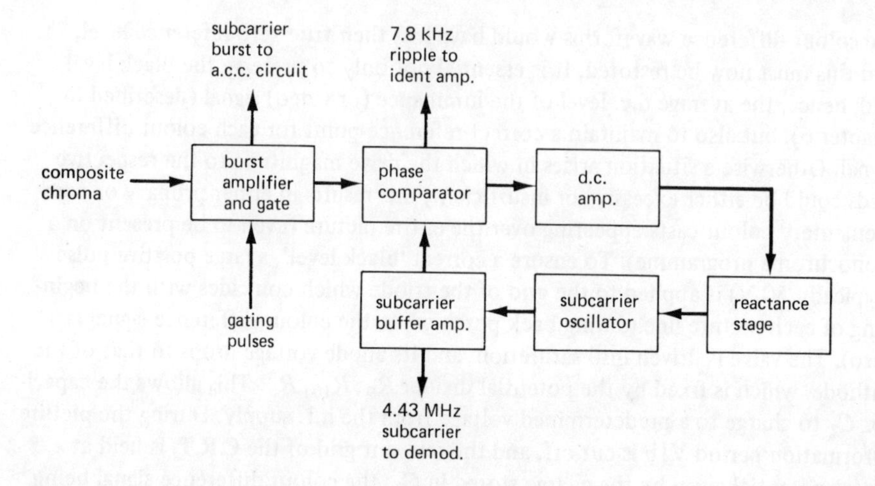

Figure 7.9 Block diagram of colour subcarrier regenerating circuit (a.p.c. loop)

during the back porch period—that is, between the line synchronising pulse and the beginning of picture information.

The purpose of the colour subcarrier regenerating circuit is to control the frequency of a crystal oscillator; a block diagram of the associated arrangement is shown in figure 7.9. The burst gating circuit conducts only during the application of a delayed pulse which coincides with the period of the burst. Usually the pulses are fed from the line output transformer, and are due to the line flyback that occurs in time with the synchronising pulse. Therefore they must be suitably delayed to arrive during the back porch period. The precise method used to shape and delay them varies from one receiver model to another. Some manufacturers use a combination of resistors and capacitors; others use a tuned circuit which is shocked into self-oscillation (ringing) by the initial pulse, the frequency of the oscillation being such that the second half-cycle occurs at the required time to open the burst gate. It is also possible to use the line synchronising pulse derived from the sync. separator and suitably delayed.

The diagram in figure 7.10 shows a simplified arrangement of the subcarrier oscillator and its control circuit. TR1 performs the dual function of a *burst gate* and a *burst amplifier.* The base bias on this transistor is not fixed, and it is normally non-conducting except for the brief period of the positive-going, delayed line flyback pulse. The delay takes place in another part of this receiver by means of a resistance–capacitance network. The diodes, D1 and D2, ensure that the magnitude of the pulse does not exceed the 25 V supply by clipping any excess; resistors R_1 and R_2 act as a base potential divider. The leading edge of the gating pulse is delayed, and in this receiver the length of the pulse is also controlled to ensure that the conduction of TR1 does not extend beyond the duration of the burst. The

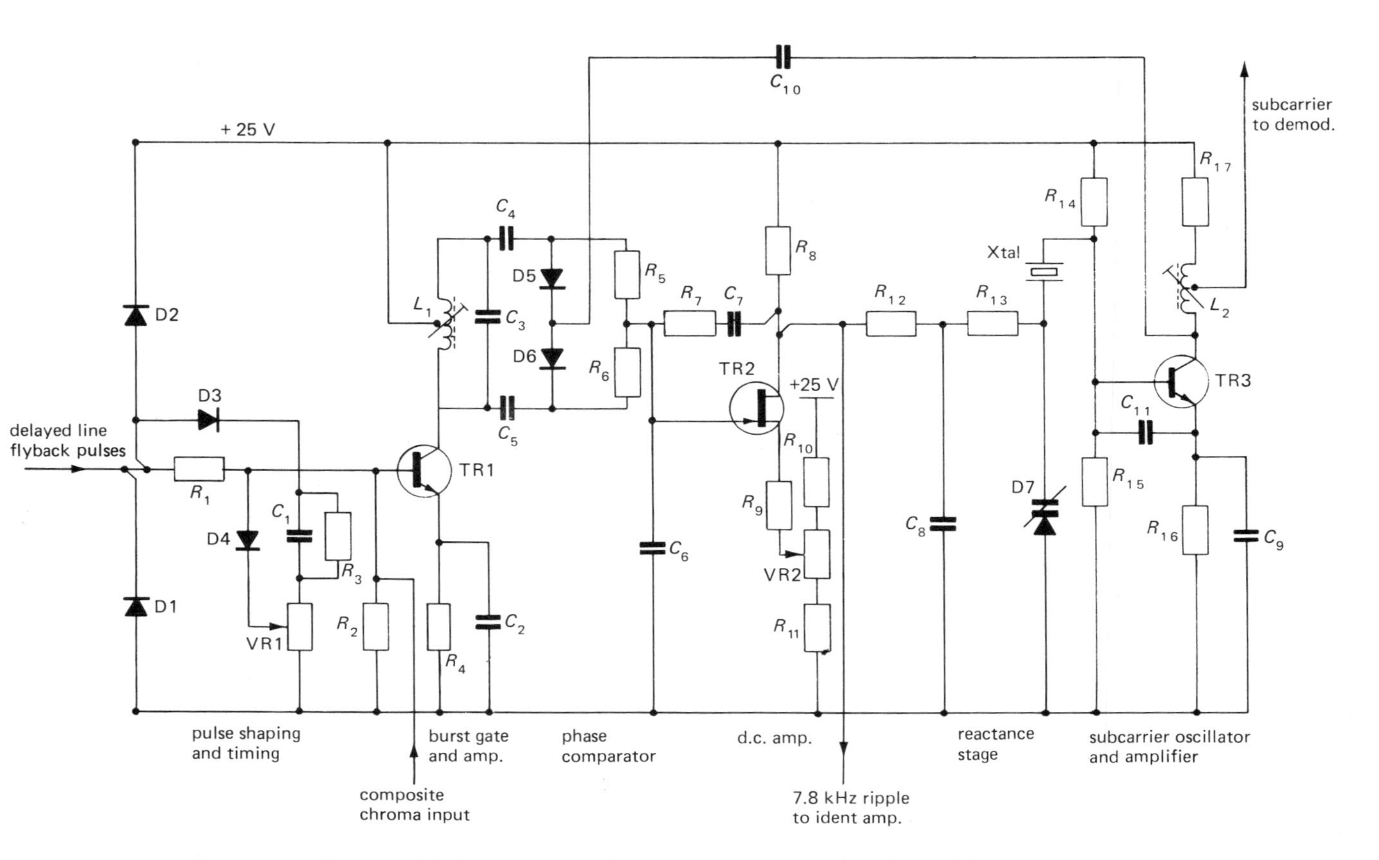

Figure 7.10 Burst gate and automatic phase control loop (Thorn)

timing circuit consists of D3, C_1, D4 and VR1. Capacitor C_1 charges from the gating pulse via D3 and VR1; during that period of time, the voltage drop across VR1 holds D4 biased OFF, and the conduction of TR1 takes place as described. Once C_1 is fully charged, the current through VR1 and the resultant p.d. are removed; D4 then conducts to divert the remainder of the gating pulse to earth, when the transistor switches off.

All gating circuits must ensure that the gate is reliably 'shut off' when not required. This is done here by means of a charge stored in C_2, which applies a positive potential to the emitter of TR1 of sufficient magnitude to prevent the remainder of the chroma signal (or spurious noise) from turning the transistor on.

The tuned circuit, L_1, C_3, resonates at 4.43 MHz, which selects the burst signal and rejects the relatively low frequency gating waveform. With an oscilloscope connected to the collector of TR1 the tuning of L_1 can be adjusted to obtain a burst of maximum amplitude. The gated burst is then applied to the *discriminator circuit*, which consists of D5 and D6 (*burst discriminator diodes*) together with C_4, C_5, R_5, R_6 and C_6.

The burst switches the discriminator diodes on, and the coupling capacitors C_4 and C_5 become charged to equal but opposite polarity potentials. The voltage at the output from the discriminator (junction of R_5 and R_6) is then zero by virtue of circuit symmetry. However, a sample of the subcarrier oscillator output is fed back from TR3 via C_{10} to the junction of the diodes. Now the circuit compares the phase and, indirectly, the frequency of the oscillator with the incoming burst. If the regenerated subcarrier passes through zero at the time of maximum burst amplitude, which is when D5 and D6 conduct, the original symmetry of the arrangement is preserved and there is still no output from the discriminator. Should the phase of the oscillator output change, then the signal fed back biases one of the diodes more than the other, and the charges on C_4 and C_5 become imbalanced. Either a positive or a negative polarity output now appears across C_6. If the oscillator frequency is incorrect, then similar action takes place, because the subcarrier zero would be out of step and would be drifting with respect to the received burst. In both cases the resultant d.c. or slowly changing d.c. (of either polarity) is amplified and used to adjust the oscillator via a reactance stage until the error basically disappears.

For simplicity, we have assumed that the transmitted burst remains constant, but the reader will recall that the PAL system uses a swinging burst, which means that its phase is alternating for every picture line. Naturally, this means that the discriminator always produces an output which also alternates in step with the phase of the burst. Obviously, the oscillator output, even if it were meant to follow the burst faithfully, would be out of step at the beginning of each line. A useful compromise solution exists: the subcarrier oscillator runs at a constant phase relationship with the original transmitter subcarrier oscillator, and it must ignore the regular, line-by-line alternations of the burst. However, the 'swinging' phase is also detected and employed in a separate circuit to identify the switched and the unswitched lines of chrominance information.

The alternating discriminator output must be smoothed before it is used to control the frequency of the oscillator; the smoothing must be carefully arranged to allow some 'swinging' discriminator signal to be directed to the PAL identifying circuitry (so-called *ident*).

In the circuit in figure 7.10 filtering takes place in two stages. The field effect transistor, TR2, is a so-called d.c. amplifier, but, in fact, it has to handle the alternating component as well. Its amplitude is first reduced by a.c. negative feedback introduced by R_7 and C_7. The resultant 7.8 kHz ripple is then sent to the ident amplifier, and is considered in the next section of the book. After the ident take-off point, final filtering can take place by means of R_{12} and C_8.

The reactance stage is formed by the varicap diode D7, whose capacitance is altered by the applied control voltage. Resistor R_{13} prevents the 4.43 MHz signal from being shunted by the remainder of the circuit. The capacitance of the diode is in series with the self-capacitance of the crystal (Xtal), allowing the resonant frequency of the circuit to be changed. The quartz crystal is connected to a transducer (see section 7.4); the frequency of vibration of the crystal depends on its physical dimensions and offers a great stability of operation. These vibrations are converted into electrical signals, and the whole arrangement behaves as if it were a tuned circuit.

The oscillator is based on the Colpitts circuit; the familiar capacitor 'tap' is now at the junction of C_9 and C_{11}. The output is amplified by means of the tuned circuit L_2 in parallel with stray capacitance; the inductance is adjusted (with the aid of an oscilloscope) to produce maximum subcarrier amplitude.

The frequency error that may be corrected is limited to a few hundred Hz, which means that if the free-running oscillator frequency (without the burst input) is too far from the required 4.43 MHz, the circuit will not be able to synchronise at all when the burst is applied. This could happen, for example, when the programme changes from monochrome to colour, or another channel has been selected, or simply when the receiver has just been switched on. To ensure correct operation, the amount of d.c. applied to the reactance stage can be adjusted, as if to simulate the action of the error signal. In the circuit in figure 7.10 this is done by controlling the amount of d.c. bias of the FET by means of VR2, while the original burst signal is temporarily reduced in amplitude or even removed. When this adjustment is being carried out, the makers' instructions must be followed, some requiring the adjustment to aim at a specified voltage; in other circuits the colour killer may have to be overridden (see next section) and the display on the screen observed. If the oscillator frequency only differs slightly from the colour burst, then the picture will be displayed by the luminance signal, but the colours will slowly drift across the screen, creating an effect of continuously changing hue. Should the oscillator be considerably out of adjustment, then the display breaks up into a drifting and meaningless coloured pattern, often obliterating the picture completely.

The output from the oscillator is usually fed to a buffer stage (an emitter follower) to limit the loading on the circuit; sometimes, as in this case, a transformer coupling (tap on L_2) is used to achieve the required impedance matching to the driven circuits.

The complete circuit formed by the burst phase discriminator, the d.c. amplifier and the oscillator is closed by the feedback connection from the output back to the input (via C_{10} in figure 7.10). This is often referred to as the *a.p.c. loop* (*a*utomatic *p*hase *c*ontrol loop).

In most receivers automatic chroma control (a.c.c.) voltage (see also section 7.3) is derived from the burst gate amplifier. Part of its output is rectified, sometimes by means of a voltage doubler, smoothed and fed to the first chroma amplifier to alter the d.c. bias and, consequently, the gain. The colour burst is used for this purpose because its amplitude is representative of the actual strength of the chrominance signal. As with the a.g.c., we need a control voltage which is independent of picture content.

7.8 The Ident and the Colour Killer Circuits

In the PAL system the modulated V component carrying the (R′ – Y′) information is switched through 180° at the beginning of every picture line. In order to achieve synchronous demodulation of the signal, it is necessary to carry out the phase reversal of either the subcarrier or the V component in step with the transmitter. This can only be done if the transmitted switched and unswitched lines are identified in the receiver. Since the coder switching creates the swinging or switched phase burst, the phase detector in the a.p.c. loop generates the necessary identifying signal, as described in the previous section. At that point its amplitude is low and the shape resembles a charge–discharge curve of a capacitor deliberately produced by the first stage of inadequate smoothing within the a.p.c. loop. The signal at this point is called the *ident ripple*, but in this form it cannot be used in the receiver PAL switch. The frequency of this ripple is equal to half-line frequency, referred to for brevity as 7.8 kHz (the 625-line scan frequency is 15 625 Hz). The phase discriminator produces a positive-going error voltage during one line and a negative output during the following line, thus giving one cycle of the ident for every two lines.

The ident ripple is fed to an amplifier which changes the waveform into a large amplitude sinewave and then into rectangular pulses. This is usually accomplished by means of a tuned circuit amplifier, often with a degree of positive feedback which can drive it so hard as to automatically clip the sinewave into an almost rectangular (square) wave. In most cases further shaping of the ident sinewave is carried out by means of diodes which follow the amplifier.

An example of an ident amplifier circuit is shown in figure 7.11. The resonant circuit formed by L_1, C_2, C_3 is tuned to 7.8 kHz (usually to obtain maximum display of the resultant sinewave on the screen of an oscilloscope connected to the collector of TR1). The waveform is then fed via C_4 to the pulse shaping circuit of D1 and TR2 with the associated components. Since the amplitude of the sinewave output is in excess of what is needed to saturate transistor TR2, a square waveform is developed across the emitter resistor R_7. Diode D1 clips the negative half-cycles of the input waveform. The ident pulses are fed to the PAL switch circuit, which we shall discuss in section 7.9.

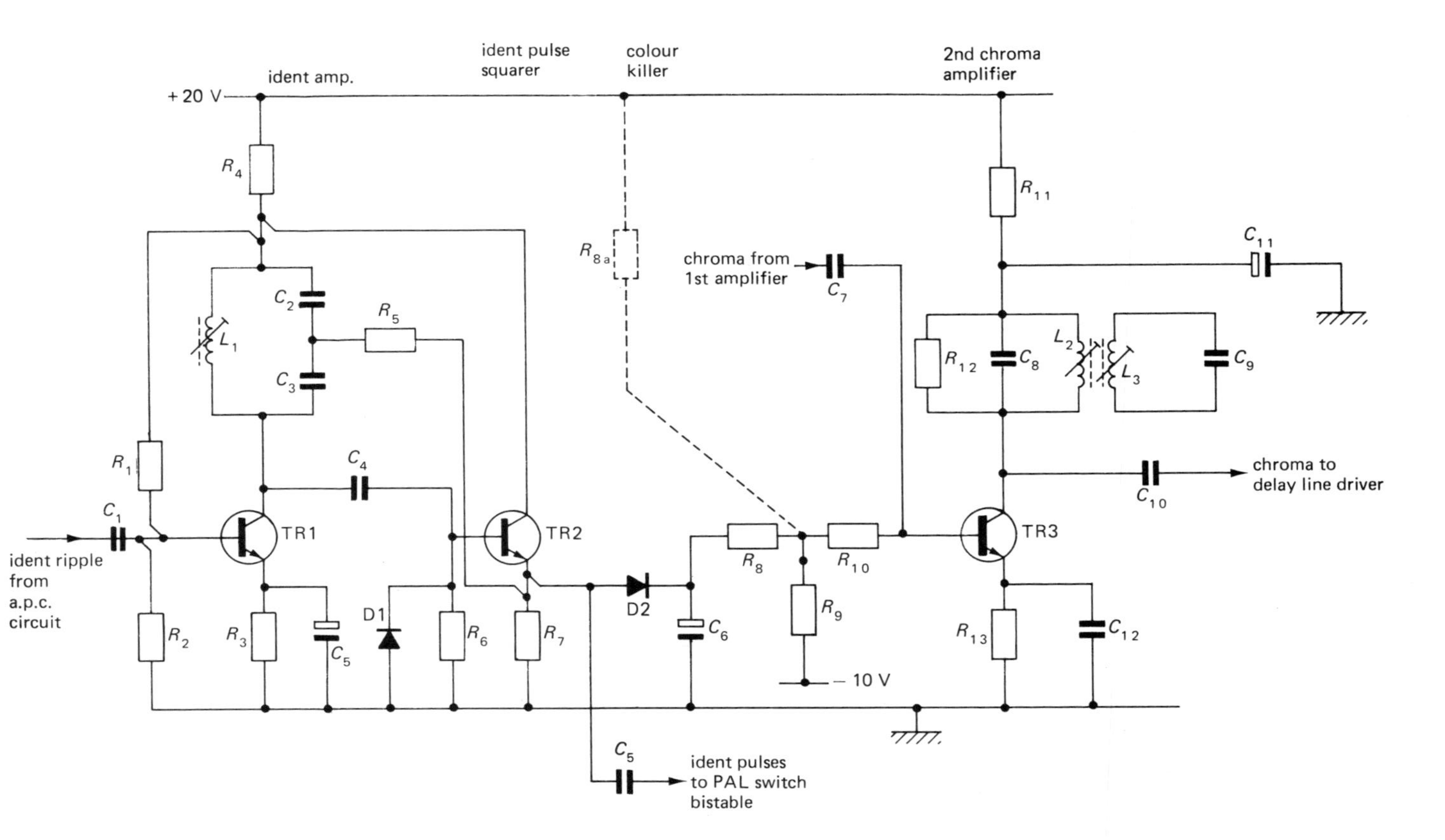

Figure 7.11 Ident amplifier, colour killer and associated second chroma amplifier (GEC)

We now look at the *colour killer* arrangement. Its chief purpose is to shut down the main chroma path during a monochrome transmission. This action prevents spurious 'coloured' noise appearing on the screen which spoils the quality of what is meant to be a black and white programme. (Normally any noise present would be swamped by the chrominance signal.) Since colour transmission is always accompanied by the burst, the colour killer can be operated, either directly or indirectly, from the burst processing circuit. The most common method is to use the ident signal to provide the correct d.c. bias for one of the chroma amplifiers. If the ident is present, the controlled stage operates normally; if it is absent, the colour killer drives the chroma amplifier into cut-off.

The circuit in figure 7.11 shows a simple colour killer arrangement. In the absence of the ident signal, the second chroma amplifier transistor, TR3, is biased OFF by a voltage derived from the *negative* supply rail via R_9. When ident pulses appear at the emitter of TR2, they charge the electrolytic capacitor C_6 via D2 to a positive voltage. Transistor TR3 is correctly biased by the potential divider R_8 and R_9, and normal amplification takes place. The reader will observe that the circuit of the chroma amplifier otherwise resembles one of the arrangements described in chapter 5 on i.f. amplifiers.

Some manufacturers use a rather more involved colour killer circuit, with an intermediate transistor which is used as a switch, either to remove or to apply the correct d.c. bias to one of the chroma stages. Occasionally the 'unkilling' bias is derived from the PAL switch, which is, however, linked with the ident circuit and thus the basic idea remains unchanged.

We have mentioned that the colour killer operates whenever the ident signal is missing, which brings us to the second purpose of the arrangement. Certain fault conditions within the decoder will delete the ident; this could cause the picture to become unviewable, but as the colour killer operates, the receiver is left with an acceptable monochrome display. By developing this idea further and by connecting the colour killer to the PAL switch, it is possible to spare the viewer the unpleasant effects of faults over a considerable part of the decoder by always leaving him with a good black and white picture.

It is frequently helpful in tracing decoder faults to override the colour killer by restoring a more or less correct d.c. bias to the chroma amplifier in question. If we then view the display on the screen, the probable area where the fault lies should reveal itself. Most makers' manuals in the section devoted to decoder adjustments indicate the easiest method of overriding the action of the colour killer. In principle, all that is needed is to complete the open link in the d.c. resistor bias chain to the controlled chroma amplifier. In the circuit shown in figure 7.11 this is indicated by means of the resistor R_{8a} (shown in broken line) connected between the positive 20 V line and the junction of R_9, R_8, R_{10}; this forms a bias potential divider consisting of R_{8a} and R_9. The value of the resistor is usually given in the service manual, but could also be obtained from potential divider calculations aiming at the normally expected voltages on the controlled transistor. We shall return to the problem of decoder fault finding later in this chapter.

7.9 The PAL Switch

The reader will recall from chapter 2 that the line-by-line switching of the V component of the composite chroma at the transmitter must be repeated in the receiver. If this were not done, as may happen under certain fault conditions in the decoder, the synchronous detector would produce an output of +(R′ – Y′) during one line and –(R′ – Y′) during the next, while the (B′ – Y′) component would not be affected. Since each colour has specific values (and polarities) of the two colour difference signals, we would see on the screen a line of the correct colour followed by a line of a completely different colour. This would be the case of very severe, and most objectionable, Hanover blinds (see section 7.4). The colours that would be especially affected are those with a high (R′ – Y′) content–for example, red or green; on the other hand, the effect on the blue or yellow will be far less.

In the synchronous demodulator it is the mutual relationship between the V signal and the reintroduced subcarrier which must be maintained; it is possible to switch either the chrominance signal or the subcarrier. The latter is only a simple sinewave; it is relatively easy to change its polarity, and that is the method most commonly adopted. The switching of the rather complex chroma signal can introduce distortion, resulting in a degraded picture quality.

The PAL switch is under the command of the ident signal to ensure that the correct line is being switched. Two approaches are found as follows: the first uses suitably amplified and shaped ident pulses to operate the PAL switch directly; the second uses line flyback pulses to initiate the switching action, while the ident signal effects a form of interlocking only to ensure that the correct line is being switched.

In figure 7.12 the PAL switch operates on the V signal by reversing its phase every other line to counteract the transmitted reversals. The positive half-cycle of the large amplitude ident sinewave biases D1 ON, and the chrominance signal is

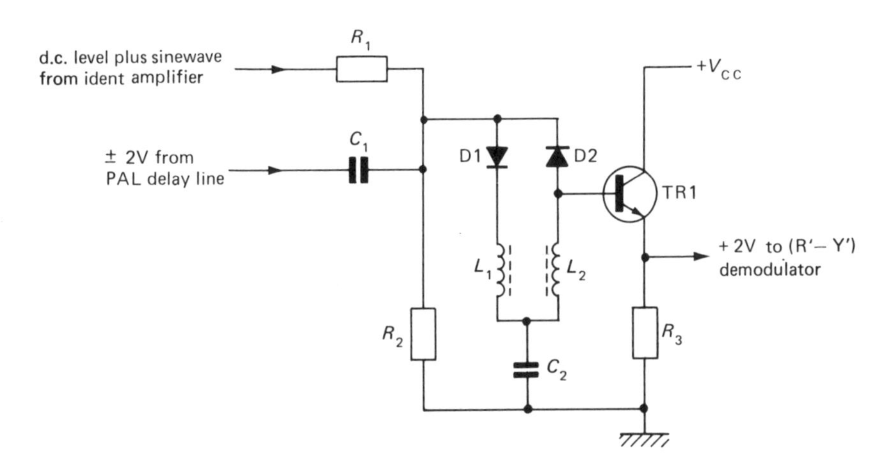

Figure 7.12 PAL switch of colour difference signal (Decca)

fed into the primary winding, L_1, of a transformer. In turn, a reversed polarity signal is induced in the secondary, L_2, which is then communicated to TR1. During the unswitched transmitted line, the ident circuit produces the negative half-cycle of the sinewave which biases D2 ON. This allows the V component to be applied directly to the transistor. D.C. base biasing is obtained from the collector of the ident amplifier via the potential divider R_1, R_2. Capacitor C_2 blocks the d.c. bias while completing the a.c. signal path through the transformer.

A problem with an 'ident-only' type of PAL switch is that it could switch over during the actual picture line, giving incorrect colours on one side of the screen and correct ones on the other, with a visible vertical band of obvious disturbance. This may happen if the ident sinewave, or the resultant pulse, is unduly delayed with respect to the beginning of the video line. A double beam oscilloscope could be useful to observe this, and to carry out such adjustment as may be recommended by the makers.

Another type of PAL switch is shown in figure 7.13. In this case it is the subcarrier feed to the (R′ – Y′) demodulator which is being reversed every other line. The two transistors, TR1 and TR2, with their associated components form a *bistable* switching circuit. In operation, when one of the transistors is saturated, then the other one is cut off owing to the cross coupling from the collectors to the bases via R_8 and R_9. This state of affairs exists until a suitable negative pulse is applied to the conducting transistor. The pulse momentarily cancels the forward bias on the base of, say, TR1 so that this transistor is cut off and its collector voltage rises. This rise will, in turn, be fed to the base of TR2, which will then be driven into saturation, its collector voltage dropping almost to zero. Therefore, even if the original negative pulse is removed, TR1 remains in the cut-off condition because its base supply voltage at the other end of R_8 is held at zero. The next line flyback pulse reverses the situation, and so on. The diodes, D3 and D4, connected to the transistor bases are called the *steering diodes* and ensure that the negative switching pulse is always directed to the saturated transistor. If a transistor is already cut off, then the pulse has no effect on it.

In effect, the bistable divides the line pulse frequency by 2; its output frequency is then equal to that of the ident.

The bistable line-by-line switching produces square wave outputs at the two collectors; however, the waveforms are exactly 180° out of phase with each other. The actual PAL switch is shown in the upper part of figure 7.13. The bistable outputs are used to bias diodes D1 and D2 alternately ON. The anodes of the diodes are connected to a fixed d.c. potential as given by the divider R_2, R_3 between the +h.t. and earth. The cathodes are joined via current limiting resistors R_4, R_5 and the windings of the transformer, L_2 and L_4, to the collectors of the bistable. When TR1 is bottomed, its collector is almost at earth potential, and D1 is biased ON. At the same time, D2 is switched off because the collector voltage of TR2 is almost at h.t. During the next pulse the state of conduction of the two diodes is reversed.

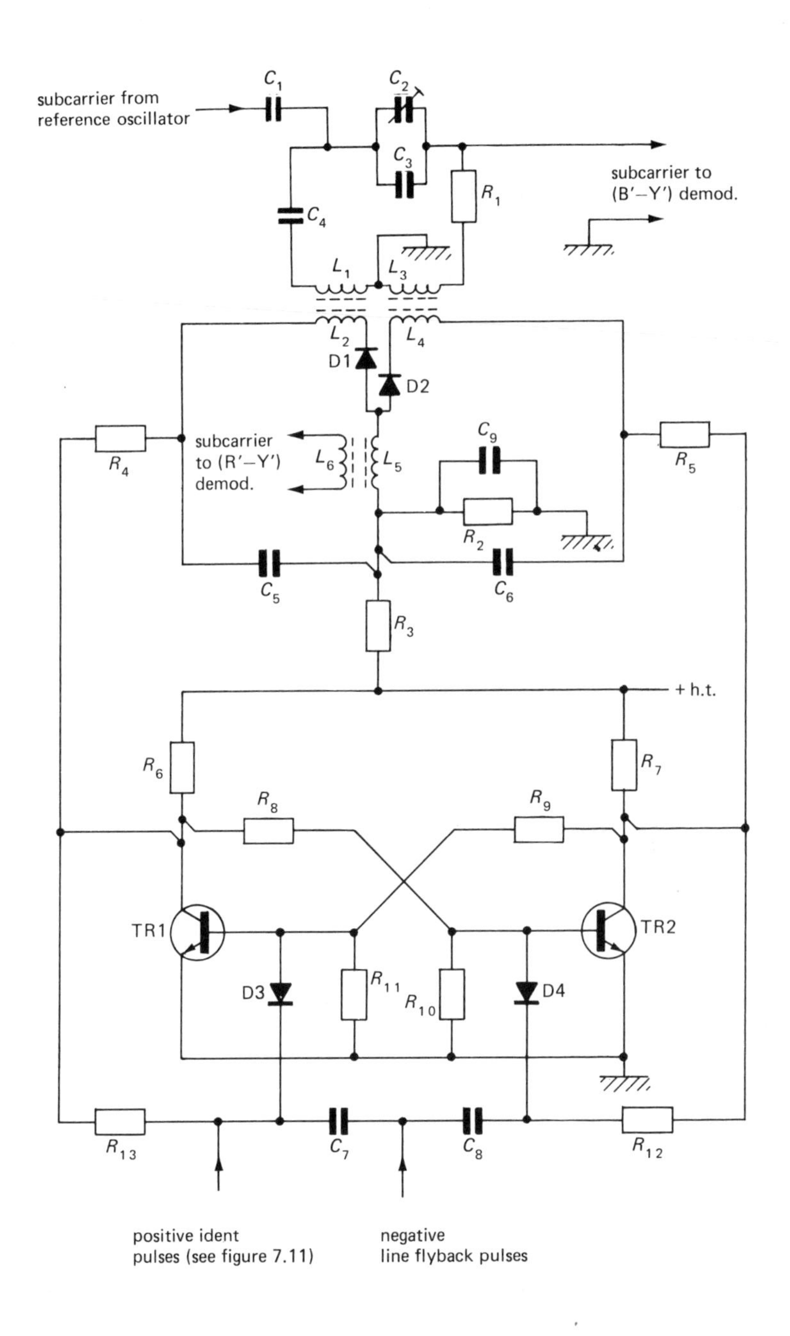

Figure 7.13 PAL switch in the subcarrier path to the (R′ – Y′) demodulator

The subcarrier from the oscillator is fed to another winding of the transformer, L_1, via C_1 and C_4. This induces signal voltages in L_2 and L_4. When D1 conducts, it directs the subcarrier from L_2 to the primary (L_5) of another transformer, the secondary of which (L_6) provides the supply to the (R′ – Y′) demodulator. When D2 conducts, it connects L_4 to L_5, but now the sense of the winding, L_4, is such that the polarity of the signal is in opposition to that in L_2; consequently the output to the demodulator is also of a reversed polarity. Capacitors C_9, C_5, C_6 provide the necessary d.c. blocking while completing the a.c. signal path within the circuit.

When the receiver is first switched on, or even when changing channels, there is a good chance that the PAL switch diodes (D1 and D2 in figure 7.13) could be in exact opposition to the transmitted reversal of the V signal because the bistable switch may have started in the 'wrong' phase. It is only a matter of chance which of the two transistors will be driven into saturation first. If this were allowed to continue, the line-by-line switching would produce a reverse polarity of the V signal which means wrong colours would be displayed on the screen. To ensure that the PAL switch is always in phase with the transmitter, the ident pulse which occurs every other line (figure 7.11) is fed to one side of the circuit at the junction of R_{13} and C_7. The *positive* ident will thus cancel out the negative line flyback pulse, rendering it ineffective as far as TR1 is concerned, and the bistable misses one operation. The next line pulse causes it to switch because the ident is at zero, and from there on, normal in-phase action continues.

The diagram in figure 7.13 also includes a method of adjusting the 90° (quadrature) phase shift of the subcarrier to the (B′ – Y′) demodulator. From the description of the coder at the transmitter (chapter 2) the reader will remember that the two feeds are in quadrature with each other, and, for correct synchronous demodulation, the same phase relationship must be maintained in the receiver. The adjustment available in this particular circuit is the trimmer capacitor C_2. The signal from the parallel circuit, C_2, C_3 combines with the one from L_3 via R_1 to provide the required phase shift. In practice, C_2 is adjusted until the maximum signal amplitude is obtained at the output from the (B′ – Y′) demodulator (as viewed on the oscilloscope).

7.10 Integrated Circuits in Colour Decoders

The complexity of the decoder lends itself to circuit integration. Designers try to achieve a compromise between several factors, including reducing the number of individual circuits, maintaining reasonable reliability and low repair costs. A wide variety of circuits are available which combine different decoding functions.

As yet, it is not possible to provide in I.C. form either large values of capacitance (in excess of, say, 20 pF) or of inductance. These and certain other components are connected externally, and some difficulty may be experienced in identifying their functions. If the reader is well acquainted with the requirements and layout of the non-I.C. decoder, then a large number of similar features can be recognised. In order to minimise the number of external components, designers occasionally use circuit techniques which differ from their transistor counterparts. For example, the ident signal may no longer need a tuned circuit to produce a large amplitude signal, because a level detector, although quite complex, can be built into an I.C.

The diagram in figure 7.14 shows a complete colour decoder which uses four integrated circuits. Some external controls and other important components have been included in a simplified manner to allow the reader to follow the arrangement more easily and also to recognise the more familiar functions.

I.C.1 (TBA 560) is a combined luminance and chrominance amplifier. The luminance signal from the detector is a.c. coupled at pin 3, and d.c. restoration takes place by means of a black level clamp which is driven by line pulses; the associated reference charging capacitor is connected to pin 4 (see chapter 6 regarding the principle of such a clamp circuit). Contrast and saturation tracking is maintained by the common connection to both the luminance amplifier and the chrominance amplifier. The external chroma bandpass filter between pins 1 and 15 is designed to shape the frequency response of the chrominance amplifiers which follow. In common with its transistor counterparts, the response complements that of the vision i.f. in the vicinity of 35.07 MHz. The reader will recall that the chroma subcarrier is usually placed on the slope of the i.f. response curve (figure 5.11); thus the lower sideband content is not equal to that of the upper sideband. Many manufacturers compensate for this unbalance in the chroma amplifiers.

The first chroma amplifier in I.C.1 is controlled by the a.c.c. voltage at pin 14, and derived from the burst (I.C.2).

Burst blanking removes the colour burst from the main chroma path, while allowing the burst to be directed to the burst gate. The gain of the second chroma amplifier stage is controlled manually by the saturation (or *contrast*) control at pin 13 and electrically linked (tracking) with the contrast control (pin 2).

The output from the delay line driver, pin 9, goes to the PAL delay line via a phase adjustment; the direct signal is applied via an amplitude adjustment to the 'add' and 'subtract' network. The resultant U and V signals are fed to I.C.3 (TBA 990). A removable link (pin 9, I.C.1) is included in the delay line circuit, allowing the decoder to revert to the PAL-S mode for various adjustments (the direct path only is now available).

I.C.2 (TBA 540) provides most of the services required for demodulation. The gated burst, whose phase has been adjusted while the decoder delay line was disconnected, is fed to the phase detector (discriminator). After filtering between pins 13 and 14, the resultant d.c. is applied to the reactance stage and then to the subcarrier oscillator. An external crystal and a trimmer capacitor are provided between pins 15, 1 and 2 for manual adjustments.

The burst phase detector yields the ident signal from which the colour killer circuit operates. In the absence of the ident, pin 7 of I.C.2 is at chassis potential; this earths the slider of the saturation control (pin 13, I.C.1), reducing the chroma amplifier output to zero. To override the colour killer, the link between the two I.C.s is removed, allowing the chroma amplifiers in I.C.1 to be biased via the saturation control only.

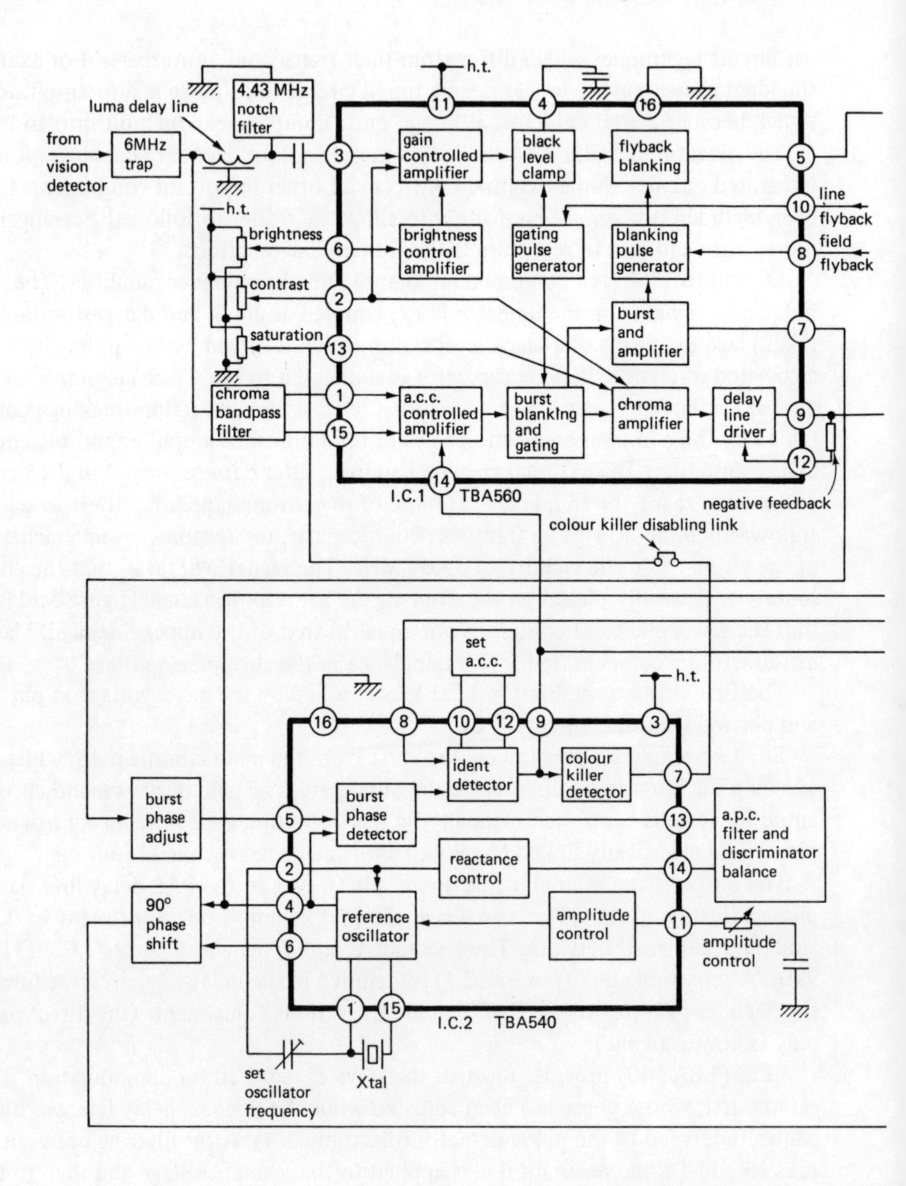

Figure 7.14 Colour decoder incorporating integrated circuits (Mullard

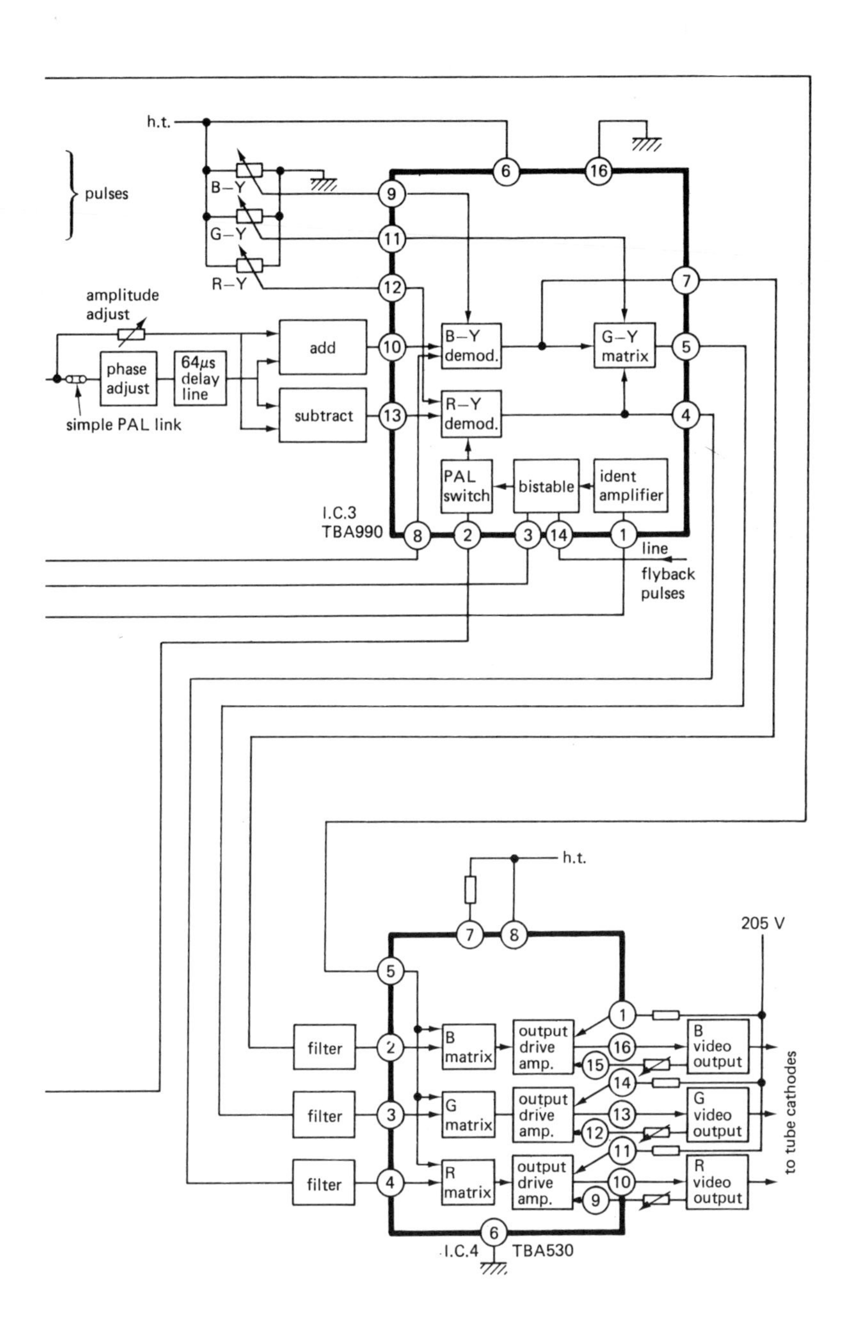
h.t.
pulses
B–Y
G–Y
R–Y
amplitude adjust
phase adjust
64μs delay line
simple PAL link
add
subtract
B–Y demod.
R–Y demod.
G–Y matrix
PAL switch
bistable
ident amplifier
I.C.3 TBA990
line flyback pulses
h.t.
205 V
filter
filter
filter
B matrix
G matrix
R matrix
output drive amp.
output drive amp.
output drive amp.
B video output
G video output
R video output
to tube cathodes
I.C.4 TBA530

I.C.3 is basically a synchronous demodulator together with the (G′ – Y′) matrix. The subcarrier outputs from I.C.2 (pins 4 and 6) are first phase shifted through 90° with respect to each other because of quadrature modulation requirement. One feed is then sent directly to the (B′ – Y′) demodulator (pin 8) and the other goes to the (R′ – Y′) detector via the PAL switch (pin 2). The switch is driven by a bistable which, in turn, is operated by the line flyback pulses and is interlocked with the ident signal. The colour difference outputs can be regulated at pins 9, 11 and 12.

Finally, I.C.4 provides the R, G, B matrix and drive to the video output amplifiers. The three colour difference signals from the preceding I.C. pass through filters, allowing only chroma signals up to 1 MHz to be included, and therefore remove the subcarrier and demodulator switching frequencies. The luminance signal is obtained from pin 5, I.C.1. The video driver stages have large amounts of feedback, not only to stabilise the output amplifiers gain, but also to cope with variations in the video output h.t. supply (205 V). The amount of blue drive is fixed, but both the green and the red are variable.

7.11 Fault-Finding in Colour Decoders

Faults in the decoder could be grouped into three major categories: no colour, incorrect colours, unlocked colours.

Before one attempts to locate a decoder fault, it is essential to establish a good monochrome picture. This should be possible in most cases, except when I.C.s are used extensively, where, as we have just seen, the luminance and the chrominance paths use the same devices. We shall deal with I.C. fault-finding later.

The points to look for are good signal strength (insufficient signal can give very weak colours or none at all); good synchronisation (burst gating could be otherwise affected); correct grey scale (colour casts can create an impression either of colour when none should be present or of wrong colours).

The easiest method of fault-finding in the decoder is by means of a suitable oscilloscope which indicates the presence and the magnitude of waveforms. The oscilloscope must have adequate bandwidth to display chroma signals which are centred on 4.43 MHz. Ideally, its frequency response should extend to 10 MHz, while one with a 5 MHz bandwidth is barely adequate. The instrument, especially of the lower bandwidth type, may load the circuit under test; this would disturb the circuits and may result in additional faulty symptoms or incorrect readings. A capacitive probe is often recommended to reduce the disturbance, but it must be remembered that the V/cm calibration of the oscilloscope is then not applicable, because some of the available signal voltage is dropped across the probe, leaving perhaps only 10 per cent for the display. As the amplitudes of the chroma signals in the initial stages are low, the oscilloscope must have high sensitivity, at least 50 mV/cm or better. Where I.C.s are used, an oscilloscope is invaluable.

No colour

Override the colour killer and observe the effect on the picture, as this helps to identify the possible fault area in the decoder:

(1) *Colour still absent.* Suspect: chroma amplifiers from the chrominance take-off point to the delay line driver; alternatively the subcarrier oscillator and its buffer stage (absence of the subcarrier shuts down both demodulators *and* removes the ident).
(2) *Unsynchronised colour.* Suspect: incorrect oscillator frequency caused by a fault within the a.p.c. loop, including burst amplifiers and gate, or misadjustments of available controls.
(3) *Colour returns to the display.* Suspect: the colour killer itself or the ident amplifier(s), although in the latter case incorrect colours may appear because of the PAL switch being out of step (see below).

Incorrect colours

If at all possible, observe either a colour bar display or a test card, and note how much the colours deviate from their standard sequence.

(1) *Primary colour fault* causing an overall colour cast on the screen: excessive drive to one gun (primary colour cast) or insufficient drive (complementary colour cast)—either can be easily observed in the 'white' bar; for example, a magenta cast is due to the absence of green. Suspect: primary colour amplifiers (after the R, G, B matrix) and black level clamps, incorrect d.c. levels in directly coupled amplifiers (such as I.C.s), faults in brightness control circuits affecting one gun only. This type of fault should have been discovered while checking grey scale tracking (see the introduction to this section).
(2) *Missing colour difference signal.* Wrong colours result since the addition of the Y' signal to the colour difference signals is now incorrect. This causes the tube to be driven by incorrect signal voltages. The colours most affected are those which would have a high content of the 'missing' signal. There are also slight differences between the displayed colours if the fault occurs before or after the $(G' - Y')$ matrix, since this signal is then formed either incorrectly or correctly. If possible, the reader should experimentally disable the various signals, using an otherwise *correctly adjusted receiver* and record his impressions of the colour bar sequence for future reference.

The following effects refer to absent colour difference signals; where there is a significant colour change the correct colour is given in brackets. The white and black bars are not affected by this type of fault.

(a) *Missing* $(R' - Y')$. The picture appears predominantly blue and green; the colour bar sequence is: white, yellow, pale mauve (cyan), green, blue (magenta), dark

green (red), blue, black. Suspect: no V signal from the delay line, (R′ – Y′) demodulator, subcarrier feed to the demodulator, certain faults in the PAL switch, (R′ – Y′) amplifiers.
(b) *Missing (B′ – Y′).* The picture appears predominantly purple and pale green; the colour bar display is: white, very pale yellow (yellow), pale green (cyan), greenish cyan (green), reddish magenta (magenta), red, black (blue), black. Suspect: no U signal from the delay line, (B′ – Y′) demodulator, subcarrier feed to the demodulator, (B′ – Y′) amplifiers.
(c) *Missing (G′ – Y′).* Symptoms on the screen are most difficult to observe, since this is the lowest amplitude signal and it has least effect. Basically all colours become somewhat dull (including the green). Suspect: (G′ – Y′) matrix and (G′ – Y′) amplifiers.
(3) *Faulty PAL switch* (normally the effects are visible only if the colour killer is overridden–see 'No colour' faults).

(a) *PAL switch out of step* (with the transmitter). Applicable to circuits which employ a bistable. Wrong colours may appear since the fault causes the polarity of (R′ – Y′) to be inverted. In a picture this produces so-called 'red' grass and 'green' faces, as confirmed by the resultant colour bar sequence: white, greenish yellow (instead of yellow), pale magenta (cyan), *reddish orange (green)*, bluish cyan (magenta), *dark green (red)*, blue, black. The fault symptoms may either appear or disappear upon changing channels, as there is an equal chance of the bistable starting up either correctly or incorrectly without ident synchronisation. Suspect: ident circuits, ident feed to the bistable.
(b) *PAL switch stopped.* Very severe *Hanover blinds*, which from a considerable viewing distance may appear as wrong colours. Suspect: bistable not switching owing to a faulty circuit or lack of line pulses; fault in the ident circuits where a bistable is not used. NOTE: The bistable output is at 7.8 kHz; consequently either this or the ident can be coupled via a low value capacitor–say 0.01 μF–to the audio output stage. A high-pitched sound will be heard from the speaker if the signal is present.

I.C. fault-finding

(1) Identify the fault area and the associated I.C.s from the fault symptoms.
(2) Check all d.c. voltages associated with the I.C. These devices are usually directly coupled throughout; if the d.c. readings are incorrect, the signal path will be affected.
(3) Check all a.c. signals (waveforms) and compare them with the makers' manual.
(4) Check that the circuits preceding (or following) the I.C. are functioning normally. If necessary, disconnect the feed to or from the I.C. and check the effect on the relevant instrument readings.
(5) Observe the following precautions:

(a) Whenever possible, connect instrument probes away from the pins of the I.C., thus avoiding accidental short circuits.
(b) Resistance checks on an I.C. are meaningless and should not be carried out unless specifically mentioned by the manufacturer.
(c) Avoid overheating the device with a soldering iron.
(d) Before replacing the I.C., ensure that the supply voltage is correct and that there are no external component short circuits.

8 *TV Sound*

8.1 Principle of Intercarrier Sound

The TV sound signal is transmitted at a carrier frequency 6 MHz above the vision carrier frequency. The mixer stage in the tuner produces two i.f. frequencies corresponding to the two carriers: 33.5 MHz, sound signal; 39.5 MHz, vision. Both i.f.s are amplified in the common i.f. strip until the sound information is finally separated in the form of a beat frequency signal—that is, 39.5 MHz – 33.5 MHz = 6 MHz. The frequency difference between the two carriers is then called the *intercarrier frequency*. The intercarrier signal is generated by a non-linear device, such as a simple diode detector which acts as a mixer. As shown in the diagrams in figure 8.1, the video detector itself is often used to produce the 6 MHz beat frequency (the chrominance signal can also be obtained on the same principle).

An example of an intercarrier signal take-off arrangement is given in a simple circuit shown in figure 8.1(b). The vision detector diode is followed by a 6 MHz tuned circuit L_2, C_2 in series with the feed to the video amplifier to provide a rejection of sound from the video stages. At the same time L_2, L_3 form a transformer so that the intercarrier signal is induced in the secondary winding L_3. Since L_3, C_3 are tuned to 6 MHz, a large amplitude signal is communicated to the 6 MHz amplifiers.

In some designs the intercarrier signal is taken off at a later point in the video amplifier stages. The disadvantage of such a method is that sound/vision or sound/chrominance beat frequency patterns can interfere with the picture. Conversely, there could also be a greater possibility of vision-on-sound interference (so called vision-buzz). A separate diode is sometimes used to free the vision detector from handling the 33.5 MHz information of any significant amplitude and to lessen the chance of interference patterns.

An alternative method to the intercarrier principle would be to amplify the 33.5 MHz signal separately. Such an arrangement would introduce unnecessary circuit complexity and, in the event of the tuner oscillator drift, the sound carrier could fall outside its own i.f. response curve. Should that happen, the resultant audio output could become either noisy or distorted. The intercarrier signal is always at the same frequency; consequently good quality sound is possible provided both the signal strength and the receiver i.f. alignment are satisfactory. The same conditions apply in order to prevent the sound signal interfering with the picture.

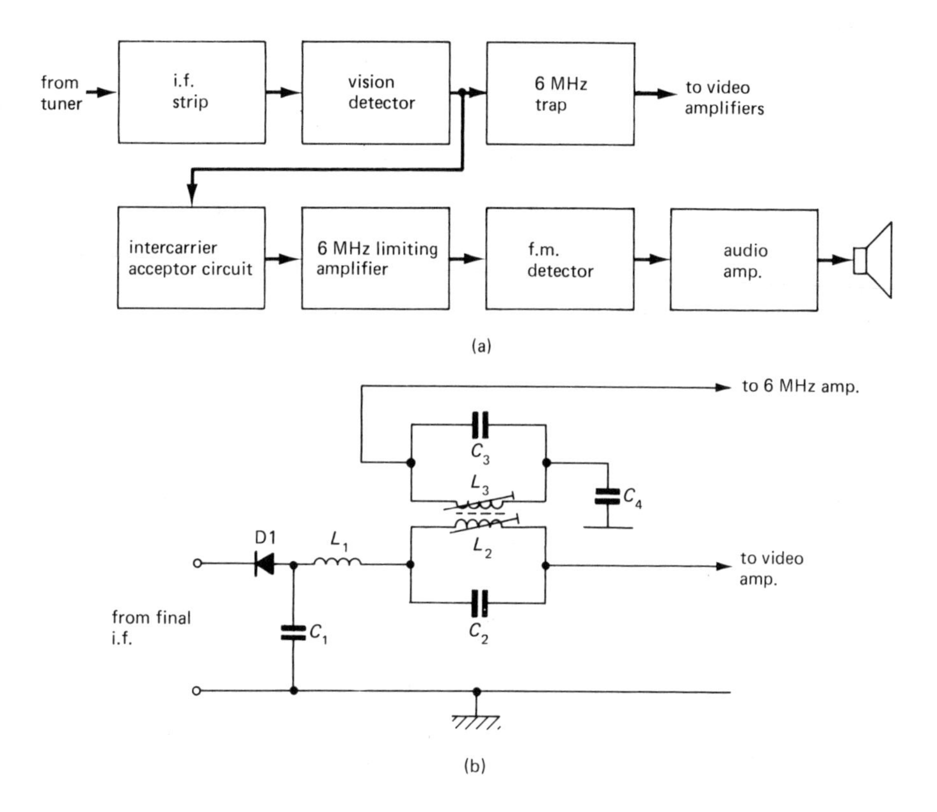

Figure 8.1 Intercarrier sound: (a) block diagram of TV sound as part of a receiver; (b) intercarrier take-off arrangement at vision detector

8.2 Intercarrier Amplifiers and Sound Demodulators

The TV sound carrier is frequency modulated, and therefore it is relatively unaffected by amplitude modulated interference signals. Interference may come primarily from the vision signal, which, on peak white, could cause accidental amplitude modulation of the sound signal. The reader will recall that the luminance signal is not allowed to modulate the vision carrier below 18 per cent of the carrier amplitude; this ensures that a sufficiently large intercarrier signal is always present. However, vision buzz would become evident if the correct relation between the amplitudes of the two carriers were lost in the receiver.

The sound system will enjoy freedom from interference provided the intercarrier amplifier offers so much gain that the peaks of the amplified signal can be 'sliced off'. Most of the interfering amplitude modulation would then be eliminated and the sound detector would be presented with a 'pure' f.m. signal. Such amplifiers are called *limiting amplifiers* because they restrict the amplitude of their output signal. Where transistors are used in the intercarrier section, a diode may be connected across the 6 MHz circuit in order to clip any excessive amplitude. Alternatively, the biasing of the tuned circuit amplifiers is such that self-limiting takes place within the stage.

Ceramic filters, which replace tuned circuits, can be found at the input to the intercarrier amplifiers. Their behaviour is similar to that of a crystal vibrating at a specified frequency. To improve the selectivity and amplifier gain, two ceramic filters following each other may be used. For TV applications the filters are tuned to 6 MHz; they are also used in f.m. radio receivers, where they are tuned to 10.7 MHz. Ceramic filters offer a narrow bandwidth, which results in excellent selectivity; they have good stability and present no alignment problems.

Modern designs use integrated circuits to provide amplification of the 6 MHz signal. The same I.C. may also contain the sound detector as well as an audio pre-amplifier.

Where the detection of the audio signal is done by discrete circuits, two main methods are used–namely a *ratio detector* or a *slope detector.*

A circuit diagram of a *ratio detector* is given in figure 8.2. The detector transformer, L_1, L_2, is tuned to 6 MHz and the tertiary winding, L_3, injects an additional voltage to the centre tap of L_2. When the incoming signal remains at its

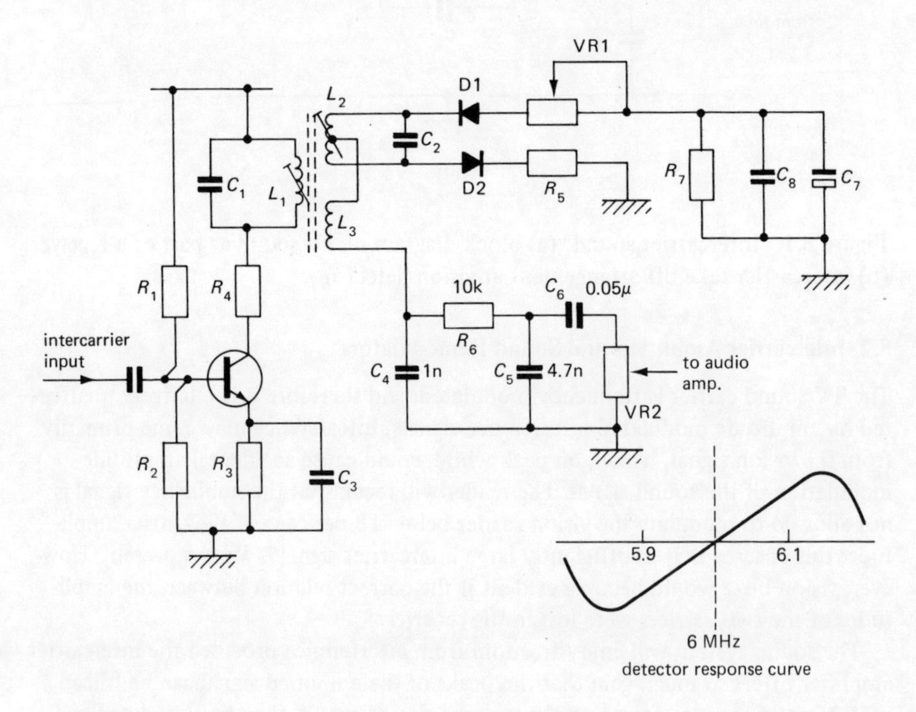

Figure 8.2 F.M. ratio detector circuit (Rank). The S-curve represents the response of the circuit–an audio output signal is produced as the intercarrier frequency deviates from 6 MHz

carrier frequency, the two diodes conduct by equal amounts and the reservoir capacitor, C_7, charges to a d.c. voltage dependent upon the signal strength. Because of the balance under this condition, the potential across VR2 is zero. When the modulating signal causes the intercarrier frequency to deviate from 6 MHz, the balance of the circuit is upset. This results in a current flow during one half-cycle through D2, chassis, C_4, L_3 and back to L_2; during the next half-cycle of the audio signal the current is in the opposite direction via D1, L_3, C_4, chassis and back to D1. The audio output is developed across C_4 and its parallel network of R_6, C_5, C_6 and the volume control VR2.

In common with f.m. radio broadcasting, TV sound is subjected to *pre-emphasis* at the transmitter. It means that high frequency audio signals are boosted by comparison with middle and low frequencies. The emphasised high frequency information swamps most of the subsequent noise, which also tends to have a predominantly high frequency content. In the receiver the signal is given *de-emphasis* or treble cut. The original high frequency audio signals are restored to their proper

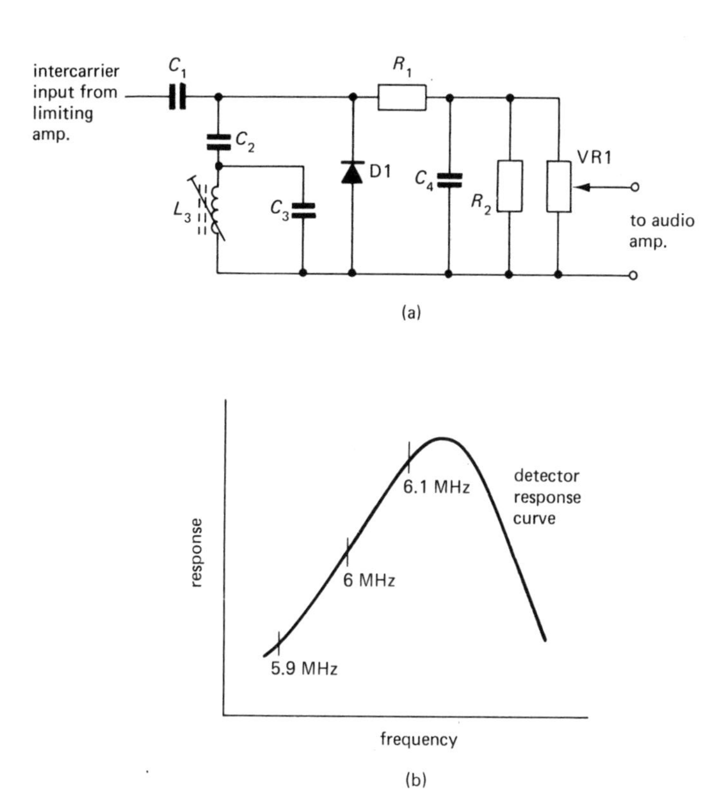

Figure 8.3 F.M. diode slope detector. The tuning adjustment (L_3) places 6 MHz on the slope of response curve (b)

level, which automatically reduces the effects of noise. A simple filter, consisting of a resistor and a capacitor, achieves the required effect. Pre- and de-emphasis are expressed in terms of the time constant of such a filter; in the UK it is 50 μs. In the circuit in figure 8.2 the filter consists of R_6 and C_5, whose time constant is 10 kΩ × 4.7 nF = 47 μs, which is near enough to the specified value. Of course, other combinations of component values are possible, but their method of connection with respect to the signal path remains as shown in the diagram.

Rejection of a.m. signals relies on the detector circuit being perfectly balanced. Therefore the total resistance of the circuit associated with D1 must be the same as that of D2. In figure 8.2 variable resistor VR1 compensates for imbalance caused by differing forward resistances of the diodes and also for the tolerance of R_5. The adjustment of VR1 aims at a minimum buzz on sound. The relatively large capacitance of C_7 also helps to suppress any a.m. interference, while C_8 is a 6 MHz filter capacitor.

There are many versions of the ratio detector circuit, but the fundamental principle remains the same for all of them. Often the centre tap in the secondary of the detector transformer is not in the winding, but is formed by the junction of two capacitors (in place of C_2 in figure 8.2).

A *slope detector circuit* is shown in figure 8.3(a). The intercarrier signal is applied to the tuned circuit formed by L_3 and C_3. The response of this circuit is such that 6 MHz is on the slope of the resultant curve–see figure 8.3(b). As the modulated intercarrier frequency swings within narrow limits (maximum ±50 kHz), the output from the arrangement varies accordingly and provides the audio signal; de-emphasis components are R_1 and C_4. The alignment of this type of circuit can be made fairly simple by tuning L_3 for a maximum undistorted output.

The final method of detection to be considered is *quadrature detection.* This is usually incorporated in an I.C., which often includes a multistage limiting amplifier and an audio preamplifier. A block diagram, illustrating the principle of quadrature detection, is given in figure 8.4(a). Transistors TR1 and TR2 produce separate pulsed outputs from the amplified and limited intercarrier signal inputs. When the frequency is exactly 6 MHz, the two pulses are 90° out of phase (or in *quadrature*) with each other; capacitor C at the input to TR2 provides the necessary phase shift.

The quadrature condition is maintained at the resonant frequency of 6 MHz by the tuned circuit L_1, C_1. Transistors TR1 and TR2 are driven between saturation and cut-off, and the resultant outputs are fed to the third transistor circuit, TR3. The conduction of TR3 depends upon the degree of *coincidence* between its two inputs. As the modulation deviates the intercarrier frequency from the centre value of 6 MHz, the presence of the tuned circuit alters the original 90° phase shift. The net result is that the two sets of pulses fed to TR3 coincide either for longer or shorter periods of time. In effect, the input to TR3 is in the form of variable width pulses which switch the transistor on, either for shorter or for longer time intervals. The resultant output from TR3, after suitable filtering, produces an output that varies in sympathy with the original frequency modulation.

An example of a complete circuit used in a TV receiver is given in figure 8.4(b).

A signal from the video detector is fed to the 6 MHz ceramic filter (C.F.), from which the intercarrier frequency of a very narrow bandwidth is applied to pin 2 of IC1. After several stages of limiting, the frequency modulated signal enters the quadrature detector. Components L_1 and C_1 form the external tuned circuit–

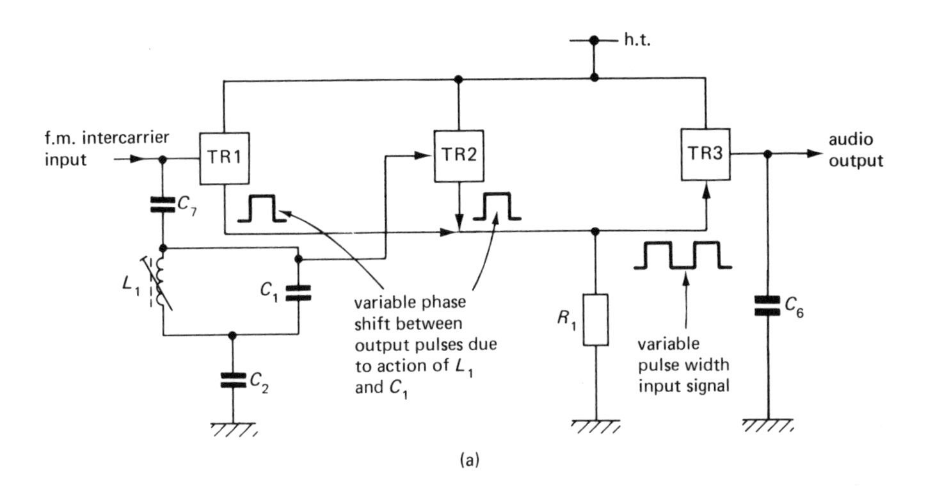

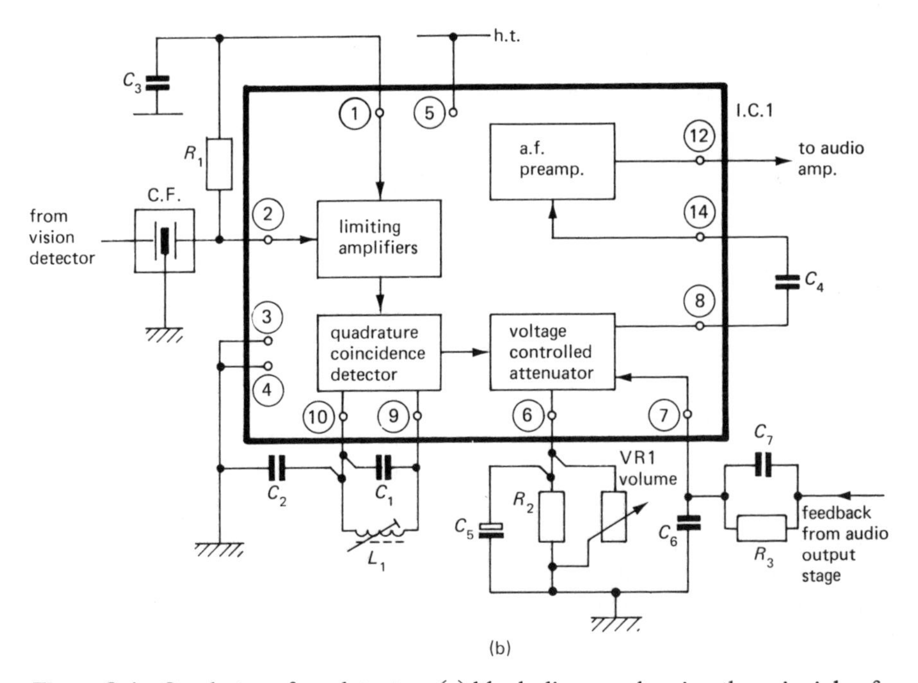

Figure 8.4 Quadrature f.m. detector: (a) block diagram showing the principle of quadrature coincidence detector; (b) complete practical circuit using I.C. type MC 1358Q (Thorn)

decoupled to chassis via C_2. The *quadrature coil*, L_1, can be simply adjusted for maximum amplitude, undistorted audio output.

The volume control, VR1, acts indirectly by varying the d.c. voltage fed to a separate attenuator stage in the I.C. The signal is then coupled via C_4 to a built-in preamplifier circuit. Finally, the audio output is fed to a power amplifier which drives a loudspeaker.

De-emphasis components are often less obvious when I.C.s are used, because the required arrangement may be part of the integrated circuit itself. In figure 8.4(b) capacitor C_6, in conjunction with the frequency dependent feedback network R_3, C_7 at pin 7, offers the necessary compensation.

8.3 Audio Frequency Amplifiers

In this part of the receiver the power level of the detected audio signal is raised from a fraction of a milliwatt to perhaps several watts in order to drive the loudspeaker.

Where valves are used elsewhere in the receiver, it is not uncommon to find a valve audio output stage. Since a power output valve may require an input voltage swing of several volts, a transistor or a valve preamplifier is necessary.

Audio output amplifiers may be divided into two groups according to their d.c. biasing: class A and class B. In *class A* amplifiers, with no signal applied, the valve or the transistor is biased so as to carry a relatively high output current, known as the *quiescent* current. When the audio signal is added to the bias current, the output current increases further during one half-cycle, and reduces below the quiescent value in the next half-cycle. To prevent signal distortion, the biasing must be carefully arranged to ensure that, with the maximum signal amplitude, the device is neither driven into saturation nor into cut off. Class A output amplifiers favour valve circuits because the quiescent current is only in the region of, say, 20 or 30 mA. In transistorised power output stages the 'standing' current may be of several hundred mA. This tends to create excessive heat dissipation; it also requires a high power rated transistor and leads to a large current drain from the power supplies. Incidentally, class A amplifiers are used in most other parts of a TV receiver, where signal amplitudes are usually low.

Class B amplifiers require two transistors (valve circuits of this kind are not used in TV receivers). They are arranged so that, theoretically at least, one transistor conducts during positive half-cycles of the audio signal, and the other transistor amplifies the negative half-cycles. Often one of them is a p–n–p device and the second one an n–p–n; the arrangement is called a *complementary symmetry* output stage (see figure 8.5). In this the p–n–p transistor deals with the negative-going part of the waveform; the n–p–n device amplifies the positive half-cycles of the input signal. In practice, it is not possible to rely on the transistors to be switched on by the signal alone, because a certain minimum value of base voltage must be reached before conduction can begin. Unless a d.c. bias of approximately 0.2 V for germanium and 0.6 V for silicon devices is provided, the output will be seriously distorted. This type of distortion is called *crossover distortion*, for it occurs when one

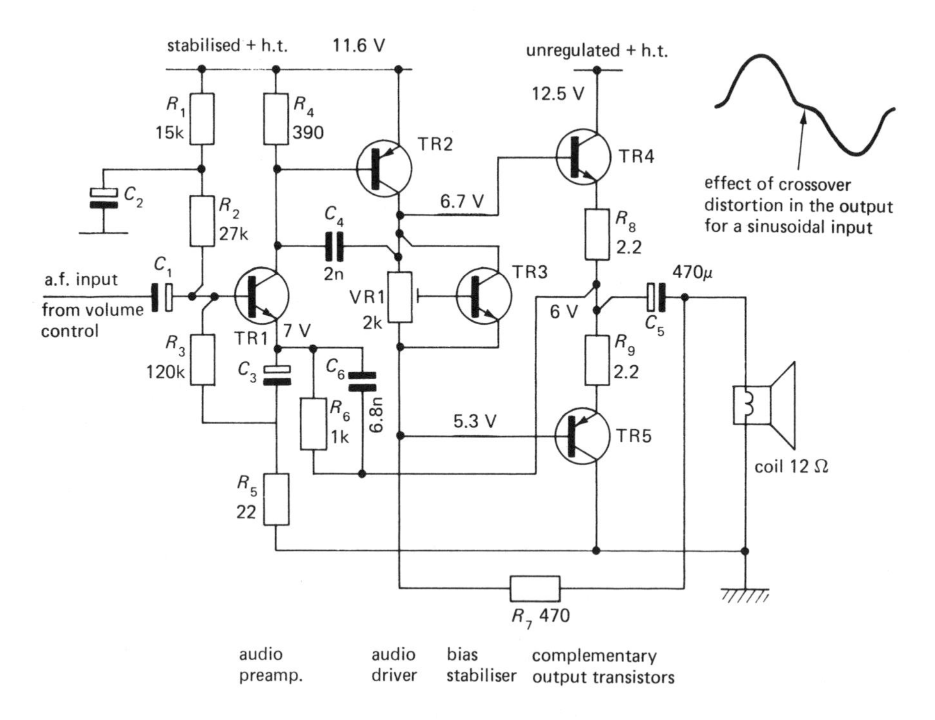

Figure 8.5 Complementary symmetry audio output stage. Voltage levels show how the d.c. output voltage is maintained at about half the h.t. supply potential (Thorn)

transistor is approaching zero current and the other one is due to take over. The effect of crossover distortion on a sinusoidal waveform is shown in figure 8.5. The reader will appreciate that the presence of such a defect would be particularly noticeable at a low volume of sound.

An example of a class B complementary output stage is given in figure 8.5. An important feature to note is that the whole stage is directly coupled, except for the loudspeaker, which is fed via capacitor C_5. Therefore the transistors not only act as amplifying devices, but also provide correct d.c. biasing for one another. Because of the direct connection between the individual stages, even slight drift in d.c. voltage in one of them could spread to the remaining stages and the biasing throughout the whole amplifier would be seriously upset. To prevent this happening, d.c. feedback is used to stabilise the operating point of the circuit.

In figure 8.5 transistor TR1 is responsible for maintaining correct overall bias for the stage. The base of TR1 is connected to a relatively conventional potential divider R_1, R_2, R_3, R_5, but its emitter is effectively returned, via R_6, to the junction of the output pair TR4 and TR5. Because of the symmetry in the output circuit, the voltage at the junction of R_8 and R_9 is approximately one-half that of the supply voltage. Transistor TR1, however 'senses' the value of the voltage at the junction of TR4 and TR5; any change in this potential is communicated to the emitter of TR1 to adjust its conduction. If, for example, the 'centre point' output voltage increased above 6 V, TR1 would be forced to conduct less. As a result the collector voltage would rise and, because of direct connection to the base of TR2, this transistor would also conduct less. In turn, the base voltage on TR4 would fall; consequently, the emitter voltage of TR4 would be brought back to its nominal 6 V.

Transistor TR3 in figure 8.5 acts as a variable resistor in the base bias circuit of the output transistors. As mentioned earlier, the output stage requires a certain amount of base bias in order to prevent crossover distortion. However, if full advantage is to be taken of class B operation, the d.c. biasing must be kept to a minimum. In this circuit the adjustment is carried out by means of VR1, which varies the conductivity of TR3 and, in effect, alters the biasing of TR4 and TR5. The setting of VR1 is critical–if the quiescent bias is too high, the transistors can become overheated and destroyed. The manufacturer's adjustment procedure for this particular receiver is to connect a milliammeter in the h.t. feed supplying the output transistors, and then adjust VR1 until the current is 4 mA (volume control set to minimum).

Class B audio amplifiers in other receivers may be without a variable bias control. There are also circuits which use output transistors of the same polarity (either both p–n–p or both n–p–n) instead of complementary types.

The audio signal path in the circuit in figure 8.5 can be followed in a similar manner to the change in the d.c. levels discussed above. The reader is invited to work out for himself what happens when the positive half-cycle increases the conduction of TR1 and the negative half-cycle reduces it. Ultimately, the positive-going portion of the input waveform causes the current in TR4 to rise, while cutting off TR5; the load current then flows through TR4, C_5, the loudspeaker and the chassis –charging C_5 in the process. During the negative half-cycle the signal current in the speaker changes its direction as TR5 conducts by discharging capacitor C_5.

Apart from the d.c. feedback, there is also a.c. negative feedback, which carries out a number of functions, including preventing instability, improving the frequency response, reducing distortion, etc. This is achieved in figure 8.5 by: C_3, R_5, R_3, C_6, C_4 and, finally, R_7.

Full understanding of the operation of this and similar types of directly coupled circuits is important in fault-finding. The effects of a fault in one part of the circuit can affect the biasing of the remaining sections, which makes the interpretation of voltmeter readings difficult. Incidentally, a circuit somewhat similar to that in figure 8.5 may also be used in the field time base of a TV receiver.

We shall not consider details of class A audio amplifiers, since they follow the usual circuit techniques where either a single valve or a transistor is used.

The diagram in figure 8.6 shows a circuit of an audio output stage which employs an I.C. This may be preceded by a demodulator I.C. of the type illustrated in figure 8.4. Integrated circuit amplifiers are capable of very high gains; consequently, their frequency response must be carefully tailored to obtain the desired bandwidth and to prevent instability. The shaping of the amplifier frequency response in figure 8.6 is by means of the negative feedback networks: R_6, C_6 from the output pin 6 to pin 9 (via C_7) and R_3, C_3 from pin 16 to the input pin 1. Components R_4, C_4 (pin 16) and C_5 (pin 15) are also used to determine the low frequency and the high frequency response, respectively. The d.c. bias of the preamplifier is obtained via R_1, R_2 (decoupled by C_2) between pins 1 and 2. The amplifier output (pin 6) is shunted by a compensating network C_7, R_7; it ensures that the effective loudspeaker load is resistive rather than inductive. The latter arrangement provides for a better response to rapidly changing signals (*transients*); it also offers protection to the output transistors in the I.C. against overvoltages which could occur if the effect of the speaker inductance was not compensated.

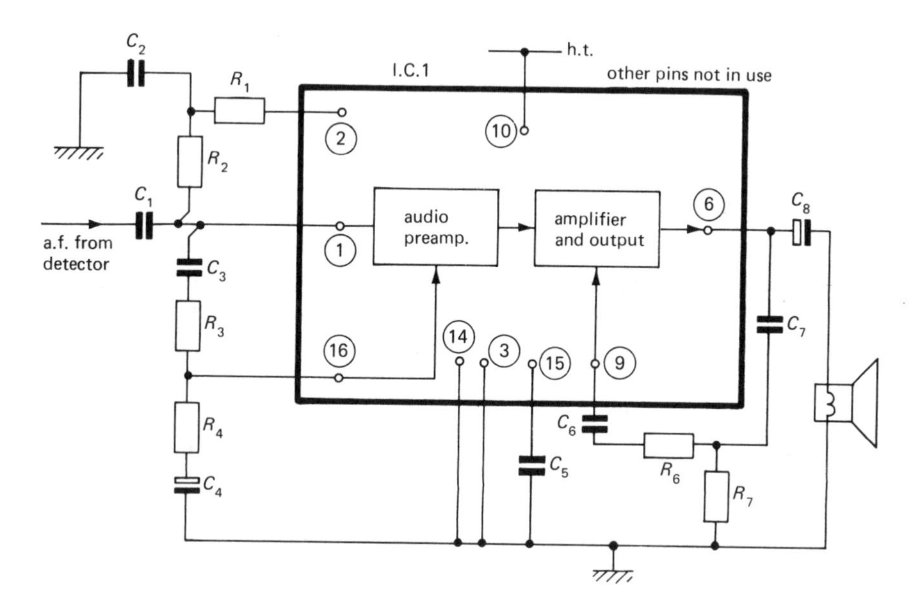

Figure 8.6 Audio output amplifier using I.C. type SN 76013/ND/7 (GEC)

9 Synchronising Pulse Separator and Field Time Base Circuits

9.1 Separation of Synchronising Pulses from Composite Video

The synchronising pulse separator circuit produces a train of relatively large amplitude rectangular pulses from the composite video waveform. The input to the circuit is from one of the video (luminance) amplifier stages. The separated pulses, after further processing, are then available for synchronising the receiver time bases with the scanning action of the transmitter camera tube.

The process of separation relies on the fact that the synchronising pulses in the composite signal always begin at a fixed reference point of the video waveform—the black level. Therefore the required circuit is a voltage level detector. The separator circuit is made non-conducting during picture information, but it is driven into saturation when the video waveform enters the 'blacker than black' region. Interference signals can also be present in the synchronising pulse part of the waveform; therefore some circuits may have additional features to render the separator relatively immune to unwanted signals.

A simple transistorised circuit is shown in figure 9.1. Negative-going video signal (hence, positive-going pulses) from the video driver stage is fed to the separator transistor, TR1, via C_1 and R_1. When the transistor is driven into saturation by the synchronising pulse portion of the input waveform, the collector voltage falls, and negative-going output pulses are generated. At the same time the input coupling capacitor, C_1, is charged to the peak value of the incoming waveform. After the synchronising pulse interval, the amplitude of the input video waveform falls, which effectively places the charged capacitor across the base–emitter junction of TR1. The negative potential on the right-hand plate of C_1 causes the transistor to be cut off, and the collector output voltage rises to h.t. This state is maintained during the active picture period as the capacitor discharges via both the preceding amplifier stage and the high value resistors R_2 and R_3. When the next pulse interval arrives, C_1 is recharged and the process is repeated. Resistors R_2 and R_3 provide a potential divider which applies a small forward bias to the base of TR1. This, by itself, is insufficient to turn the transistor on, but it ensures a much quicker changeover from cut-off into saturation when the pulses are applied.

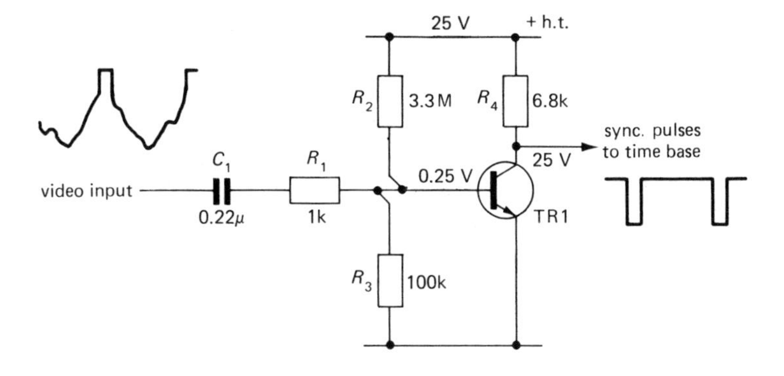

Figure 9.1 Transistorised sync. separator circuit

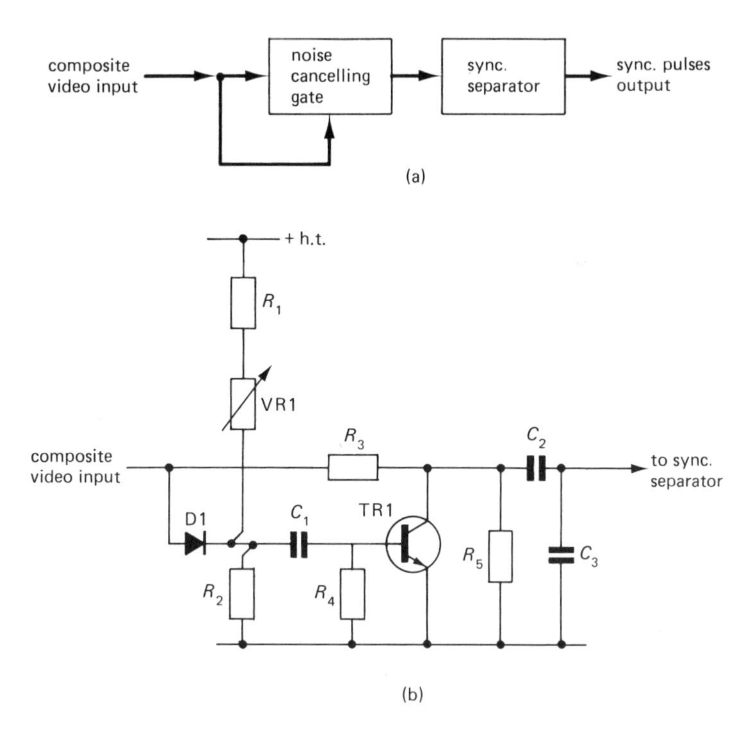

Figure 9.2 Noise gated sync. separator: (a) block diagram–interference pulses on video waveform are used to operate the gate circuit; (b) practical circuit; D1 and VR1 set the threshold of interference level and TR1 is the noise cancelling gate (Rank)

Negative-going output pulses are produced by the circuit in figure 9.1. Should positive synchronising pulses be required, either an inverting amplifier might be included or the take-off point would be chosen from a suitable luminance (video) amplifying stage. Alternatively, a p–n–p transistor sync. separator circuit may be found in an arrangement of n–p–n transistors.

A number of receivers use a *noise gated synchronising separator.* An ordinary separator circuit is likely to amplify interference signals and produce spurious pulses at its output. It will be shown later, however, that both the line and the field time base oscillators are not easily affected by such unwanted signals. For that reason most receivers do not take any special precautions against noise in the separator stage.

Noise gating, or noise filtering, is based on the fact that the amplitude of interference pulses is much greater than the amplitude of synchronising pulses. A block diagram of a noise cancelled synchronising separator is shown in figure 9.2(a)—the gate renders the actual separator circuit inoperative during interference periods. The practical circuit may be either contained in an I.C. or can use discrete components as given in figure 9.2(b). In this example both the synchronising pulses and the accompanying interference are *positive*-going. The gating diode, D1, is normally reverse biased from the potential divider formed by R_1, VR1 and R_2. The amount of bias is adjusted by VR1, so that the synchronising pulses cannot, by themselves, turn the diode on, and the pulses continue to the separator via C_2. Large amplitude negative-going interference pulses bias D1 ON, which, in turn, switches the gating transistor, TR1, on. In effect, a momentary short circuit is placed across the signal path to the separator and the noise is removed. Capacitor C_3 is of a relatively low value (330 pF); it removes the high frequency chrominance information, which could also extend into the synchronising pulse level. The presence of C_3 alone

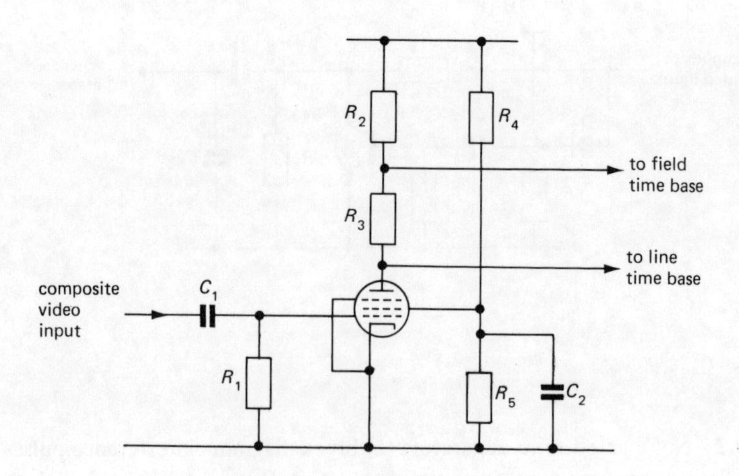

Figure 9.3 Valve sync. separator circuit

would have helped in reducing the amount of interference reaching the synchronising separator. In many receivers such a simple capacitor filter provides all the necessary noise cancellation.

A valve synchronising separator circuit is shown in figure 9.3. The principle of operation is otherwise similar to that described for its transistor counterpart. The video input has positive-going synchronising pulses which drive the valve into saturation and charge C_1 at the same time. For the remainder of the waveform the charge stored in the capacitor maintains the pentode in a cut-off state until the arrival of the next pulse. Therefore the outputs from the anode circuit are in the form of negative-going pulses. The screen voltage of the valve, set by R_4, R_5, is relatively low, which ensures that the cut-off state is also maintained under no-signal conditions. The valve is usually part of a triode–pentode arrangement; the other section may be used in one of the time bases.

Finally, the synchronising pulse separator circuit can be incorporated into an I.C. The device would, naturally, provide other functions, as will be shown later in this chapter (see figure 9.14).

The output pulses from the sync. separator must be of the correct magnitude, shape and timing. These conditions are fulfilled by the values of the components in the circuit. Should a fault develop in this area, the separator might, for example, switch during picture information, or it could produce pulses of insufficient amplitude. Such defects would cause lack of proper synchronisation, or lead to the synchronisation being upset by the picture content. Both the line and the field time bases could be affected by a faulty sync. separator. However, the flywheel circuit in the line time base can sometimes mask the effect of separator faults on horizontal synchronisation (see section 10.1).

9.2 Processing of Field and Line Synchronising Pulses

The output from the sync. separator is in the form of either positive- or negative-going pulses: the line, field and equalising pulses are all present in succession. They must now be sorted out in order to make the two time bases respond only to their respective synchronising information. That is done by means of simple resistor–capacitor networks. Depending upon the values of the components and their method of connection, these circuits can be called either *integrating* or *differentiating circuits.* The names are derived from mathematical operations, because such networks can, apart from their TV applications, perform the mathematical equivalents of either integration or differentiation in, for example, certain computer circuits.

Integrating circuits

An example of such a circuit is given in figure 9.4(a). Capacitor C is allowed to charge via resistor R from the driving circuit. The graph in figure 9.4(b) shows how the voltage across the capacitor rises if the input, V_{IN}, is maintained constant for a relatively long time. Ultimately, the capacitor would become fully charged

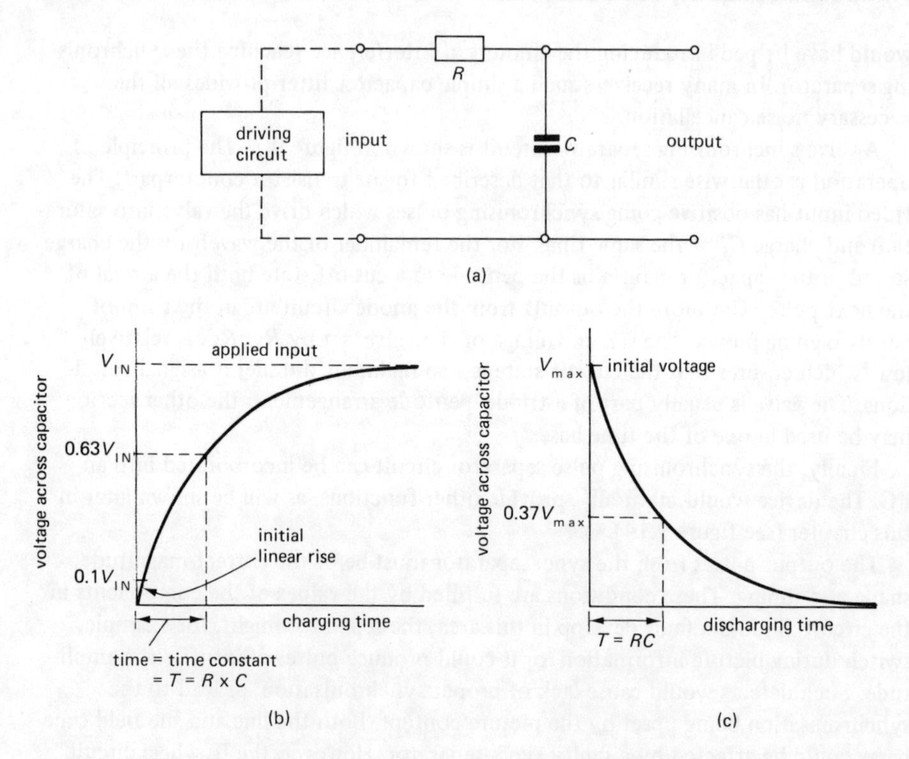

Figure 9.4 Pulse integrating circuit: (a) basic circuit–input waveform charges C via R; (b) graph of voltage rise across capacitor during charging (assuming constant d.c. level at the input to the circuit); (c) graph of voltage fall across capacitor during discharging

and the output voltage would be equal to the input. The speed with which the potential across C rises depends upon the value of the series resistance, R, and the capacitance, C. The product $R \times C$ is called the *time constant* of the circuit (T). From the mathematical properties of the charging graph it can be proved that after a time interval equal to the time constant the voltage on the capacitor reaches 63 per cent of the maximum driving voltage. The initial shape of the graph up to, say, 10 per cent of the maximum is reasonably linear, after which the graph becomes increasingly curved.

If the circuit is fed with synchronising pulses, the voltage, V_{IN}, is sustained only for the duration of each pulse. Therefore the time constant of the integrator is deliberately made long by comparison with the duration of the input pulse. Now the capacitor voltage will rise only along the initial portion of the charging curve in figure 9.4(b).

To sum up: an integrating circuit has a *long time constant* when compared with

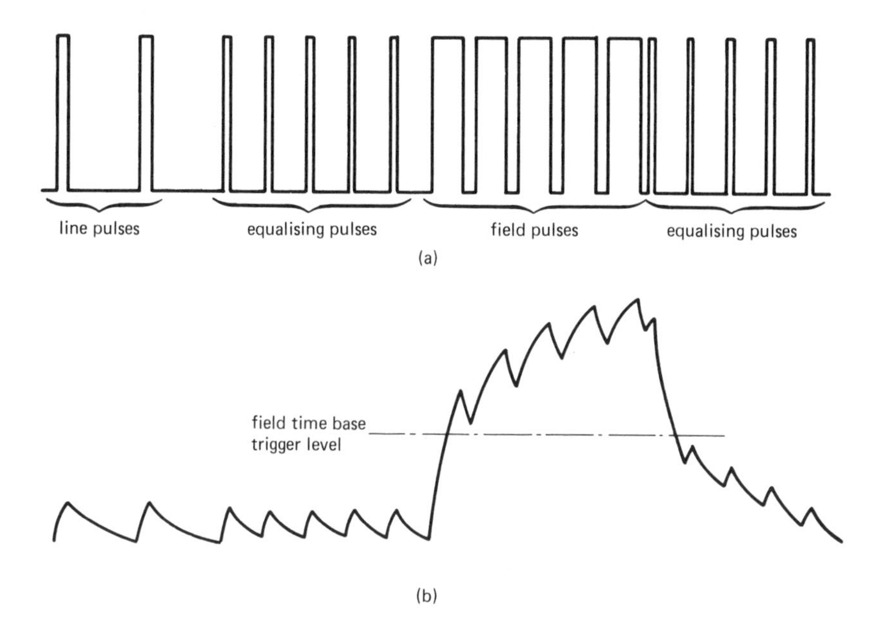

Figure 9.5 Formation of a field time base trigger pulse (see also figure 3.8): (a) output from sync. pulse separator; (b) output from field pulse integrating circuit

the duration of the applied pulse, *and* the output is taken from across the capacitor. This type of circuit changes the shape of a rectangular, or square, pulse input waveform into a sawtooth type of output waveform. An integrating circuit is used to produce a field time base trigger pulse from the incoming field synchronising sequence. Another application is in the generation of sawtooth scan waveforms and certain convergence correction waveforms in the line and field time bases.

The diagram in figure 9.5 explains how the field synchronising sequence of pulses is changed into a field trigger pulse by the action of an integrating circuit. The pulses produced by the sync. separator are shown in figure 9.5(a); these are then fed to an integrator. The resultant output waveform across the integrating capacitor is illustrated in figure 9.5(b).

In a practical version of the circuit in figure 9.3 the value of resistance can be 18 kΩ and that of the capacitance 2000 pF; the resultant time constant $T = R \times C =$ 18 000 Ω × 2000 pF = 36 μs. This is relatively long compared with the duration of the various input pulses. Therefore, turning to figure 9.5, the equalising pulses (2.3 μs) charge the capacitor only to a low voltage because of the long time constant of the circuit. During the intervals between successive equalising pulses the capacitor loses most of its voltage. When the broad field pulses (27.3 μs) are applied, the charging period is longer than the discharge time; consequently, the capacitor output voltage is in the form of 'serrations' of increasing amplitude. The intervals between the equalising pulses which follow the field pulses allow the capacitor to discharge to its initial voltage level. The actual calculations of the amplitudes of the charging and discharging steps in the waveform are beyond the scope of this book.

When the build-up of the pulse voltage across the capacitor reaches a predetermined value, the field time base oscillator is synchronised to initiate field flyback. A possible trigger level is suitably marked in figure 9.5(b). The first part of the waveform also indicates that line synchronising pulses (4.7 μs) charge the integrating capacitor to a relatively low potential. This is below the field oscillator trigger level and has no effect. Similarly, short duration interference pulses cannot produce sufficient output to upset field synchronisation.

The need for equalising pulses can also be seen from figure 9.5(b). These are necessary to maintain good *interlace* of the two fields, which form a picture (frame). The first field ends on a complete line (this corresponds to the waveform in figure 9.5) so that the integrating capacitor can discharge almost completely before the field synchronising sequence begins. The second field ends on *half-line*; the reader will observe that the remaining capacitor voltage would then have a higher value at the start of the next field pulse sequence. The slight differences between the two fields would cause the fixed trigger level to be reached later on the first field than on the next, resulting in poor interlace. Equalising pulses 'reset' the capacitor voltage always to the same value, giving the broad pulses an equal starting potential at the end of either field.

Differentiating circuits

An example of a simple differentiating circuit is given in figure 9.6. It consists of a capacitor and a resistor, but now the output is taken from across the resistor. This means that the output waveform is proportional to the capacitor *current.* When a pulse is applied to the input, a large initial current flows, its value being limited only by the resistance in the circuit. As the charge on the capacitor builds up, the current becomes less until, finally, it reaches zero. The complete process follows a graph which is identical *in shape* with that in figure 9.4(c).

In a differentiating circuit the value of the time constant ($R \times C$) must be ***small*** when compared with the width of the pulse applied to it (readers are reminded that for an integrator the time constant must be longer than the pulse width). A short time constant ensures that the charging is completed rapidly, and the output falls to zero for the greater part of the input pulse.

When the pulse is removed, the fully charged capacitor behaves like a storage battery and it discharges through R and the driving circuit. However, the flow of the discharge current is in the opposite direction to that of the charging process—hence a negative-going pulse in the output, as indicated in figure 9.6(c).

To sum up: a differentiating circuit has a *short time constant* when compared with the duration of the applied pulse, *and* the output is taken from across the resistor. The shape of a rectangular (or square) pulse input waveform is changed into two sharp 'spikes' at the output, which coincide with the leading and the trail-

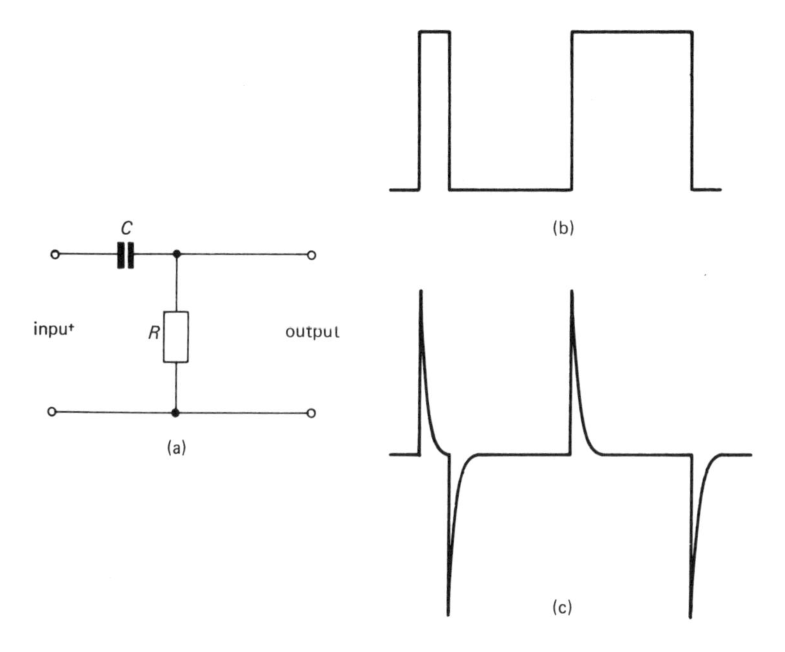

Figure 9.6 Pulse differentiating circuit: (a) basic circuit–input waveform charges *C* via *R*; compared with figure 9.4, output is now across resistor, and the time constant is much shorter; (b) example of an input waveform–one narrow pulse followed by a wide pulse (part of field synchronising sequence); (c) output waveform as a result of differentiating the input pulses

ing edges of the original pulse. Differentiating circuits are chiefly used to produce sharp pulses in order to *switch* another circuit–for example, a time base oscillator or a bistable in the PAL colour decoder.

Since the output from a differentiating circuit is relatively independent of the input pulse width, this type of circuit produces line time base trigger pulses from the narrow equalising pulses, from the broad field pulses and, of course, from the line synchronising pulses themselves. This effect is illustrated in figure 9.6(b) and (c), where two different input pulses result in identically shaped output pulses.

9.3 Receiver Time Base Requirements

The purpose of the time base circuit is to supply a *current waveform* to the deflecting coils whose resultant magnetic field will deflect the electron beam in the tube. A correctly synchronised 'linear' picture should be scanned on the screen. Good *linearity* suggests that the picture is free from distortion, as is evident when simple geometric forms are displayed. In colour receivers the time base must provide additional waveforms for convergence and other raster correction purposes.

The current waveform in the scan coils is principally a *sawtooth*—the relatively long rise in the waveform generates the scan and the rapid fall initiates the flyback. A simple integrating circuit, of the type shown in figure 9.4, can be modified to produce a sawtooth waveform. Charging the capacitor via the resistor at a relatively slow rate gives the scan portion of the waveform. At the end of the scan period the capacitor must be rapidly discharged by means of an electronic switch in the form of a time base oscillator connected across the capacitor. The discharge time is very brief because the resistance of the switch is low (for example, a saturated transistor or a valve), giving a very short time constant. The resultant sawtooth waveform is further amplified before being fed to the scan coils.

The presence of the inductance of the scan coils and, to some extent, that of the transformer can also modify the applied waveform. If a rectangular voltage pulse is fed to an inductor, the resultant current will rise relatively slowly because of the back-e.m.f. in the coil. The graph of the *current* growth is then identical in shape with the voltage rise across a capacitor [figure 9.4(b)]. The rate of increase again depends upon the time constant, which for an inductive circuit is given by L/R (L is the inductance of the coil and R its resistance). This type of circuit can also act as an integrator, provided the time constant is long compared with the duration of the applied voltage pulse.

The length of the *field scan* is approximately 1/50 s, or 20 ms (ignoring the duration of flyback); therefore the time constant of an integrating circuit which produces the field sawtooth waveform would have to be, say, 100 ms. This is easily achieved with a capacitor and a resistor of suitable value. The alternative method of using the scan coils to generate a sawtooth waveform is not possible. The inductance of the coils is, for example, 50 mH, and their resistance is in the region of, say, 50 Ω, which gives a time constant of only 1 ms.

The situation is considerably different in the line time base, since the scan consists of two periods of 26 μs each. Therefore the time constant of the scan coils is sufficiently long to generate the required sawtooth, as will be shown in chapter 10.

9.4 S-correction and Scan Linearity

So far it has been assumed that the *scan current* waveform is in the form of a perfectly linear sawtooth. The practical waveform is somewhat different because a wide angle tube has an almost flat screen. Consequently, the *deflection centre* (the point at which the beam begins to deflect) does not correspond to the centre of the screen curvature. A 'flat' screen implies that its radius of curvature is relatively long; therefore equal changes in the *deflection angle*—based on the deflection centre—do not result in correspondingly equal distances on the screen. This effect is illustrated in figure 9.7(a)—the distances covered by the beam near the centre of the screen are shorter than those away from the centre. A linear sawtooth would produce a picture which was elongated at the edges. To correct for that, the rate of change of scan current must be slowed down towards the extremities of the scan. The resultant current waveform in the coils resembles a somewhat slanted letter

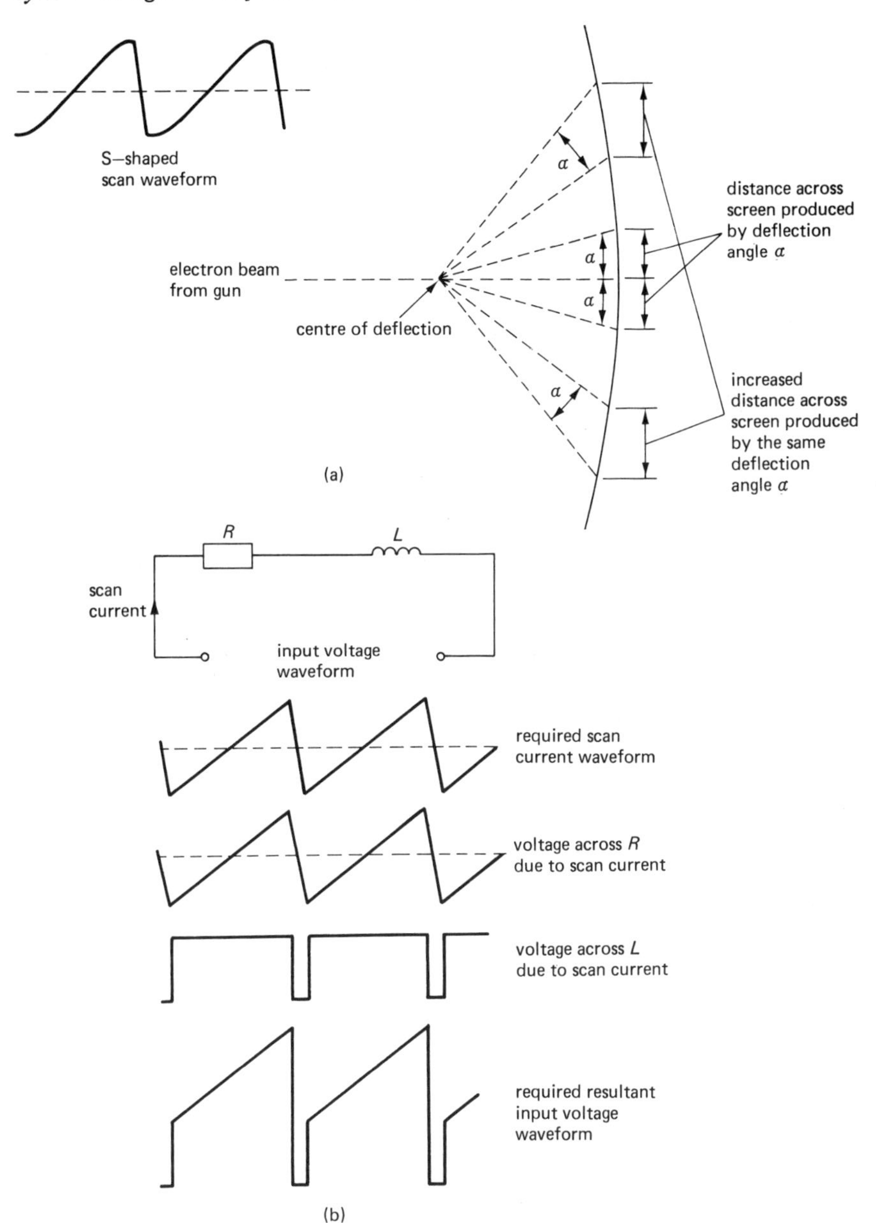

Figure 9.7 Some of the factors to be considered in the formation of scan drive waveforms: (a) S-correction, without which the picture would be stretched near the edges of the screen; scan waveform is flattened towards its peaks, thus reducing progressively the amount of deflection and producing a 'linear' picture; (b) 'trapezium' drive waveform compensates for the combined inductance and resistance of scan coils

'S', and the correction which must be applied to the original sawtooth is called *S-correction.* In field time base circuit S-correction is a part of *picture linearity* correction arrangements within the field amplifier stage.

There are other factors which could cause picture distortion; consequently, the amplifier drive waveforms must be further modified in order to obtain a linear display on the screen. The effect of the time constant of the scan coils on the shape of the current waveform has already been discussed. If the coils have a high inductance (long time constant), the drive waveform would have to be rectangular; if the inductance is very small (very short time constant), the required voltage waveform ought to be a sawtooth. Practical coils may need a trapezium-shaped voltage waveform which, as shown in figure 9.7(b), is the sum of a rectangular waveform and a sawtooth. When the scan coils are coupled to the output stage via a transformer, waveform distortion can again be introduced. This is caused by the saturation of the core owing to the combined effect of the no-signal direct current and the superimposed peak of the scan waveform. The field output stage is a power amplifier driven alternately between cut-off and near-saturation. Such a large signal swing could easily introduce distortion if the bias conditions of the output valve or transistor were incorrect. In fact, some degree of waveform distortion under these circumstances is inevitable.

Linearity controls are provided to compensate for the effects that the above factors have on the scan waveform. Suitable shaping circuits predistort the desired drive waveform, so that the effect of subsequent distortion can be cancelled out. Linearity correction in the field time base is achieved by means of negative feedback. Capacitors and resistors are often used in the feedback path, which further modify the shape of the signal before reintroducing it to the input. The complete arrangement acts as a waveform adding network—the basic sawtooth is developed by partly charging a capacitor and then adding suitable linearity correction waveforms. Their actual shape and amplitude are adjusted by linearity controls until a correct picture is displayed on the screen.

9.5 Valve Field Time Base Circuits

Since the approach to circuit design differs between valve and transistor receivers, it is necessary to consider the two types separately. The block diagram in figure 9.8 shows the basic functions of a typical circuit where valves are employed.

The trigger pulse shaping circuit develops a synchronising pulse for the field oscillator from the pulses arriving from the sync. separator. Basically this consists of some form of an integrating capacitor circuit, as described in section 9.2. Additional relative complexity at this stage ensures correct interlace; any unwanted pulses that could otherwise cause premature flyback are often removed by an *interlace filter.* An example of this is given in the diagram in figure 9.10 as part of a complete time base circuit which will be discussed later.

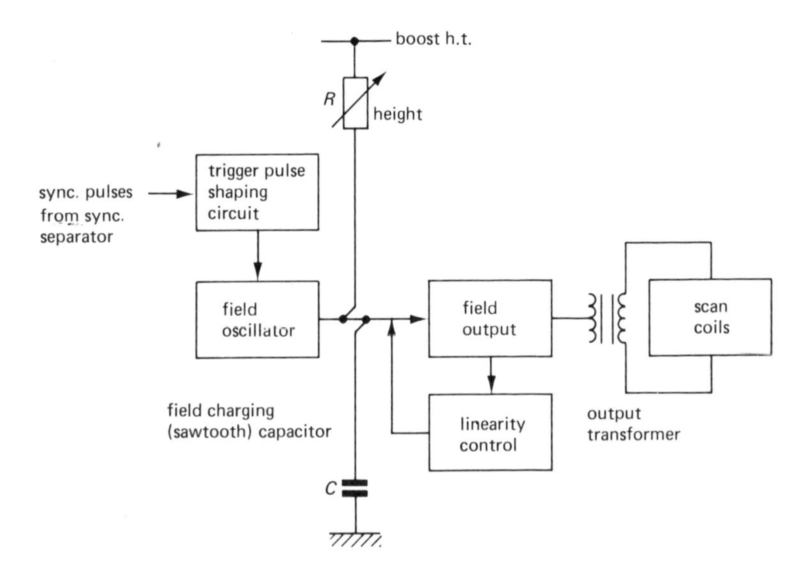

Figure 9.8 Block diagram of a valve field time base. The oscillator acts as a switch allowing C to charge via R (long time constant), and to discharge via the oscillator itself (short time constant)

Valve time base oscillators

The field oscillator produces pulses which allow the field charging (sawtooth) capacitor to charge during the scan period, and then to rapidly discharge it to initiate the flyback. The circuit behaves like a changeover switch connecting the capacitor alternately to the supply line or to chassis. In valve field time bases the most common form is a multivibrator or, occasionally, a blocking oscillator.

The two basic forms of *multivibrator* are: anode to grid cross-coupled and a cathode coupled circuit. The circuits in figure 9.9 show the fundamental arrangements of each. The more frequently encountered cross-coupled multivibrator is shown in figure 9.9(a). The oscillator is free-running as the valves are driven alternately between cut-off and saturation. The oscillator starts up when the circuit is first switched on, since the inevitable inequality between the two valves combined with the anode–grid coupling drives, for example, V2 into saturation while V1 becomes cut off. During this stage, the cross-coupling capacitor C_1 charges from h.t. via R_1 and the grid–cathode of V2. The situation is then reversed; V1 becomes saturated and V2 cuts off—now C_2 charges from h.t. via R_2 and the grid–cathode of V1. At the same time the previously charged capacitor C_1 is discharging through the saturated valve, V1, and resistor R_4. In effect, the negative plate of C_1 is connected between the grid of V2 and the chassis; therefore the valve remains cut off. The rate of discharge depends on the time constant R_4C_1, so that after a cer-

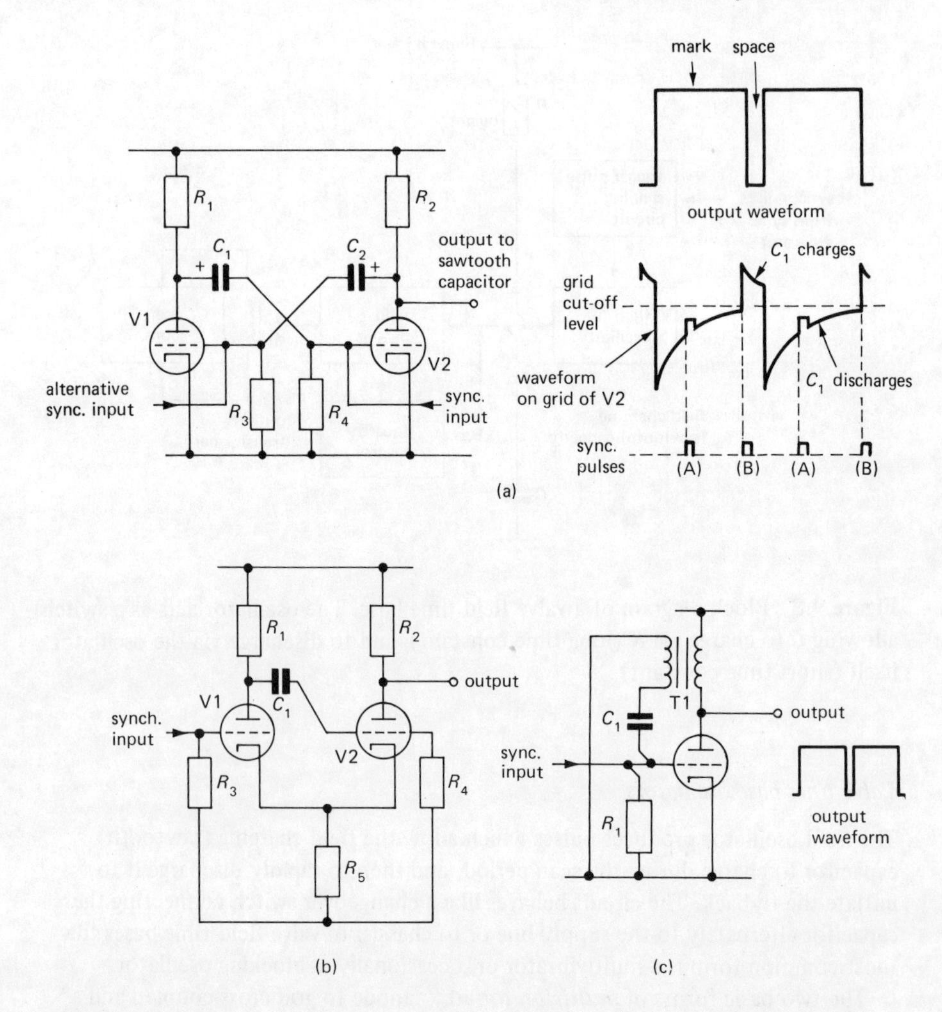

Figure 9.9 Basic forms of field time base oscillators: (a) cross-coupled multi-vibrator; (b) cathode-coupled multivibrator; (c) blocking oscillator. NOTE: All waveforms are idealised. Effect of sync. pulse timing is shown in (a): premature pulse at (A) cannot trigger the oscillator

tain time interval the voltage on C_1 can no longer maintain the negative grid bias and V2 switches on. The negative plate of the previously charged capacitor C_2 is now connected to the grid of V1 and this valve is driven into cut-off. The time constant R_3C_2 governs the timing of the second changeover as C_2 gradually discharges through R_3 and V2.

The basic multivibrator output waveform is rectangular in shape, owing to the anode voltage switching between the h.t. level (cut off) and almost chassis potential (saturation). The mark space ratio and the frequency of the output waveform depend upon the time constants of the cross-coupling components. Therefore, if one of the resistors is made variable, the oscillator frequency can be altered. Synchronisation of the multivibrator can be achieved by applying a suitable pulse to one of the grids. A positive-going trigger pulse applied to V2 while the valve is cut off overrides the negative bias of C_1 and switches the valve on. Alternatively, a negative-going synchronising pulse could be applied to the grid of V1 to take the valve out of saturation into cut-off, which would, in turn, switch V2 on. The timing of the trigger pulse is important, as indicated on the oscillator waveforms in figure 9.9(a). Synchronising pulse A is applied too soon, so that it cannot overcome the cut-off bias on the grid of V2 and synchronisation does not take place. Pulse B, however, is timed to occur just before the 'free-running' changeover point; consequently V2 can be triggered into conduction. This system ensures that the oscillator is relatively immune to interference pulses occurring during the scan, but it also means that the multivibrator free-running frequency must be close to, although *slightly* lower than, the required value.

A *cathode-coupled multivibrator* is shown in figure 9.9(b). There is only one cross-coupling capacitor (C_1), the other feedback connection being between the cathodes. When C_1 is charging from the h.t. via R_1, triode V2 is driven into saturation. Its high current produces a voltage drop across R_5 of such a magnitude as to keep the other valve, V1, cut off. When the capacitor is fully charged, the current in V2 falls, the cathode bias reduces and V1 starts to conduct again. This, however, effectively places the negative plate of C_1 on the grid of V2, which, in turn, switches the valve off. The reduction in the current of the second triode further reduces the cathode bias; V1 conducts more and is driven into saturation. The capacitor is now discharging via V1, R_5 and R_4 until a point is reached when its negative charge is no longer effective and V2 is again allowed to conduct; recharging C_1 repeats the above process. A negative-going synchronising pulse can be applied to the grid of V1 to take it out of saturation.

Another form of pulse generator occasionally found in field time base circuits is the *blocking oscillator* shown in figure 9.9(c). This requires a transformer, which could be an added disadvantage, since at an operating frequency of 50 Hz the size and cost may not be justified. When first switched on, the valve conducts heavily because of lack of cathode bias—the rising anode current induces an e.m.f. in the secondary winding of the transformer. The direction of the e.m.f. is such that it charges capacitor C_1 via the grid–cathode circuit of the valve. Once the valve reaches saturation and its output voltage is nearly at chassis potential, no further increase in the anode current is possible and the e.m.f. in the secondary winding falls to zero. In effect, this places C_1 between the grid and the cathode—the negative potential on the capacitor cuts the valve off and the anode voltage rises to h.t. The capacitor slowly discharges via R_1 until the negative bias is sufficiently low to cause the valve to start conducting again. A positive-going synchronising pulse can be applied to the grid to overcome the cut-off bias and trigger the oscillator.

Practical field time base circuit

The circuits discussed in figure 9.9 were the basic forms of the theoretical arrangements. Their practical versions often differ to such an extent that it can be difficult to recognise the original principle. In older designs a simple multivibrator was followed by the power output stage. The tendency in more modern circuits is to combine the multivibrator with the output amplifier and thus eliminate one valve. A triode–pentode is used, the triode section acting as one part of the multivibrator, while the pentode serves the dual purpose of being the second part of the oscillator as well as the field output amplifier.

An example of a complete field time base circuit is given in figure 9.10. Negative-going pulses from the sync. separator are fed to the integrating circuit R_1, C_1 which develops the field trigger signal (waveform A–the 'serrations' from figure 9.5 could be seen if the oscilloscope time base were set to display line frequency pulses). The *interlace filter* is formed by R_2, D1, C_2 and C_3. Its function is to establish a voltage level at which the field synchronising pulse initiates flyback. The *interlace diode*, D1, clips the pulse developed across C_1 and allows it to be coupled to the oscillator via C_3. When the diode conducts, a negative trigger pulse is fed to the grid of V2 which cuts the valve off.

The two valves (actually a triode–pentode valve) form a multivibrator as well as the power output amplifier. Because of the presence of the amplifying stage, the circuit departs from that shown in figure 9.9. In the practical circuit in figure 9.10 V1 is driven alternately between cut-off (scan) and saturation (flyback) by means of the cross-coupling from the anode of V2. The coupling via C_9 and the network of R_{10}, R_{11}, R_{12}, R_{13}, C_{10} and C_{11} ensures the correct timing of the oscillator waveform. Manual frequency control is obtained by VR1–the *field hold* control. However, the second valve, V2, is no longer switched in the typical multivibrator fashion. Instead, its grid is fed with a suitably shaped drive waveform which the pentode amplifies and delivers the current to the *field output transformer*, T1.

The drive waveform starts as a sawtooth (waveform C); this is due to the field charging (sawtooth) capacitor C_5, which is charged from the boosted h.t. line via R_9 and VR2. The capacitor charges when V1 is maintained in its cut-off state by the negative portion of waveform B. The field charging circuit is an integrating circuit with a relatively long time constant. Assuming that VR2 is set to its mid-value of 1 MΩ, then the time constant $T = (R_9 + \text{VR2}) \times C_5 = 1.47\ \text{M}\Omega \times 47\ \text{nF} = 69$ ms, which is longer than the scan time of approximately 20 ms. The boosted h.t. originates from the line output stage; this voltage is much higher than the normal h.t. and it can be anything between 400 V and 1000 V. The use of the boosted h.t. supply has a number of advantages. Firstly, C_5 can charge to a much higher voltage and still follow the more or less linear portion of the charging graph (figure 9.4). Secondly, a high potential across C_5 will suffice to drive the output valve fully and without any intermediate amplification. Finally, the boosted h.t. has a relatively high output impedance, so that simple voltage stabilisation by means of a *voltage-dependent resistor*, VDR2, is satisfactory. Regulating the supply at this

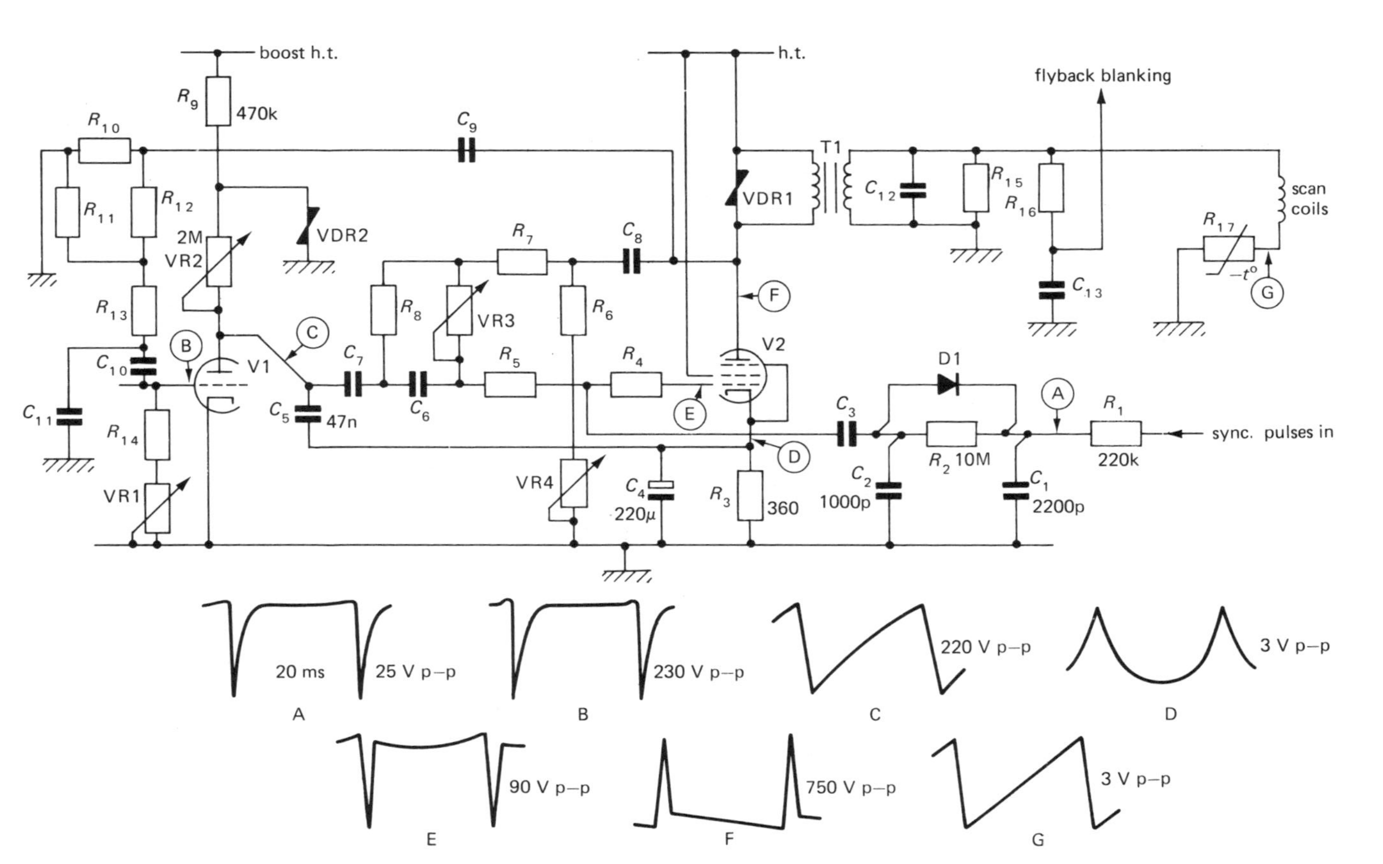

Figure 9.10 Valve field time base circuit diagram. Expected waveforms are shown by their corresponding letters in the diagram (Rank)

point helps to maintain constant height of the picture. The magnitude of the voltage across C_5 governs the drive to—and the output from—the power amplifier; therefore VR2 is the *height control.*

The sawtooth as developed so far is further modified to ensure picture linearity (see section 9.4). The modifications to the waveform produced across C_5 are carried out by the linearity correction network. Since it is impossible to design a fixed circuit to provide the required waveform shaping, two variable controls are incorporated: VR3 (so called '*top linearity*') and VR4 ('*overall linearity*'). The correction is applied by means of negative feedback from the anode of V2 to its grid via C_8, R_7, R_8, C_6, VR3, R_4, and R_5. These components alter the shape of the waveform fed back, which is then added to the basic sawtooth coupled in via C_7. A second linearity correction circuit, formed by C_8, R_6 and VR4, is effectively placed across the output from the valve to modify the waveform applied to the output transformer T1.

The field charging capacitor, C_5 in figure 9.10, is connected to the cathode bypass capacitor C_4 in preference to a direct connection to the chassis. This introduces yet another form of linearity correction. At first glance it appears that C_4 bypasses the cathode resistor R_3 in the usual way. In fact, another waveform is generated across the capacitor which, together with the slope resistance of V2, forms a long time constant integrating circuit. The basically *sawtooth current* flowing in the output stage produces a voltage waveform called a *parabola* (waveform D) This parabolic waveform is, in effect, added to the input to improve the linearity.

Waveform E applied to the grid of V2 consists of a suitably shaped 'sawtooth' and a large negative pulse of a relatively short duration. This pulse cuts V2 off to initiate the flyback, which, in turn, causes a high amplitude positive pulse to appear at the anode of the valve (waveform F). The effect is produced by the sudden interruption of the scan current and by the energy stored in the scan coils. If this energy is not quickly dissipated, the output circuit can be shocked into self-oscillations (ringing) that could interfere with the flyback as well as continue into the next scan. Damping is provided by R_{15} and to some extent by VDR1 across the primary of the output transformer. The voltage-dependent resistor reduces the amplitude of the flyback pulse, because the resistance of the device falls rapidly as the voltage across it increases. If the pulse magnitude was not restricted, the insulation of the associated components, chiefly that of the capacitors and the transformer, could be damaged. The flyback pulse across the secondary of the transformer is shaped via R_{16} and C_{13}, and then applied to the video amplifier to provide *field flyback blanking* (see chapter 6).

The field scan coils are connected to the secondary of the output transformer via a thermistor, R_{17}. Physically the thermistor is attached to the coils to sense their temperature. As the temperature increases, the resistance of the scan coils rises, which tends to reduce the amplitude of the scan current and, hence, the picture height. The resistance of R_{17} falls, owing to the *n*egative *t*emperature *c*oefficient (n.t.c.) of the device, so that the effective resistance of the entire circuit remains constant. Waveform G across the thermistor represents the current in the coils, although it is not normally given in the makers' manual.

9.6 Transistorised Field Time Base Circuits

The block diagram in figure 9.11 shows the circuit arrangement of a typical transistorised field time base. Some of the functions can be easily recognised by comparison with the valve counterpart discussed in the previous section. Now the oscillator tends to be an independent unit followed by a multistage amplifier. In most cases there is no output transformer; the scan coils are coupled via a suitable capacitor to the output stage.

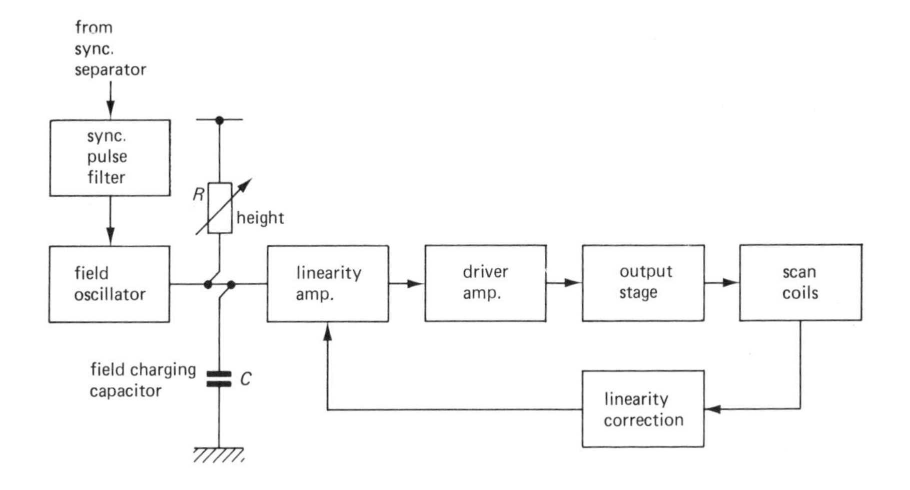

Figure 9.11 Block diagram of a transistorised field time base

Field oscillator

The field oscillator in some receiver designs may be either a cross-coupled multivibrator or a blocking oscillator. The principles of operation are similar to those of their valve versions described previously; any differences are primarily due to the d.c. biasing requirements of transistors. However, a somewhat modified multivibrator circuit is shown in figure 9.12(a). When first switched on, the cross-coupling capacitor C_1 charges up via R_2 and the base–emitter junction of TR2. The charging current drives TR2 into saturation, which, in turn, causes TR1 to cut off owing to the lack of forward bias. When C_1 is fully charged, TR2 is left without any forward base bias and the transistor switches off. Transistor TR1 turns on (its bias restored via R_5), which allows C_1 to discharge through R_3, VR1 ('hold' control) and TR1. The capacitor again recharges via R_2 and TR2, so that the above sequence is repeated. The oscillator output is in the form of short duration, positive-going pulses developed across R_5. The charging time constant associated with C_1 is short, which corresponds to the flyback, while the discharge time constant is long to allow for

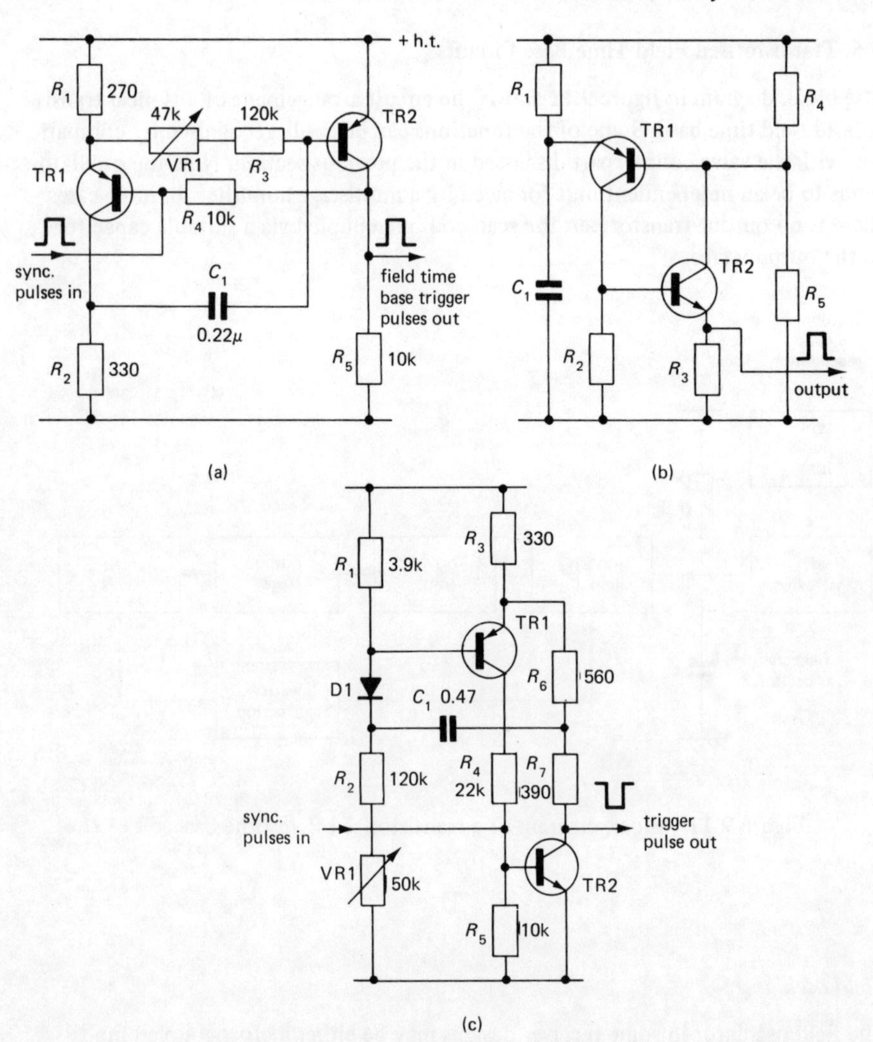

Figure 9.12 Field time base multivibrator circuits. These arrangements trigger the field charging capacitor charge/discharge circuits: (a) field time base multivibrator (Thorn); (b) complementary symmetry multivibrator (basic circuit); (c) complementary symmetry multivibrator (Thorn)

the duration of the scan. Positive-going synchronising pulses switch TR1 off, the charging of C_1 begins and the resultant output pulse initiates field flyback.

The fundamental principle of another form of oscillator is illustrated in figure 9.12(b). Transistors TR1 and TR2 are complementary types and their method of interconnection resembles that of a two-transistor equivalent of a thyristor (see section 14.6). When the circuit is first switched on, C_1 charges via R_1; therefore the emitter of TR1 is close to the chassis potential and the transistor remains cut off. Similarly, TR2 is non-conducting, because of the absence of any forward base bias. The charging process of C_1 finally raises the emitter voltage of TR1 above the base potential derived from R_4, R_5 and the transistor turns on. This switches TR2 on, and the two transistors are driven rapidly into saturation as they provide base bias for each other. The resultant current produces a positive-going output pulse across R_3. Capacitor C_1 is discharged through the conducting transistors; consequently the emitter voltage of TR1 falls below the base voltage, both TR1 and TR2 cut off and the capacitor charging process starts again. Synchronisation can be obtained by the application of a 'turn-on' pulse to the base of onc of the transistors. Field hold control could be provided by making R_1 variable.

The circuit shown in figure 9.12(c) is based on the above principle. Initially, however, both transistors are switched on, owing to their respective biasing arrangements. During that time capacitor C_1 charges from h.t. via R_3, the forward biased base–emitter junction of TR1, D1, R_7 and TR2 (the path of least resistance). When C_1 is fully charged, its negative plate is in effect connected to the emitter of TR1, which causes the transistor to turn off; the base bias to TR2 is now interrupted and this transistor also switches off. Capacitor C_1 slowly discharges via R_2, VR1 (field hold control), the h.t. decoupling capacitors (not shown in the diagram), R_3 and R_6. Eventually the capacitor voltage falls to a value where D1 becomes forward biased, so that the conduction of TR1 and TR2, as well as the recharging of C_1 start again. The output from the oscillator is in the form of short duration negative-going pulses that occur during the conduction periods of the transistors. A positive-going synchronising pulse applied to the base of TR2 switches the transistor on, which allows the charging of C_1 to take place.

A number of receivers use a form of *thyristor*, known as a *silicon controlled switch* (SCS 1), in the field oscillator as shown in figure 9.13. (The principle of operation of the thyristor is described in detail in chapter 14.) In this circuit the thyristor has two gates available: the *cathode gate* (connected to R_5) and the *anode gate* (connected to the junction of R_1 and R_2). In operation the SCS is not conducting until the anode is positive with respect to the anode gate. In the circuit in figure 9.13 the anode gate potential is fixed by the divider R_1, R_2, but the anode voltage is variable because of the repeated charging and discharging of capacitor C_1. As C_1 charges slowly via VR1 (field hold) and R_3, the anode voltage is low and the SCS is held off by its gate potential. However, a point is reached when the SCS is triggered into conduction as the anode voltage rises above the gate potential. During this period the timing capacitor C_1 is rapidly discharged through the SCS (and so is the field charging capacitor C_2—to initiate flyback). Positive-going synchronising pulses are applied to the anode of SCS 1 to advance its 'firing' point; diode D1 prevents the synchronising pulses being diverted via R_4 and C_1 to chassis. Alternatives to the arrangement described here do exist, but their operation is always based on the provision of a suitably timed change in the voltage level between the anode and the anode gate of the SCS.

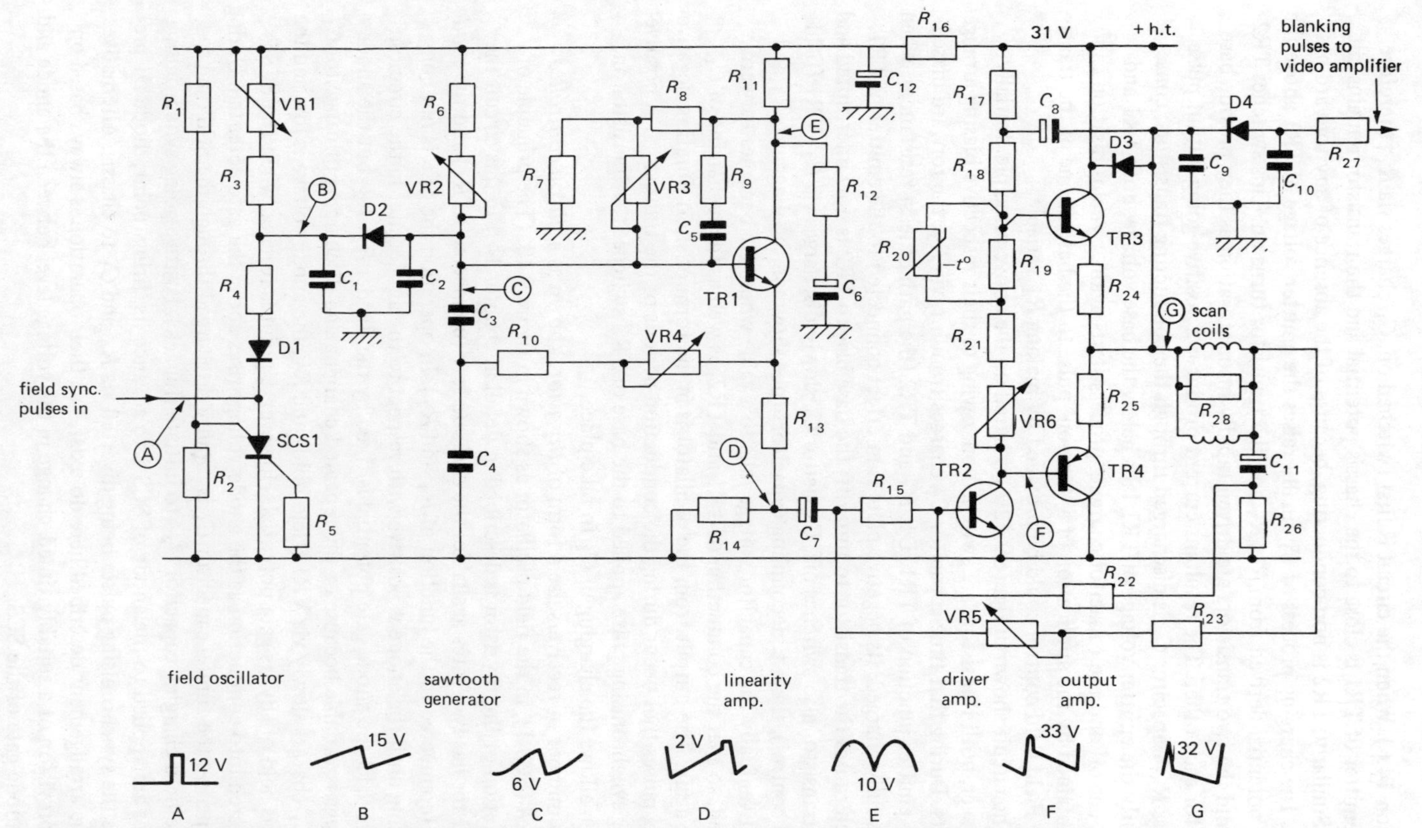

Figure 9.13 Transistorised field time base circuit diagram. Expected waveforms and their peak-to-peak values are shown encircled. (Philips)

Some field oscillator circuits have no vertical hold adjustment, although a suitable modification can be sometimes introduced if a field hold control is required.

Practical field amplifier circuit

At first sight the field time base amplifier circuit shown in figure 9.13 tends to resemble an audio output stage (see chapter 8). The loudspeaker is now replaced by the scan coils! The reader is asked to refer to section 8.3 (and figure 8.5) and note the discussion on crossover distortion and bias stabilisation in this type of circuit. Additional features of the amplifier in figure 9.13 dictated by the requirements of the field time base will now be discussed.

The field waveform is initially developed across the field charging capacitor C_2 connected to the h.t. through VR2 (*height control*) and R_6. A linearity correction waveform is fed back from the emitter of TR1 via R_{10} and VR4 ('*top linearity control*'); it is shaped by the integrating capacitor C_4 and coupled in via C_3. A parabolic waveform appears at the collector of TR1 because of the action of the integrating circuit R_{12}, C_6. A sample of that waveform is also fed back via R_9, C_5, R_8 and VR3 ('*bottom linearity control*').

Field scan takes place when SCS 1 is switched off and C_2 is charging. The time constants associated with C_1 and C_2, respectively, are so arranged that the charging potential across C_2 is lower than that on C_1 and the *oscillator isolating diode* D2 is biased OFF. To initiate flyback, SCS 1 is switched on, D2 becomes forward biased and C_2 is rapidly discharged.

The output from the linearity amplifier, TR1, is coupled to the driver transistor TR2. Stabilisation of its d.c. operating point is achieved by feeding the base bias current from the junction of the output transistors, TR3 and TR4, via the scan coils and the voltage adjustment VR5. Additional linearity correction takes place by feeding back, via R_{22}, a sample of the output waveform developed across the low value resistor R_{26} (1.2 Ω in this design). There is also a feedback connection from the output through the so-called *bootstrap capacitor* C_8.

D.C. biasing of the output transistors, TR3 and TR4, is accomplished by the combined effect of TR2, VR6 (eliminates crossover distortion), R_{21}, bias stabilising thermistor R_{20}, R_{19}, R_{18} and R_{17}. During the first half of the scan the drive waveform makes TR3 conduct while TR4 is switched off. The second half of the field scan is completed by the conduction of TR4, but the transition from one output transistor to the other is gradual in order to avoid crossover distortion. If this were present, non-linearity would appear in the middle part of the picture. However, any adjustments to the bias resistors (VR5 and VR6 in figure 9.13) must be carried out according to the manufacturers' instructions. There could otherwise be a danger of overdriving and possibly destroying the output transistors.

Field flyback is caused by the rapid discharge of C_2, which switches off TR1, whose negative-going output pulse turns off the driver (TR2) and, finally, the output transistor TR4 cuts off as well. The sudden collapse of the magnetic field in the scan coils induces a large back e.m.f.; its polarity tries to keep the current flowing towards the junction of R_{24} and R_{25} (that is, in the same direction as at the end of the scan). The magnitude of such a pulse could be damaging to the output transistors; however, the flyback pulse turns D3 on, so that the back e.m.f. from the scan coils is clipped at a level just above the supply voltage (see the output waveform). The flyback pulse is also fed via Zener diode D4 and the shaping components C_{10} and R_{27} to the video amplifier to provide retrace blanking, as described in chapter 6.

A number of receivers use a single transistor output stage. The scan coils are then fed via a double-wound transformer or an autotransformer, or directly from the collector decoupled from the h.t. by means of a choke. A suitable diode circuit is again included to damp the flyback pulse to a safe value, and the action of the whole output stage is very similar to that of its valve counterpart.

9.7 Integrated Circuits in the Field Time Base

I.C.s can be used to provide synchronising pulse separation, the field time base oscillator, the driver and the field output amplifiers. External components are necessary to provide, in addition to d.c. biasing and decoupling, such functions as integration of the composite field trigger pulse, field charging capacitors, height and hold controls, linearity correction feedback circuits, etc. There are, of course, a number of combinations of different functions in any particular type of I.C. to suit the designer's approach.

An example of an integrated field time base is shown in figure 9.14, where two devices are used to complete the circuit. The reader is advised to refer to the other arrangements where discrete components were used in order to recognise many similarities in the operation of the circuits.

Integrated circuit I.C. 1 in figure 9.14, apart from being shared with the line time base to be discussed in chapter 10, includes the synchronising pulse separator and the field oscillator. Composite video is fed in at pin 9 (I.C. 1) via C_1, where the capacitor together with resistor R_1 is part of the sync. separator time constant (compare with C_1 and R_3 in figure 9.1). A low value capacitor C_2 (470 pF) shunts any high frequency content of the video, e.g. colour burst and noise. The sync. separator supplies its output pulses to the line oscillator (top half of I.C. 1 in the diagram) and to the field oscillator (bottom half of the circuit). The integration of the field trigger pulse and the timing of the free-running frequency is determined by C_3 and VR1 (field hold control) connected to pin 12.

The field sawtooth is developed by charging C_4 and C_5 from h.t.2 via the height control circuit–VR2 and R_2. The resultant waveform, including linearity feedback connected to the junction of the two capacitors, is applied to pin 7 of I.C. 2. The latter is a power amplifier which drives the field scan coils coupled to pin 1 via C_8. The waveform developed across the small value resistor R_5 is then fed back via R_4 and VR3 (linearity control) to the input in a similar way to that shown in figure 9.13 (see components R_{26} and R_{22}, also VR4 and R_{10}). Diode D1 between the field charging capacitors (C_4 and C_5) and the field oscillator is the *isolating*

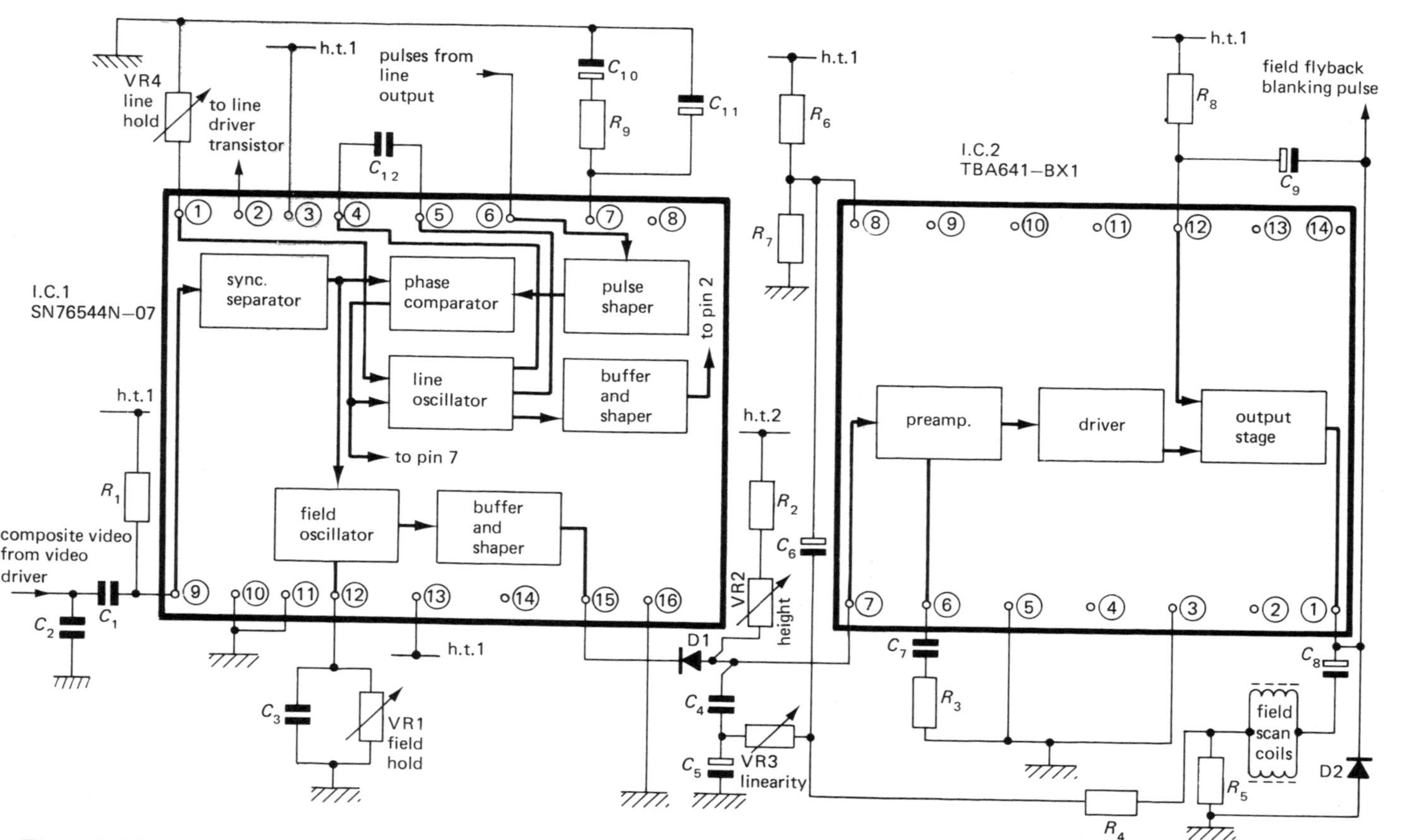

Figure 9.14 Complete field time base using I.C.s. Circuits include part of line time base and the sync. separator. (Thorn)

diode (equivalent to D2 in figure 9.13). Diode D2 connected across the output from I.C. 2 provides positive-going pulses for field flyback blanking (flyback blanking pulses could also be obtained from pin 14 of I.C. 1). Capacitor C_9 is the '*bootstrap capacitor*' (equivalent to C_8 in figure 9.13); resistors R_6 and R_7–pin 8– provide d.c. biasing for the I.C.; C_6 is a decoupling capacitor. R_3 and C_7 form a filter to prevent possible high frequency instability in the amplifier.

9.8 Class D (Switched Mode) Field Time Base

The previously discussed class A and class B field output stages dissipate large amounts of power in the output devices. This is particularly high in colour receivers employing 110° tubes, because of their additional convergence circuitry. These power requirements make it difficult to provide a suitable I.C. for such a receiver.

A switched mode field time base reduces the power consumption of the stage and, consequently, most of the circuitry can be accommodated in an I.C. (for example, Mullard TDA 2600). This is achieved by using the output transistors as switches which are pulsed either ON (high current, low voltage, low power loss) or OFF (no current, no power loss). The drive waveform to the output devices is a train of 150 kHz pulses of continuously changing mark space ratio. The resultant output pulses are converted into the scan sawtooth by an external filter consisting of a suitable choke and capacitor (see the discussion on integrating circuits). External energy recovery diodes are provided to maintain the flow of current in the scan coils when the output transistors are cut off by their input pulses.

The width of the high frequency drive pulses is varied at the required field rate– that is, 50 Hz. The I.C. contains a 150 kHz triangle waveform generator, the output of which is modulated by a separately derived 50 Hz sawtooth. Their timing components are external to the I.C. This type of modulation is known as *pulse width modulation*: at the beginning of the scan the 'on' period is short, while the 'off' interval is long; as the scan progresses, the 'on' period becomes longer at the expense of the 'off' interval.

9.9 Fault-Finding in Sync. Separator and Field Time Base Circuits

Faults in the synchronising pulse separator circuits affect the quality of time base synchronisation. Because of the interaction between the various sections of a receiver, it is possible that a fault outside the separator circuit could produce similar symptoms. The necessary process of eliminating the effect of 'external' stages could be helped, for example, by observing the quality of the picture (and sound!), the effect of contrast control, time base 'hold' controls, etc. Preset adjustments should, however, be carried out in accordance with the makers' instructions where applicable.

In principle, faults in the sync. separator circuit produce either no pulse output, or pulses of insufficient amplitude or of a distorted shape or incorrectly timed. The resultant symptoms can be divided into the following groups: (1) loss of both line and field synchronisation, (2) loss of only one time base synchronisation, (3) line 'pulling' and (4) field 'bounce'.

To ensure both the correct biasing and the required time constants, the values of components used in the separator circuit are quite critical. In addition, the characteristics of the valves or transistors in this type of circuit are important. The only satisfactory way of checking the valves is by substitution, but transistors and I.C.s are less prone to gradual changes, and should not be removed unnecessarily. The indirect method of line time base synchronisation (*flywheel*) often causes only the field time base to be affected by sync. separator faults.

Faults in the field time base can be divided into a number of groups as shown below, although at times at least two effects may be present. (It is assumed that the possibility of faulty valves has already been considered.)

(1) *Incorrect field frequency.* Apart from faults in the sync. separator (or elsewhere in the receiver!), this could be caused by a defective field oscillator—in particular, in the trigger pulse circuit or the oscillator timing components (the latter often leads to a critical field hold.

(2) *Loss of vertical deflection* ('*frame collapse*'), resulting in a thin horizontal line across the screen. This can be caused by a failure of either the field oscillator or the field output stage (including the scan coils, or even the convergence circuitry in colour receivers). The defective section can be easily isolated with the aid of an oscilloscope. However, a voltmeter can often indicate whether the oscillator is functioning, since the resultant waveform usually develops a reverse grid (or base) bias. If the oscillator output could be fed to the audio amplifier via, say, a 0.01 μF capacitor, a low frequency harsh note from the loudspeaker would confirm the presence of a field waveform. The field output stage could be tested by applying a suitable signal to its input. In valve time bases a.c. mains waveform can be injected to the grid of the output valve through a 0.1 μF capacitor from the lower voltage end of the heater chain. Should the display open up after a fashion, then the output is functioning. Alternatively, touching the input terminal to the field amplifier with a meter test prod may result in a momentary 'bounce' of the horizontal line if the stage is not faulty. Any such short-cuts must be carried out with some foresight, especially when applied to transistor or I.C. stages, as semiconductor devices can be damaged by excessive amplitude 'test' signals. Faulty field output transistors are often caused by a defect in a different part of the circuit—particularly, in directly coupled stages, which can easily produce grossly incorrect bias conditions.

Whenever investigating a frame collapse type of fault, it is essential to reduce the brightness to a minimum; alternatively, the beam could be cut off by removing the drive to the tube, switching the guns off or similar measures. This is necessary because the energy of the tube current will be concentrated over a very small area in which the screen material could be damaged. Some receivers have a protection arrangement that automatically cuts the beam off in the absence of a field drive.

(3) *Insufficient height.* This is often associated with poor picture linearity or picture *foldover.* The fault area is difficult to predict, because both the oscillator and the field amplifier stages can be responsible. An oscilloscope is valuable in establish-

ing whether correct amplitude waveforms are present. In valve circuits the common cause of this type of fault is low h.t. (or boosted h.t.) supplying the field charging circuit; alternatively the resistors in the height control circuit often increase considerably in value.

(4) *Poor linearity.* This can be associated with some lack of height, as mentioned above. This defect often affects only part of the picture–namely cramping at the top, cramping at the bottom or compression at the centre. The latter can only be found in transistor circuits, owing to the presence of crossover distortion. Non-linearity can be caused by incorrect d.c. biasing in field amplifiers or faulty linearity correction components. In valve field output stages the cathode resistor and its bypass capacitor are a common source of trouble. The resistor fixes the d.c. bias, while the capacitor provides a correction waveform; thus an open circuited capacitor causes cramping at the bottom of the picture together with a reduction in height. Similar symptoms can occur if the capacitor (or resistor) is short circuited, since the valve would be driven into its non-linear region during the second half of scan. A high value cathode resistor causes non-linearity at the top, because of the excessive negative grid bias voltage at the beginning of the scan. Generalisations on fault symptoms in transistorised circuits are difficult, because of their relatively greater complexity. The underlying principles, however, are similar, except perhaps for the flyback circuit, where a number of components are involved–both at the field output and in the field oscillator stage.

(5) *Miscellaneous faults.*

(a) *Poor interlacing.* This exaggerates the line structure of the picture; it can be caused by a fault in the interlace filter (where present) or the pulse integrating components, allowing interference or stray line flyback pulses to trigger the field time base oscillator.

(b) *Trapezium distortion.* Either the top or the bottom edge (or both) of the picture is slanted, which causes part of the screen not to be scanned–this is due to a fault in one of the field scan coils. In colour receivers a fault in the raster correction network can be responsible (see under 'symmetry control' in chapter 11).

(c) *Pincushion distortion.* 'Bowed' top and bottom edges of the picture–deferred to chapter 11.

10 Line Time Base Circuits

10.1 Line Time Base Synchronisation

Horizontal scanning of the screen is under the control of an oscillator which is synchronised with the action of the electron beam of the camera tube at the transmitter. In the 625-line system the line repetition frequency is 15 625 Hz, and each line period of 64 μs consists of 52 μs of actual picture information (*active picture line*) and 12 μs *line blanking*. During the latter period there is a line synchronising pulse which lasts 4.7 μs. These timings are important, because they explain the different design of line time base circuits when compared with their field counterparts described in chapter 9. The common factor between them is the requirement to provide an approximately sawtooth deflection current waveform in the scan coils.

The problems involved in the synchronisation of the line oscillator will now be discussed. It was explained in chapter 2 how interference pulses can affect the synchronising pulses when negative picture modulation is used, since the amplitude of any such noise rises towards the synchronising pulse level. Synchronisation of the field time base is not seriously affected by interference, because of the presence of a trigger pulse integrating circuit. The capacitor which develops the field trigger pulse tends to 'absorb' short term spurious noise. This is not possible in the line time base, because of the nature of the synchronising pulses; here they are of relatively short duration—and so is the interference. We saw in section 9.2 that the line synchronising information is sorted out from the field pulses by means of a differentiating circuit. Such a circuit has a short time constant; therefore when a narrow interference pulse is applied to the input, a differentiated pulse will appear at the output. If the line oscillator were triggered directly from the differentiating circuit, the synchronisation of the time base would be disturbed by the unwanted 'spikes'.

Modern receivers use a form of buffer stage between the sync. separator and the line oscillator which offers considerable immunity against incorrect triggering. This method is called *flywheel synchronisation*, because its effect in the circuit is similar to that of a flywheel in a mechanical system whose speed of rotation is maintained relatively constant.

Precise line synchronisation is obviously necessary to ensure a viewable picture; however, it is also needed for correct gating of the colour burst and for the functioning of gated a.g.c. where applicable.

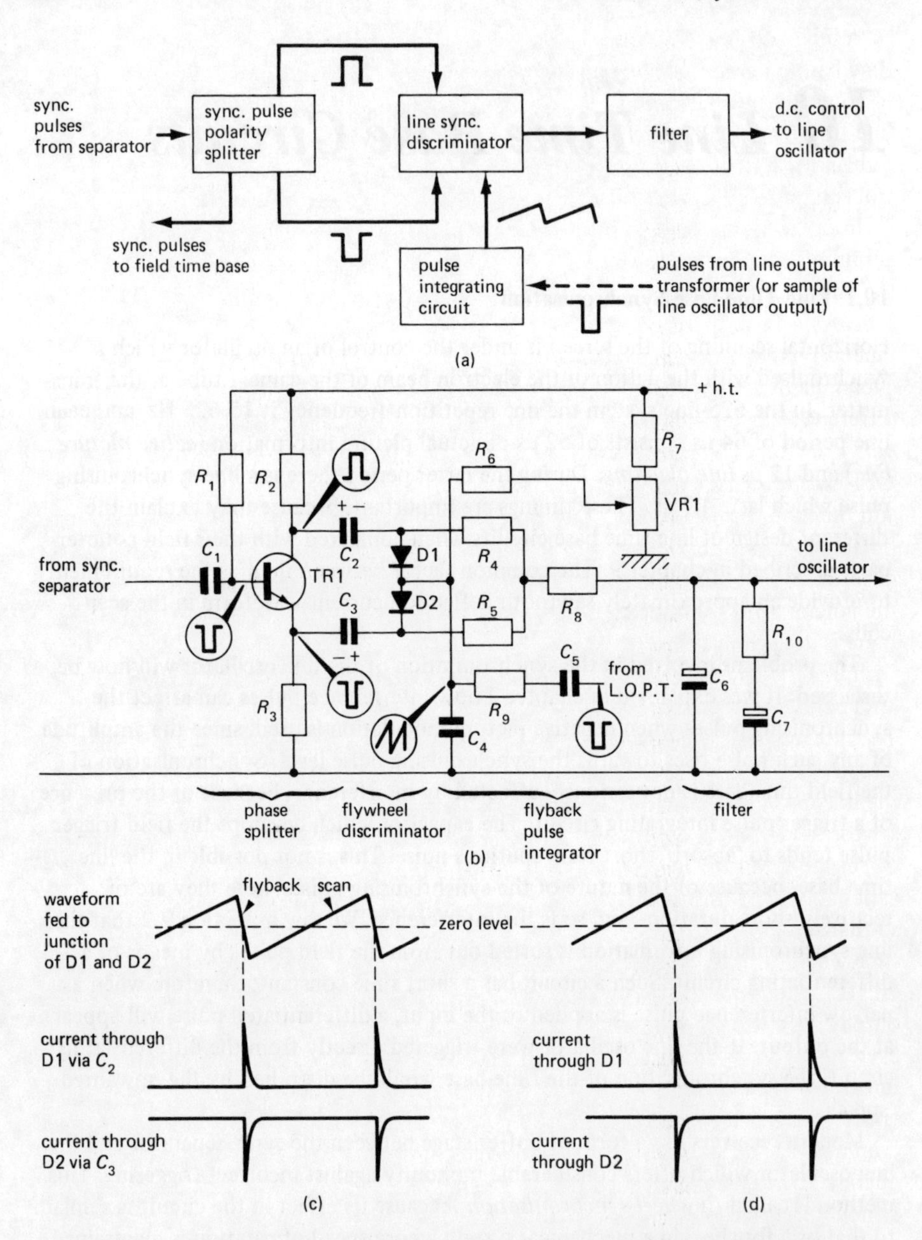

Figure 10.1 Flywheel synchronisation of line time base: (a) block diagram; (b) practical circuit diagram (Thorn); (c) and (d) diode sampling of the integrated line output waveform. In (c) synchronisation is correct–diodes sample zero level of integrated flyback waveform. In (d) synchronisation is incorrect–flyback occurs too soon, diodes sample negative part of flyback sawtooth. *Note*: Current in diodes is due to sync. pulse differentiation

A block diagram and the corresponding circuit arrangement of flywheel line time base synchronisation are shown in figure 10.1. The purpose of the circuit is to compare the phase and frequency of the synchronising pulses with those of the line time base oscillator. This is done by checking the coincidence of the actual flyback with the synchronising information. Any discrepancy generates a d.c. control voltage to adjust the oscillator frequency until the error disappears.

In figure 10.1(b) pulses from the sync. separator pass through the polarity splitter amplifier, TR1, so that equal but opposite outputs are obtained at the collector and the emitter, respectively. TR1 provides a switching waveform for the *diode line synchronising pulse discriminator* (D1 and D2). When the two diodes are simultaneously biased ON, they charge the coupling capacitors C_2 and C_3. The capacitors, in conjunction with the low resistance of the diodes, provide short time constant differentiating circuits during their charging period. *In the absence of synchronising pulses*, TR1 is switched ON, as implied by its waveforms; D1 and D2 are biased OFF by the charges on C_1 and C_2. The capacitor discharge path is via the transistor and the loop formed by R_4 and R_5. Because of the symmetry between the top and bottom half of the circuit, the current due to each capacitor is equal, and the potential differences across R_4 and R_5 are equal in value but of opposite polarity. Therefore the junction of R_4 and R_5 is at zero potential with respect to chassis—in effect, there is no output from the circuit. However, the junction of the two diodes is fed with integrated flyback pulses from C_4 in addition to a d.c. potential derived from VR1. The presence of these two inputs modifies the output from the discriminator. The integrated flyback pulses indicate the actual timing of line flyback and, indirectly, the frequency and phase of the line oscillator. Initially, the pulses from the line output transformer, at C_5, are rectangular in shape, which is changed into sawtooth by the action of the integrating circuit R_9, C_4. The coupling capacitor, C_5, ensures that the waveform has no d.c. component, so that the sawtooth also varies about the zero level [figure 10.1(c) and (d)]. The time base circuit is set up so that when the oscillator frequency is correct, both D1 and D2 conduct at the time when the integrated *flyback sawtooth* passes through zero, as shown in figure 10.1(c). Therefore the flyback waveform does not contribute anything to the output from the discriminator and the line oscillator remains unaffected. If, however, owing to an incorrect timing of the flyback, the diodes conduct when the integrated sawtooth is, for example, negative [figure 10.1(d)], the balance of the entire circuit will be upset (D1 conducts more than D2). Consequently, the discriminator will also produce a negative-going output to correct the oscillator frequency accordingly. Conversely, a positive control voltage will appear if the diodes switch during the positive portion of the integrated waveform.

The application of a d.c. voltage to the junction of D1 and D2 results in a d.c. level at the output from the discriminator. This serves as a d.c. bias for the line oscillator and any correction voltage swings above or below that fixed value. Effectively, VR1 alters the oscillator frequency and it becomes a *line hold* preset. It should be adjusted according to makers' instructions; usually this means that VR1 is set to produce a specified output voltage from the discriminator while the sync. separator has been disabled.

The output from the diode circuit is smoothed by R_8 and C_6, followed by a damping filter R_{10} and C_7 to prevent 'hunting'. 'Hunting' would cause the line oscillator frequency to rise and fall before finally settling down–for example, whenever channels were changed over.

The flywheel effect is created by the value of the smoothing filter components. Any short-term changes in the output from the discriminator are simply absorbed by capacitor C_6 in figure 10.1(b). Similarly, variations in the line synchronising pattern during the field synchronising sequence are also swamped by the filter. Only long-term changes can alter the charge on the filter capacitors sufficiently to affect the line oscillator.

A number of variations on the above circuit are possible. In some receivers polarity splitting can be performed by a centre tapped transformer; in others there may be two antiphase feeds from the line output transformer and only single polarity synchronising pulses.

In the above-mentioned circuits three separate signals were needed. A somewhat simpler alternative is shown in figure 10.2; there synchronising pulses are applied directly from the separator. The underlying principle is similar to that of the previous arrangement, but the actual mechanism of its operation is different. The coupling capacitor, C_1, having been previously charged, is quickly discharged by the negative-going synchronising pulse when the two diodes are switched on. If at the same time the flyback sawtooth passes through zero [see figure 10.1(c)], C_1 becomes fully discharged; should the sawtooth be, respectively, either positive or negative, the coupling capacitor will acquire a net charge of the resultant polarity. This charge is then transferred to the output, where, after smoothing, it becomes a d.c. control voltage.

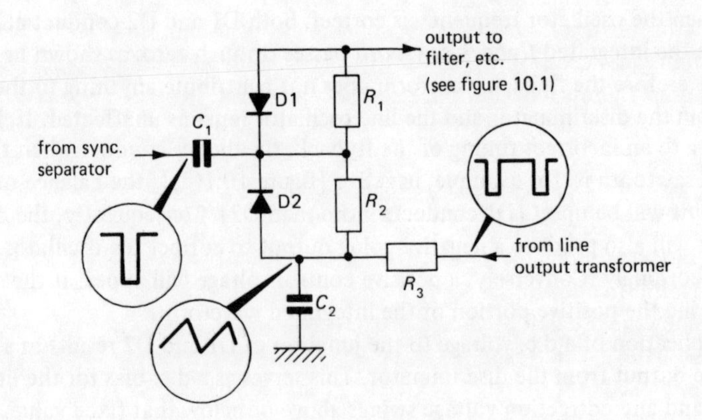

Figure 10.2 Line sync. pulse discriminator (alternative arrangement)–polarity splitting no longer needed

There are several versions of this type of arrangement. For example, in some the sawtooth feed may be applied at the output terminal; in others provision for adjusting the control voltage ('line hold') can be included.

The flywheel discriminator can also be contained inside an I.C. An example of such a circuit was given in the previous chapter, where it was combined with the sync. separator, the line oscillator and other functions. The reader is asked to refer to figure 9.14. Flyback pulses from the line output transformer are fed in at pin 6 and, after suitable shaping, they are communicated to the phase comparator, which performs the function of the discriminator. Here the pulses are not integrated as they were in the previous circuits, since the comparator 'looks' at the displacement between the edges of the synchronising and the flyback pulses. The output is then smoothed by the filter capacitor, C_{11}, and the 'anti-hunt' network R_9, C_{10} connected externally to the I.C. (pin 7). The resultant d.c. voltage controls the line oscillator.

Other I.C. types are also possible; a different external component arrangement would then suggest an alternative design approach to circuit operation.

Semiconductor flywheel circuits are now commonly used. Slightly older receivers employed a synchronising pulse coincidence detector in the form of a valve—for example, a triode-pentode. The synchronising pulses could be applied to the grid of the triode while the integrated flyback waveform was fed to the pentode section. The two outputs at the anodes would then be added, but the shape and the average value of the resultant waveform would depend upon the degree of coincidence between the respective inputs. After a filter the average value would become the oscillator control voltage.

The frequency stability of the flywheel-controlled line time base can have its disadvantages when a video recorder is to be connected to a TV receiver. Owing to slight variations in the speed of the tape/head mechanism, there is a tendency for the recorded synchronising pulses to occur at somewhat irregular intervals. A stable line oscillator might not be able to respond quickly to those variations; therefore modifications are necessary to such receivers as may be recommended by the makers. Some TV sets have built-in alternative flywheel filter components for this purpose.

10.2 The Line Oscillator

Both the blocking oscillator and a multivibrator are used to provide the line time base waveform. In addition, most modern receivers designed in Britain tend to favour a sinewave oscillator, usually of the Hartley type (the exception being found in I.C. time bases). The reason for this is the improved frequency stability of a tuned circuit oscillator over one of the other types. That, of course, further enhances the flywheel effect already introduced by the discriminator described in section 10.1.

At first sight it may be puzzling how it is possible to use a sinewave in order to obtain a sawtooth scanning waveform. What happens in practice is that the sinusoidal shape can be easily converted into a more or less rectangular form, which is, as we shall see, the required drive for the line output stage (especially in transistorised circuits).

The frequency of the line oscillator must be controlled automatically by means of a d.c. voltage derived from the flywheel discriminator. If the oscillator uses a tuned circuit, then its frequency depends only on the *effective* value of its capacitance and inductance. Therefore the control voltage must change one of those properties, which implies that the effective *reactance* of the circuit has to be altered. This problem exists elsewhere in a TV receiver–namely in tuner a.f.c. and the automatic control of the colour subcarrier in the decoder. In both applications there is normally a varicap diode whose capacitance changes with the applied d.c. voltage. However, the line oscillator frequency is relatively low (15 625 Hz), and the amount of capacitance change available directly from a diode would be insufficient. A separate amplifying stage, known as the *reactance stage*, is incorporated in the sinewave line oscillator. The reactance stage is connected to the original tuned circuit in such a way as to behave like a capacitor (or sometimes as an inductor) whose effective value depends upon the amount of the applied d.c. bias.

Two examples of a sinusoidal line oscillator with its reactance stage are given in figure 10.3; circuit (a) is based on valves and circuit (b) is transistorised. Consider the valve arrangement first.

The connections to the pentode section are such that in effect it functions as if it were a double valve. The oscillator circuit is formed by the control grid and the screen, so that the electron flow to the anode is controlled by the action of the oscillator; the anode circuit functions in the usual amplifier fashion. Positive feedback is applied via C_3 to the grid to sustain the oscillations. The tapped, tunable inductor L_1 can be recognised as being the *Hartley* circuit. Its parallel capacitance is formed by C_1 together with the *effective* capacitance of triode V1. The latter effect is produced by the signal taken from the tap on the coil and fed to the cathode of V1 via C_2. The combined phase shift introduced by R_1, C_2 and R_2 is such that an angle of 90° lead is created between the input current to the triode and the voltage induced in L_1. Thus V1 behaves as if it were a capacitor; the amount of capacitance depends on the gain of the valve, which, in turn, varies with the amount of control d.c. applied to the grid.

The fundamental difference between the circuit used in the line time base and a normal sinusoidal oscillator is the shape of the output waveform. In most applications it would be expected to generate as pure a sinewave as possible. Now the output must be in the form of a pulse whose shape has to be adjusted to suit the line output stage (see anode waveform of V2). The overall shaping is done in two stages: (1) by a suitable biasing of the combined oscillator and amplifier valve V2 so that it is driven alternately between saturation and cut-off (the resultant clipping is not necessarily symmetrical); (2) by a suitable shaping network [R_5, C_5 in figure 10.3(a)] connected across the output.

The transistorised version of the line oscillator shown in figure 10.3(b) appears to be similar to its valve counterpart. The chief difference, however, lies in the

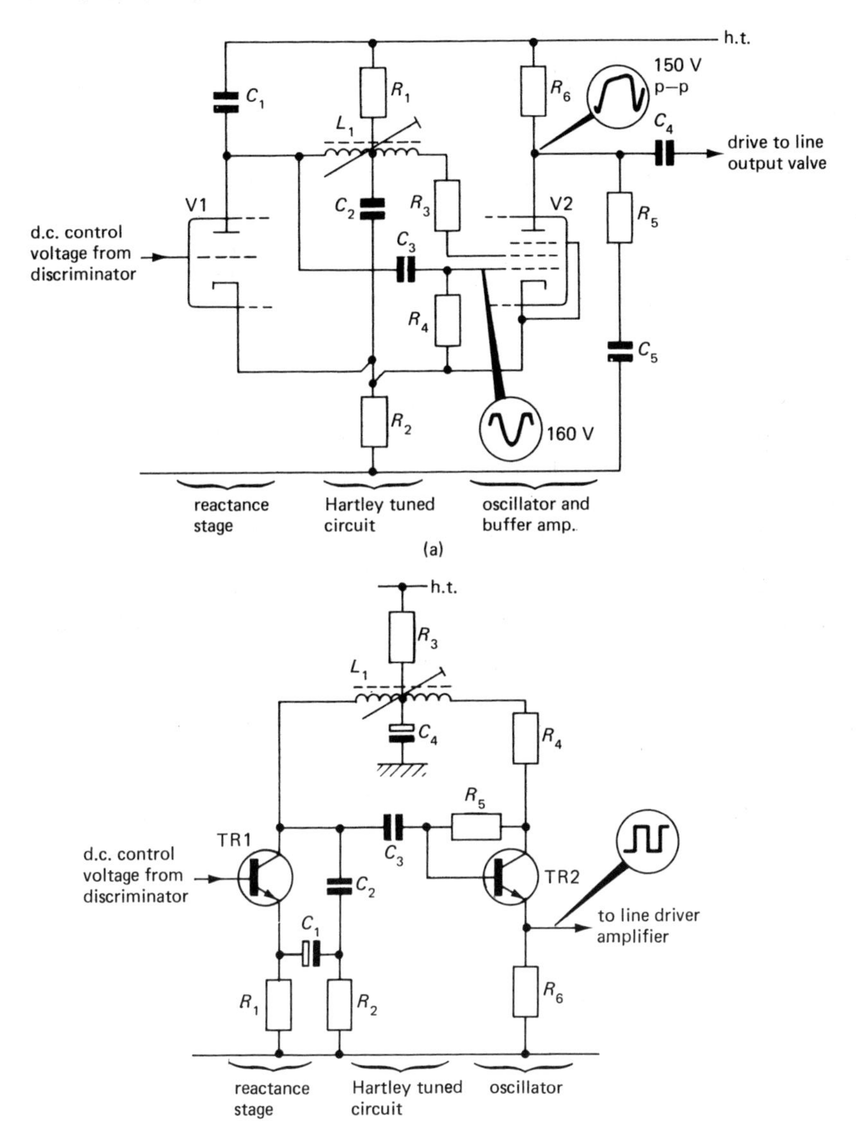

Figure 10.3 Examples of sine wave line oscillators: (a) GEC; (b) Thorn. Note differences between the output waveforms required in a transistorised circuit and in a valve circuit

actual method of introducing the 90° leading phase shift between the reactance stage (TR1) and inductor L_1. The phase shifting network consists of C_2 and R_2; the signal developed across R_2 is fed to the emitter of TR1 through the coupling capacitor, C_1. The actual amount of 'leading' current supplied by the reactance stage depends on the d.c. bias applied to the base of TR1.

The sinusoidal waveform developed within the tuned circuit drives TR2 alternately from saturation to cut-off, so that the output at the emitter is almost a square wave. The differences between the drive waveforms required by a transistorised and a valve line output stage, respectively, will be explained later in this chapter.

In some receivers the reactance stage does have a varicap diode. However, its effective capacitance must be increased, which is done by means of a transistor amplifier known as the *capacitance multiplier.*

The Hartley oscillator circuit can be tuned by means of the ferrite core of the inductor; its adjustment becomes the *line hold* preset. This is often in addition to the line hold potentiometer provided in the line sync. pulse discriminator [VR1 in figure 10.1(b)]. Neither adjustment is normally available to the viewer, since it is important that the recommended procedure be observed.

The action of the synchronising pulse discriminator opposes the effect of any slight lack of accuracy in the oscillator free-running frequency. This may create the impression that precision in carrying out the adjustments is not important because of the apparent 'play' in the system. Unfortunately, the automatic line oscillator control circuit can only correct a limited range of frequency errors. Furthermore, this range is somewhat wider when the circuit is already synchronised. If the oscillator is to be 'pulled in' from its free-running state (for example, when the receiver is first switched on, or after channel changeover) the discriminator can deal successfully with a much lower frequency error.

The adjustments, as already mentioned in section 10.1, should be performed

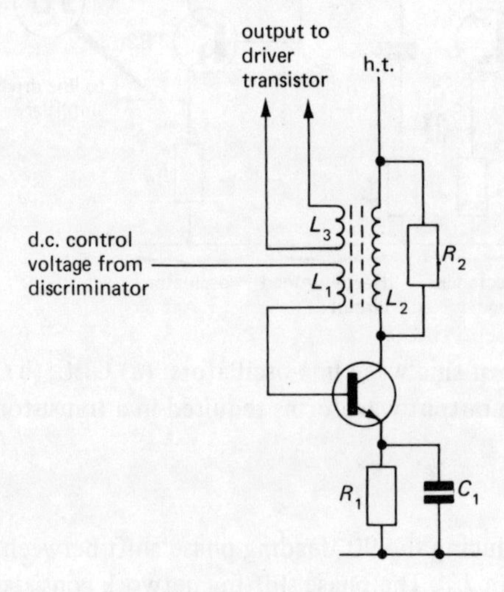

Figure 10.4 Line time base blocking oscillator. Conduction is 'blocked' by C_1, which, when fully charged, applies a positive bias to the emitter

with the automatic control feature disabled; in *some* circuits it could be done by short circuiting the discriminator diodes. The tuning of the coil can be altered until a 'floating' picture is obtained; this should lock in as soon as the action of the sync. discriminator has been restored to normal.

Where the time base uses a blocking oscillator, the circuits are very similar to their basic forms; a valve version was shown in figure 9.9, and its transistorised counterpart is given in figure 10.4. The high operating frequency gives a smaller size transformer compared with the design used in the field time base. The blocking of transistor conduction in the arrangement in figure 10.4 is brought about by C_1 in the emitter circuit. Positive feedback is applied from the collector to the base by the action of the transformer (windings L_1 and L_2). When the transistor is driven into saturation, C_1 charges to a positive voltage. Once the collector current has reached its maximum, the e.m.f. induced in L_1 falls to zero and the transistor switches off. The charge on the capacitor makes the emitter positive with respect to the base and the cut-off state is maintained until C_1 has discharged sufficiently via R_1. At this point the d.c. control voltage on the base takes over, the transistor is again driven into saturation and line flyback begins. The output from the oscillator is coupled to the next stage via tertiary winding L_3.

The inductance of the blocking oscillator transformer produces a large back e.m.f. each time the transistor switches off. In some circuits a diode may be connected across the primary winding to discharge the magnetic energy and the large voltage pulse is avoided. The switch-off pulse tends to be followed by self-oscillations (ringing) of the tuned circuit formed by the inductance of the windings and their stray capacitance. In the arrangement shown in figure 10.4 ringing is damped by resistor R_2.

Control of the blocking oscillator frequency is performed by the synchronising pulse discriminator. This time the d.c. voltage is fed to the base of the transistor to override the reverse base–emitter bias and initiate flyback.

Multivibrator circuits are not discussed here, since they are rarely used in modern receivers, because of their poor frequency stability. If encountered, the circuit will resemble one of the basic arrangements illustrated in chapter 9.

It is interesting to note that some makers who do not use a sinewave oscillator in the line time base include a tuned circuit in conjunction with their multivibrator or blocking oscillator. This is done to improve frequency stability of the oscillator, making it relatively unaffected by 'stray' trigger pulses.

A number of receivers use an I.C. line time base oscillator. This usually incorporates other functions, and the reader is asked to refer to chapter 9, figure 9.14. After the phase comparator and the filter (pin 7, figure 9.14), the d.c. control voltage is fed to the oscillator. Its frequency is governed by the external capacitor, C_{12} (pins 4 and 5), and the manual 'line hold' adjustment, VR4. The oscillator in an I.C. is often designed to generate a triangular waveform in the first instance, which is then shaped within the device to produce a rectangular output. Frequency stability of the oscillator itself can be obtained by means of internal circuitry, so that a sinewave circuit is not needed.

10.3 Transistorised Line Driver Circuits

The purpose of the driver stage is to amplify the oscillator output waveform before feeding it into the line output transistor. This type of arrangement is needed in all power amplifiers (audio, field and video output), because of their requirement for a high amplitude input current at a relatively low input voltage.

In the line time base the driver stage may also help to shape the oscillator waveform to produce rectangular output pulses.

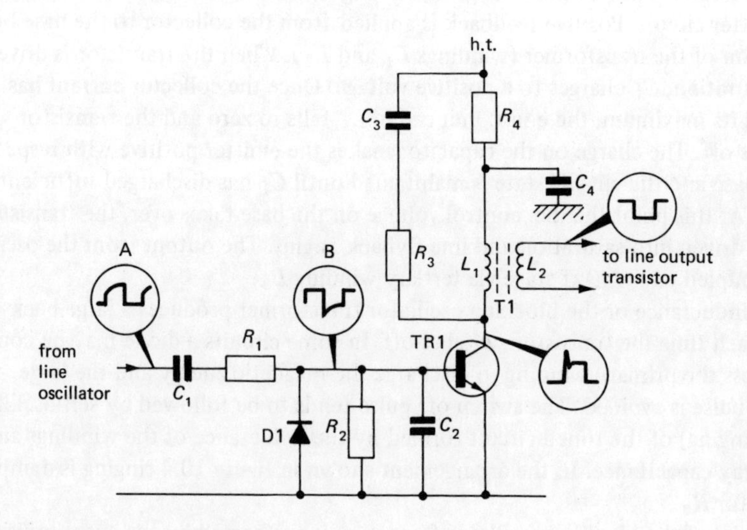

Figure 10.5 Line driver amplifier (Thorn)

A typical circuit is shown in figure 10.5. The oscillator waveform (A) is coupled via C_1 to the driver stage. The somewhat modified waveform (B) switches the driver transistor alternately on and off. The output at the collector is basically a square wave. Its large amplitude voltage swing must be stepped down by the driver transformer, T1. As shown by the waveform (C), when the transistor switches off, there is a voltage spike accompanied by 'ringing'. This is due to the initial back e.m.f. from the transformer winding and the subsequent self-oscillations in the collector circuit. Excessive ringing on the drive waveform could have a detrimental effect on the line output transistor; therefore damping is applied by means of C_3, R_3 and R_4 connected effectively across L_1. When the transistor switches on, R_4 also limits the maximum d.c. current taken from the h.t. supply.

Diode D1 across the base-emitter junction of TR1 protects the device by preventing a large negative signal being applied to the transistor input.

10.4 Equivalent Circuit of a Line Output Stage

The design and operation of the line output stage differs considerably from that of the field time base. The difference is caused by the effect of the long time constant of the line scan coils and the associated circuitry on the deflection waveforms. It was shown in section 9.2 how an *integrating circuit* produces a sawtooth current waveform in a suitable inductor. The line output transistor acts simply as a switch which either connects the scan coils across a supply (usually via a transformer) or, at the end of a line, disconnects them, to initiate flyback. When the voltage is applied to the coils, the current through them rises gradually; if the circuit time constant is relatively long, the sawtooth deflecting waveform is obtained.

The behaviour of the line output stage can be better described with reference to the simplified equivalent circuit shown in figure 10.6; L is the effective inductance and R is the effective resistance of the stage. Parallel capacitor C represents not only the inevitable stray capacitance, but also additional capacitors which tune the entire circuit.

At first, with switch S1 (the line output transistor) open, no current flows and the beam is at the centre of the screen. When the switch closes (the transistor is turned on by its input drive waveform), the current through the coil rises at a rate given by the time constant L/R; this causes the beam to be deflected from the centre to the right-hand side of the screen. At the end of the picture line, S1 opens (the output transistor is turned off) and flyback begins. The scan current cannot disappear at this instant because of the energy stored in the inductance; instead the parallel circuit formed by L and C begins to oscillate (ring). Initially, the sudden interruption of the main current path produces a very large positive voltage pulse from the scan coils towards the upper terminal of switch S1. At the same time, the first half of the beam flyback takes place as the current in the inductor falls to zero, having charged up the parallel capacitor. The second half of the flyback is produced when capacitor C discharges into the scan coils. The lower plate of the capacitor would have previously become positive; therefore the flow of current is now as shown by the broken line arrow in figure 10.6.

When the beam has reached the left-hand side of the screen, the relatively slower scan begins. The oscillations must be stopped, because both their timing and waveform (sinewave instead of sawtooth) are no longer useful. The energy which would have maintained the ringing could be dissipated in a suitable damping resistor, but this would be extremely wasteful, since the process is repeated every 64 μs. Instead the switch is closed at the end of the flyback, the h.t. supply with its negligible resistance prevents any further oscillations and the scan current falls at a linear rate. The beam is, therefore, deflected from the left-hand side to the centre of the screen and the first half of the scan is completed. During this process the energy stored in the coil is gradually transferred to the supply capacitor, where it will not be wasted– it is, in fact, available for the second half of the scan. If the switch remains closed, the current in the coils builds up in a sawtooth manner and the deflection continues from the centre to the right-hand side of the screen. Oscillatory flyback then

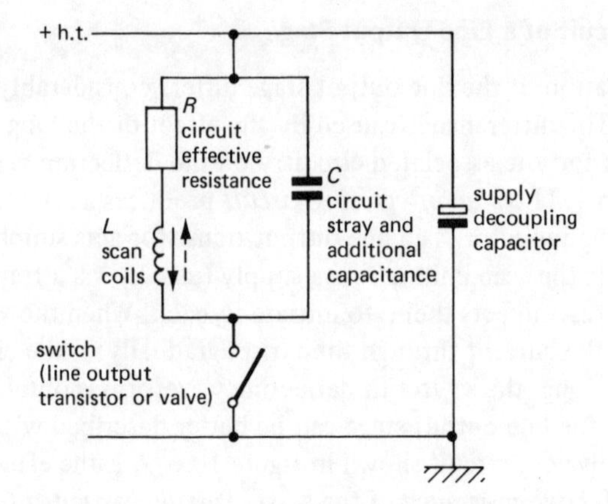

Figure 10.6 Simplified equivalent circuit of a line output stage. Continuous line arrow indicates direction of current in scan coils when the beam is deflected between screen centre and the right-hand side. Broken line arrow shows direction of current during deflection between centre and the left-hand side of the screen. When current is zero, the beam is at the centre. As switch opens (flyback), L, C form a tuned circuit; with the switch closed, R, L form an integrating circuit

follows and the process is repeated as described above.

The whole set-up resembles a kind of pendulum which is slowly moved from its vertical position to the right (part of scan); it is then released to swing back to the extreme left with the aid of the potential energy acquired previously (flyback). The speed of that return movement depends on the length (tuning) of the pendulum. When the pendulum has reached the topmost left position, a suitable brake must be applied to prevent further swings and to control the rate of descent (next scan). Energy has to be added to the system only to lift the pendulum from the vertical to the top right.

The reader will recall that the video waveform allows approximately 10 μs for line flyback. If this were not achieved, one would be faced either with width problems or with picture distortion in the left-hand side of the screen. Therefore the frequency of flyback ringing is relatively critical; this means that the value of the tuning capacitor C in figure 10.6 is critical too. This capacitance also governs the amplitude of the initial voltage pulse developed at the beginning of flyback; the lower the value of C the greater the back e.m.f. from the coil.

10.5 Simple Transistorised Line Output Stage

The practical line output stage shown in figure 10.7 is now considered. The circuit is of a relatively simple design, since it is used in a low power, battery/portable type of receiver.

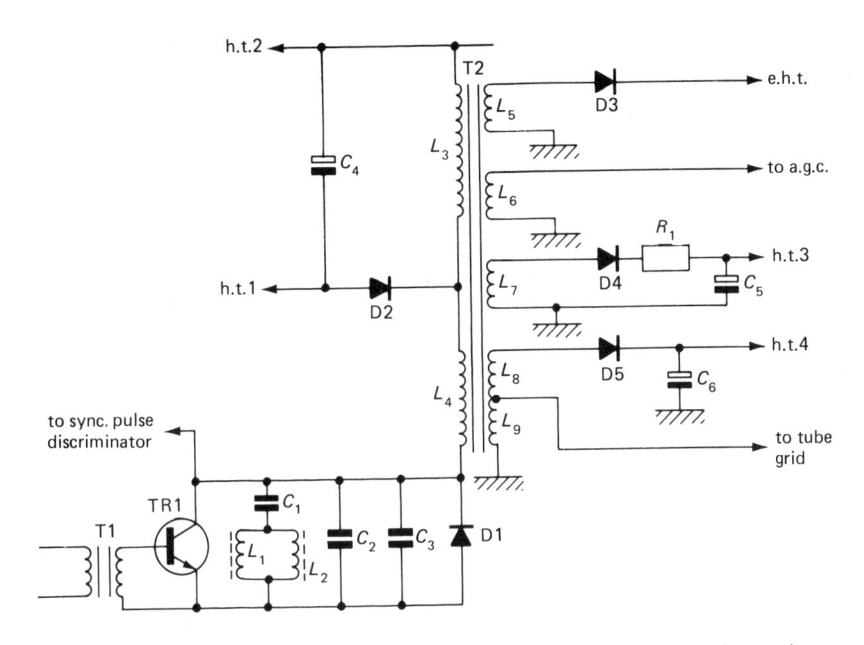

Figure 10.7 Line output stage of a battery/portable receiver (Thorn)

It will be noted that, from the scan current point of view, the parallel-connected deflecting coils L_1 and L_2 are in series with the output transistor TR1 via C_1. The receiver power supply (h.t.1) is thus used to 'top up' the energy losses in the scan circuit in addition to the provision of various h.t. levels to other sections of the receiver.

The line output transistor has no fixed d.c. bias, as it is alternately switched on and off by the base drive waveform from the driver transformer T1. When the transistor is turned on, it discharges the scan coil *S-correction* and d.c. blocking capacitor C_1, which results in a current flow through L_1 and L_2 in the 'upward' direction in the diagram in figure 10.7. At the end of scan the drive waveform switches TR1 off, but the e.m.f. in the coils tries to maintain the original direction of the current. That means that the parallel combination of C_2 and C_3, together with any stray capacitance, are now charged; the polarity of the flyback pulse is such that diode D1 is reverse biased. Flyback oscillations begin, and the beam is returned to the left-hand side of the screen. It is then that the direction of the fly-back e.m.f. reverses and diode D1 is biased ON. The low resistance of the diode damps the oscillations immediately and the current gradually decays—flowing in the 'downward' direction (figure 10.7) in the deflection coils L_1, L_2. The first half of the scan is thus produced from the energy already stored in the scan coils. The efficiency of the output stage is therefore high, and for this reason diode D1 is called the *efficiency diode*.

In many receivers, especially in those where the line output stage is fed from a relatively high supply voltage (100 V or more), the efficiency diode is not used. The line output transistor itself provides efficiency action by conducting in reverse. This type of design will be discussed in section 10.6.

The reader can now appreciate why the line drive waveform (see figure 10.5) is almost square (the mark/space ratio is approximately 1:1.3). The mark period switches the line output transistor on, while the space interval turns it off. During the transistor off-state, flyback takes place, followed by the first half of scan, when the efficiency diode conducts. The exact timings depend upon the practical design of the circuit, but lack of continuity between the changeover from one state to the other could cause picture non-linearity towards the centre of the screen.

The parallel capacitors, C_2 and C_3 in figure 10.7 tune the line scan coils during flyback thus affecting the width of the picture. Since C_2 and C_3 are equivalent to capacitor C in figure 10.6, they also govern the magnitude of the flyback pulse, and therefore, protect the output transistor against excessive voltages.

Capacitor C_1 in figure 10.7 performs two functions–firstly, blocking the d.c. from the scan coils, and, secondly, improving the linearity by introducing S-correction (see section 9.3). The latter occurs because the charging of the capacitor tends to be slightly non-linear, which, in turn, counteracts the tendency to overscan towards the edges of the raster.

As far as the scan coils are concerned, the line output transformer, T2 acts as a choke in the h.t. feed. A current flows through it during scan to 'top up' the charge on C_1 and to supply power to a number of auxiliary h.t. rails; similarly, flyback pulses are also fed to the transformer. This means that a number of a.c. outputs are available from the different windings. These outputs are rectified and smoothed to provide power supplies to other sections of the receiver. The scan part of the induced waveform is normally used to supply low voltage, high current demands; conversely, the flyback generates high voltages at relatively low currents. The latter produces e.h.t. for the final anode of the picture tube, which in a battery portable type of receiver would be approximately 10 kV.

The advantage of using the line output stage to supply various circuits in the receiver lies in the fact that the frequency of the resultant a.c. is high (15 625 Hz), making the filtering very simplified and giving a smaller transformer size when compared with a 50 Hz mains unit. In battery portable receivers the main supply is limited to 12 V d.c., but the line output transformer can supply the necessary higher voltages. For example, in figure 10.7 h.t.3 of approximately 330 V is fed to the screen grid and focus control of the tube, h.t.4 of 90 V to the video output stage, h.t.2 of 24 V to the line and field oscillators as well as to the video driver stage.

Supply h.t.2 is also known as the *boosted h.t.* This is obtained by adding to h.t.1 (11 V main receiver supply from the battery) a rectified and smoothed supply across C_4 (derived from L_3 and *boost diode* D2). If the voltage across C_4 is 13 V, then the total with respect to receiver chassis is 11 V + 13 V = 24 V. Of course, h.t.2 cannot be established until the line oscillator and the line output stage become operational. Therefore, when first switched on, the h.t.2 rail is at a reduced voltage of less than 11 V fed from h.t.1 via D2. This is sufficient to start the system up until normal action is finally reached.

Some designs do not employ a parallel efficiency diode (D1 in figure 10.6); instead the series boost diode is used to provide the energy recovery action during the first half of scan. This method is similar to the one adopted in a valve line output stage, as will be shown in section 10.7.

The line output transformer in figure 10.7 also provides a number of pulses: winding L_6 communicates pulses to the gated a.g.c. circuit; the tap on L_8, L_9 gives negative pulses to the grid of the picture tube for line flyback blanking; the flywheel line sync. pulse discriminator is supplied from the lowest point on the transformer (L_4).

10.6 High Power Transistorised Line Output Stage

In large screen, mains-driven receivers the arrangement of the line output stage is somewhat more involved. The design is even more complex in a colour TV set. This is due to a far greater power demand placed upon the line output stage, which, in turn, is handled by the output transistor. The main design features are: high level of h.t. voltage (100 V or more) and high amplitude currents (1 A or more). The available transistors are also subjected to flyback pulses of nearly 2000 V and are close to the limits of their operating capability; they have to be protected against excessive conditions of drive current and voltage.

The diagram in figure 10.8 is of the line output stage in a colour receiver which employs 110° deflection. In this type of circuit the power requirements are particularly high because of the wide angle scanning. The two line output transistors, TR1 and TR2, are connected in series, basically to share the unusually large flyback pulse in view of the relatively high h.t. (275 V in this receiver). The flyback tuning capacitors, C_3 and C_4, are of equal value, so that they ensure a balanced division of voltages across each device. The waveform from the driver stage is coupled via transformer T1, current limiting resistors R_2, R_4 and adjustable inductors L_1 and L_2 to the bases of the transistors. When TR1 and TR2 are driven hard into saturation, it is essential to their survival that they be bottomed as quickly as possible. This means that the collector–emitter voltage drops rapidly to less than one volt before the output current has risen to any extent and power dissipation in the transistors can then be kept relatively low.

A similar requirement applies when switching off, but this is not so straightforward. A heavily saturated power transistor takes a certain amount of time to turn off after the removal of its base drive. If the input signal were *suddenly* interrupted, it would make the collector current decay slowly, while the voltage across the device would rise sharply, and the power rating might be exceeded. The base inductors, L_1 and L_2, are used to reduce the base current *gradually*—when the voltage waveform from the driver reverses the coils oppose the sudden change. However, when the delaying action of the chokes finally stops, the reverse drive applied to

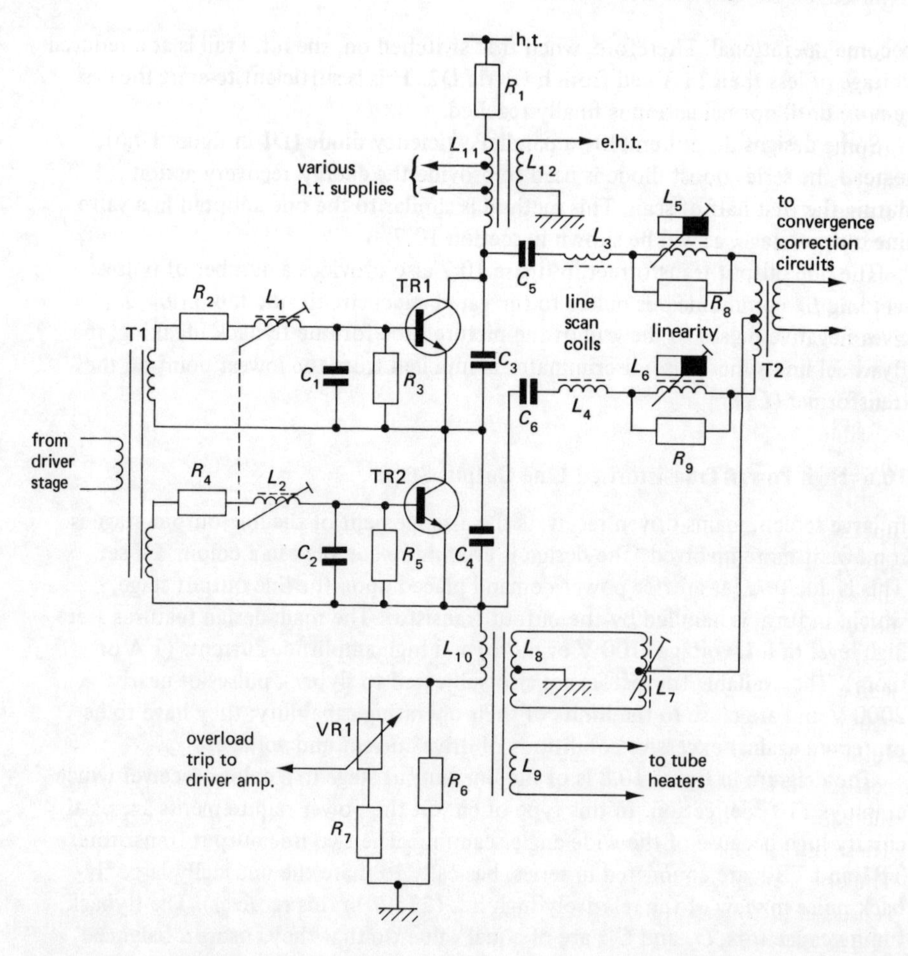

Figure 10.8 Simplified circuit diagram of line output stage of a colour receiver (Rank)

the bases rapidly turns the transistors off. In the arrangement shown in figure 10.8 the inductors are variable, to ensure balanced operating conditions of each transistor. The adjustment must be carried out according to the makers' instructions, and usually involves a measurement of an out-of-balance current in the circuit. This current is often indicated by a p.d. across a fusible resistor (not shown in figure 10.8) connected between the junction of the two output transistors and another point in the circuit. The adjustment aims at a minimum reading across the resistor (less than 1 V); incorrect settings can lead to the destruction of the transistors.

In many other designs separate base inductors are not used; their function is then taken over by the inductance of the driver transformer itself.

Line flyback takes place in the manner already described; however, the reader will notice that an efficiency diode is not used in the circuit in figure 10.8. The output transistors are allowed to conduct in *reverse*, to provide the current during the first half of scan. This action occurs when the back e.m.f. in the scan coils reverses at the end of line flyback. The emitter potential of each transistor is now positive with respect to its collector; consequently, the base is also positive in relation to the collector (via either R_5 or R_3 in parallel with the respective secondary windings of T1). Efficiency action takes place because the base–collector junctions are forward biased, allowing current to flow. It will also be noted that the 'new' base biasing from the emitter terminals turns the transistors on; it appears that the original functions of the emitter and the collector are interchanged. This is possible because the name n–p–n (or p–n–p) implies the same material structure of the device in either direction.

Efficiency action continues until all the stored magnetic energy has been used up; the scan current is then at zero and the electron beam is at the centre of the screen. Before the efficiency period ends, the drive waveform is again inverted by the action of the oscillator, ready for the 'normal' conduction of TR1 and TR2 to be resumed for the second half of scan.

The above method is often adopted when the h.t. supply to the line output stage is in excess of 100 V. The use of a transistor in place of an efficiency diode has one major disadvantage. When the transistor conducts in the conventional direction, the voltage across the device is less than 1 V, but during part of the reverse conduction period this potential difference is in the region of 5 V. The latter value consists of the base–collector forward bias and the base–emitter 'cut-off' drive fed from T1. These differences can cause scan non-linearity at the beginning of each picture line. The effect of the inequality cannot be swamped if a relatively low supply voltage is used in the line output stage.

The scan coils, L_3 and L_4, are connected to the output transistors via the S-correction, d.c. blocking capacitors, C_5 and C_6. In series with the coils are two *linearity correction inductors,* L_5 and L_6, with their permanent magnets. It will be recalled that in the field time base linearity correction took place before the waveform was fed to the output amplifier. This is no longer possible in the line output stage, which merely functions as an ON–OFF switch, and is unable to deal with anything but a rectangular (square) waveform. The linearity coils apply correction directly to the line scan current waveform. L_5 and L_6 vary their inductance as the amplitude of the current increases; this effect is caused by the permanent magnets placed close to the coils. Magnetic saturation is produced in the cores of the inductors as the flux from the magnets adds to the flux set up by the rising scan current. Consequently, the total inductance of the entire circuit falls towards the end of the line period, which, in turn, alters the shape of the waveform. This effect is particularly noticeable near the edges of the picture.

The core of a linearity coil can be adjusted too. The overall inductance will then be varied, which affects both the *width* and the *linearity.* In many receivers a variable inductor, without a magnet, is used in series with the scan coils to provide width control.

Additional inductors in the scan circuit can be shocked into self-oscillations by each flyback pulse. This ringing is damped by parallel resistors–for example, R_8 and R_9 in figure 10.8. If the oscillations were allowed to continue, the beam would tend to move to and fro at the same time as it traversed the screen. This would lead to a display of alternate dark and light vertical bands, known as *striations*, extending from the left towards the centre of the screen.

A colour receiver needs further scan correction by means of *convergence* waveforms, described in the next chapter; such waveforms are also derived from the line output transformer (L_7 and T2).

Resistor R_1 in the h.t. feed to the output stage in figure 10.8 limits the current when the transistors are switched on–especially in the event of a tube flashover.

The 'earthy' end of the stage is connected to the chassis via VR1 and the associated resistors. This arrangement senses the current in the circuit; if it is excessive a 'turn-off' bias disables the driver and the line output transistors remain in the 'off' state (assuming that the fault is not within the transistors themselves!).

The capacitors in the base circuits of the line output transistors (C_1 and C_2) are also a form of protection against tube flashovers–they divert any resultant pulses, which could otherwise damage the transistors.

10.7 Valve Line Output Stage

The behaviour of the scan coils again, is, similar to that described in the equivalent circuit in section 10.4. However, the difference between the design of transistorised and valve output stages is due to the following factors: valves do not act as low resistance switches even when fully saturated; valves cannot conduct in reverse, but are capable of withstanding relatively high voltages–especially those which occur during flyback.

A simplified circuit diagram is shown in figure 10.9. The line output valve, V1, is a high power pentode. The oscillator waveform suitably shaped is applied to the grid via the coupling capacitor, C_1. The cathode is connected directly to chassis, and its biasing is derived from the drive waveform together with a correction bias fed via R_3 and its associated circuit (R_1, R_2, VR1, VDR1, C_8). The latter stabilises the performance of the line output stage–that is, the width of the picture, the magnitude of the boosted h.t. (which will be described later) and, to some extent, even the e.h.t. This is done by means of feedback from the line output transformer via C_8. The amplitude of the line pulses at that point depends upon the amplitude of scan current which represents the picture width. At the same time, the scan current is dependent upon the pentode biasing and the resultant anode current. Since e.h.t. is derived from suitably stepped-up line flyback pulses, the feedback arrangement indirectly senses the magnitude of the e.h.t.

The voltage-dependent resistor, VDR1, rectifies the pulses from C_8. The rectification action is due to the fact that the resistance of the device is high during the

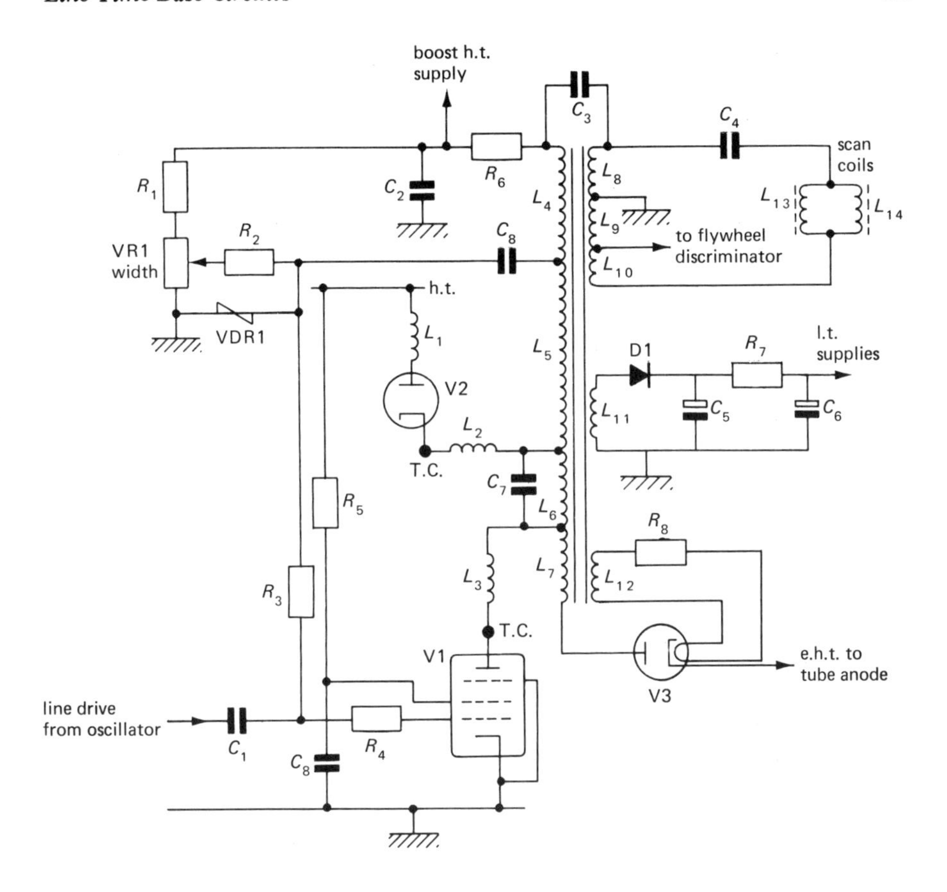

Figure 10.9 Valve line output stage, including width stabilisation circuit

low amplitude scan waveform and is low during the large amplitude, but opposite polarity, flyback waveform–exactly as in a rectifier. Its connection to the line output transformer is so arranged that a negative voltage is developed by the action of the VDR and capacitor C_8. The larger the pulse–suggesting excessive line output conditions–the more negative is the feedback control voltage. The drive to V1 is thus reduced by this voltage and the output brought back to normal. The feedback voltage is added to a positive potential obtained from the divider R_1, VR1 which allows the conduction of V1 to be preset as required.

VR1, or its equivalent, may be labelled: '*set boost h.t.*', '*set e.h.t.*' or '*width*', depending on the manufacturer. Its setting must be carried out according to the makers' instructions, since excessive drive could result in damage to the valve itself or to the line output transformer. Alternatively, e.h.t. which is above normal can also cause serious problems. In most cases the adjustment aims at a specified value of the *boosted h.t.*

The line scan coils, L_{13} and L_{14}, are fed from a section of the line output transformer between L_8, L_9 and L_{10} via the S-correction capacitor, C_4.

When the receiver is first switched on, the line output valve is connected to the h.t. supply via L_3, L_6, L_2 and diode V2. As the drive waveform turns the pentode on harder, its anode current increases. By transformer action a corresponding current flows through the scan coils, causing the beam to be deflected from the centre to the right-hand side of the screen. At the end of the line period V1 is rapidly cut off by the oscillator to initiate the flyback. A large back e.m.f. pulse appears at both the anode of V1 and the cathode of V2. For this reason the two electrodes are brought out to the top of the valve glass envelope in the form of a top cap (marked T.C. in figure 10.9). This solves the problems of adequate insulation between the adjacent pins of the valve base.

The inductance of the line output stage (the transformer and the scan coils) plus its stray and flyback tuning capacitance are again shocked into self-oscillations when V1 switches off. Beam flyback takes place in the manner described in section 10.4; during that time diode V2 is reverse biased by the back e.m.f. in the transformer. When the polarity of this e.m.f. changes at the end of flyback, V2 switches on to damp any further oscillations. The scan current gradually decays as C_3 charges via V2 and the first half of the scan is produced. The energy stored in the scan coils 'drives' the windings L_8, L_{10}, which, by transformer action, induces a voltage and current in L_4 and L_5. Since the diode allows energy recovery to take place, it is called the *efficiency diode.*

During the 'efficiency' part of the scan C_3 is charged by the sum of two voltages: h.t. supplied to the circuit by V2 and the e.m.f. induced in the transformer windings L_4, L_5. As a result, the potential on this capacitor, with respect to the chassis, is well in excess of the main receiver h.t. The supply rail fed from C_3 is called the *boosted h.t.*; the efficiency diode is also known as the *boost diode* and C_3 is referred to as the *boost capacitor.* The value of C_3 is usually between 0.22 μF and 0.47 μF. The reader is invited to compare this arrangement with the derivation of the boosted h.t. in the transistorised circuit shown in figure 10.7.

The boosted h.t. in valve circuits can have any value from 400 V to 1000 V, depending on receiver requirements. It is used to supply some of the electrodes of the picture tube (screen grids—or the so-called 'A1' and *focus* in monochrome receivers) and the field charging capacitor in valve field time base circuits. The line output valve is also fed from the boost circuit once the receiver is operating normally.

In the previous discussion it was implied that the line output valve was completely switched off during the first half of scan. In practice V1 in figure 10.9 is biased ON before the beam has reached the screen centre. Power losses in a valve circuit are higher than in the transistorised equivalent; therefore the stored energy is used up before the first half of scan is completed. The drive waveform is shaped and timed in such a way as to ensure a smooth takeover from the efficiency diode to the line output valve. Failure to achieve this can cause picture non-linearity.

E.H.T. is generated by rectifying the large flyback pulse, which is stepped up by an additional winding, L_7, known as the *e.h.t. overwinding.*

Diode V3 is the *e.h.t. rectifier*; its heater is supplied, via a current limiting resistor, from a separate winding of just a few turns on the line output transformer (L_{12}). It is not practicable to connect the heater to the main valve heater chain (see chapter 14), because the filament would then be at a relatively low voltage with respect to chassis, while the cathode would be maintained at e.h.t. Such an arrangement would create problems with the insulation between the heater and the cathode of the rectifier.

The line output transformer, in both valve and transistor circuits, supplies a number of pulses to other sections of the receiver; in figure 10.9 only a tap feeding the line sync. pulse discriminator (L_9, L_{10}) is shown. There are, however, other pulses—namely to the video amplifier (or directly to the tube) to provide line flyback blanking, black level clamp (mostly in colour receivers), a.g.c. gating (where applicable) and the colour decoder (burst gating, burst blanking, PAL switching).

10.8 Generation of E.H.T.

In most receivers e.h.t. is obtained by rectifying a stepped-up line flyback pulse in the manner illustrated in figures 10.7-10.9. The value of e.h.t. can be from 5 kV (small screen, battery portable) to 25 kV (large screen colour receivers). The output from the e.h.t. overwinding can be of the required amplitude—only rectification is then necessary; or it may be at a lower level, in which case voltage multiplication is used.

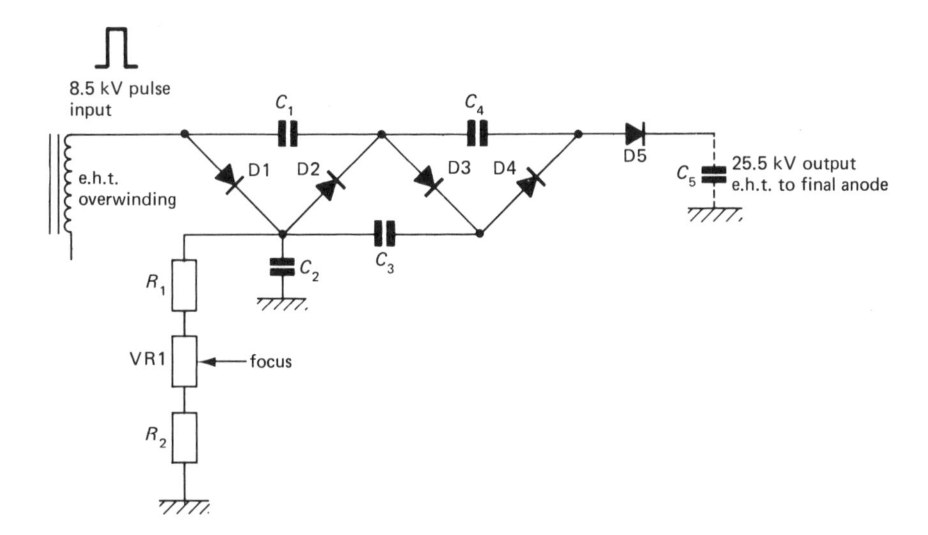

Figure 10.10 Voltage tripler. This arrangement is used in a colour receiver which requires a relatively high focus potential. Final capacitor C_5 is formed by the tube body

Semiconductor e.h.t. rectifiers are commonly used, because of their negligible heat dissipation, greater reliability and lack of X-ray radiation. X-rays can be emitted whenever electrons are accelerated through a large p.d. and then suddenly decelerated by hitting the anode.

Voltage multiplication is only economical in semiconductor circuits, since a large number of rectifier diodes are needed. The circuit shown in figure 10.10 represents a typical *voltage tripler.* The energy supplied by pulses from the overwinding is transferred to the final capacitor, C_5, via the intervening C_1, C_2, C_3 and C_4 through a 'pumping' action of the diodes; each flyback pulse initiates a 'pumping' stroke. The first pulse after switching the receiver on charges C_2 via D1 to 8.5 kV. During scan the input from the overwinding falls to zero; D1 is cut off, but D2 conducts, charging C_1 from C_2; each capacitor ends up with just over 4 kV. The next pulse recharges C_2, and during the following scan the potential on C_1 is raised again, this time to over 6 kV, and so on, until there is almost 8.5 kV on both capacitors. C_3 and C_4 charge in a similar manner; when C_1 and C_2 have been charged, pulses are steered to C_3 via D3, and then D4 charges C_4. Let us assume that the final capacitor, C_5, has not yet been charged, so that it 'sees' the series combination of C_2 and C_3 (as well as C_1 and C_4); since there is 8.5 kV on each, C_5 can initially charge to 17 kV via D5. When this voltage is reached, the last pulse in the whole charging sequence will raise the potential on C_5 to 17 kV + 8.5 kV = 25.5 kV. The whole process takes some time to arrive at its state of balance, because each time the input to the tripler falls to zero, the existing charges on the capacitors are being distributed around the circuit.

The tripler is a self-contained unit with the exception of capacitor C_5, which is normally formed by the body of the picture tube, as will be shown later.

In colour receivers a relatively high voltage is needed for the tube focus electrode and it can be conveniently tapped from the tripler. The potential across C_2 in figure 10.10 is approximately 8.5 kV, which is then reduced by means of a divider formed by R_1, VR1 and R_2. Occasionally a special VDR with a sliding contact can be employed. Some designs use an arrangement known as a *thick film* potentiometer. This could be compared to an integrated circuit, since a number of resistors (for example, R_1, VR1, R_2 in figure 10.10) are deposited in the form of a film over a ceramic substrate. These units are obviously not repairable, but they have good reliability and, unlike large value carbon resistors, their stability is high.

Some receivers have an e.h.t. of less than 20 kV, and a *voltage doubler* may be used. In that case the circuit action is similar to that of the tripler, but the section comprising C_3, D3, C_4 and D4 (figure 10.10) is omitted.

Some designs use a *diode split* e.h.t. winding instead of a tripler. The winding is split into a number of sections which are linked by diodes. Each diode rectifies the flyback pulse from its own section, while the smoothing is performed by the 'stray' capacitance that exists between adjacent layers of the winding. The final value of the e.h.t. is equal to the sum of the d.c. levels obtained from each section. Reliability of diode split line output transformer is good, because the high voltage capacitors normally used in triplers are not required.

Modern receivers use *harmonic tuning* in the generation of e.h.t. The overwinding, by virtue of its design, has a relatively high value of inductance and it also possesses the inevitable stray capacitance. Therefore each flyback pulse will shock the circuit into self-oscillations. The 'ringing' not only affects the input to the e.h.t. rectifier, but is also transformed to other sections of the line output transformer—notably to the collector of the output transistor.

At this point the reader must not confuse the *flyback tuning* with the *harmonic tuning.* The latter is primarily used to help to produce the required e.h.t. by taking advantage of the ringing of the e.h.t. overwind. Flyback tuning is necessary to time the return of the electron beam at the end of each scan.

Instead of damping the self-oscillations of the e.h.t. overwind, they are encouraged, provided the frequency of such oscillations is strictly controlled. The winding is tuned by its inductance in conjunction with suitable, low value external capacitors. The capacitors do not have to be connected to the e.h.t. coil itself—for example, C_7 in figure 10.9.

Depending upon the frequency of self-oscillations, either *third harmonic* or *fifth harmonic* tuning could be used. In this context it means that the frequency is linked ('in harmony') with the value of the flyback frequency (which, as we know, is tuned separately). The mathematics of the associated theory is beyond the scope of this book, but it can be shown that the desirable frequency of e.h.t. ringing is approximately three times the flyback frequency; hence *third harmonic tuning.* The oscillation waveform *adds* to the flyback pulse waveform in the overwind, so that the resultant is an even more 'peaky' output to the e.h.t. rectifier. This means that the input pulse is increased in magnitude without the necessity for a greater number of turns in the winding. At the same time, by transformer action, the induced ringing *subtracts* from the flyback pulse appearing in the main winding connected to the output transistor (or valve). Thus, the advantages of third harmonic tuning are: increased e.h.t. pulse often making a tripler unnecessary; reduced flyback pulse in the main winding—an important feature when transistors are used; the ringing is effectively damped at the end of flyback. The explanation of the last feature involves complex mathematics, but from a practical point of view, if the e.h.t. ringing extended into the scan period, it could again produce vertical striations (see section 10.6).

With the advent of colour receivers *fifth harmonic tuning* has been widely used. In this case the frequency of e.h.t. ringing is approximately five times the flyback frequency. The resultant e.h.t. pulse does not increase in its peak amplitude; therefore a tripler may be necessary. The principal advantage of fifth harmonic tuning is that it produces a relatively constant e.h.t., since the output pulses from the overwind become broad—almost rectangular in shape. This is particularly important in colour receivers, where the beam current is relatively high and the e.h.t. must remain stable. The disadvantage of fifth harmonic tuning is that the flyback pulse at the collector of the output transistor tends to be more 'peaky'. Careful design, however, ensures that the ratings of the output device are not exceeded.

There is no obvious way of deciding from the circuit diagram which of the two methods is used; details may only be obtained from the makers' circuit description. Harmonic tuning is sometimes made adjustable, in which case manufacturers' instructions must be consulted.

10.9 Thyristors in Line Output Stage

The chief disadvantage of a transistorised line output stage is the risk of damage to the device if either the current or the voltage limits are even momentarily exceeded. Since the transistor only acts as a switch, it might appear that a thyristor could be used instead. Thyristors have been used in power electrical engineering for many years, and devices of high voltage and current capacity are easily available. The principle of operation of a thyristor is described in chapter 14; briefly, it behaves like a rectifier, since it conducts in one direction only—that is, with its anode positive with respect to the cathode. However, unlike the rectifier, the conduction does not normally start until a control signal is applied to the third electrode called *the gate.* In that respect it resembles a transistor, which also needs a suitable bias on its third electrode—the base. The difference between the thyristor and the transistor lies in the fact that when the base drive is removed, the transistor switches off; when the thyristor gate signal is removed, the anode current continues until it is reduced to a very low value (practically zero) by the external circuit. On a.c., or unsmoothed d.c., no problem exists, since the supply waveform naturally reduces to zero. In d.c. circuits with a smoothed output this is not the case, and other means of turning the thyristor off have to be adopted.

A simplified arrangement of a line output stage with thyristors is shown in figure 10.11. Each thyristor (TH1 and TH2) is connected in parallel with a diode (D1 and D2, respectively). This is necessary to allow for both the forward and the reverse flow of current in the scan coils in the manner described in section 10.4. A line output transistor can conduct in either direction if required; thyristors can pass current only one way.

The *line output thyristor*, TH2, is in parallel with the *efficiency diode*, D2—this arrangement resembles the conventional transistorised stage shown in figure 10.7. The output pair is connected to a suitable tap on the line output transformer, which, in turn, feeds the scan coils and provides other services in the usual manner.

The other thyristor-diode pair (TH1, D1) is used to switch off the output devices and initiates the flyback. It is easier to explain the circuit action if one examines it at the instant when the pulse from the line oscillator switches the *flyback thyristor*, TH1 on (it actually occurs a couple of microseconds before the flyback is due). This action connects the tuned circuit formed by L_1 and C_1 to chassis. C_1 was originally charged up to h.t. when TH2 was conducting; now it discharges via L_1, TH1 and the output pair—D2 and TH2. Effectively, this presents a reverse bias across the output thyristor which switches off, but the discharge of C_1 continues via D2. The change in the flow of current is communicated, via the output transformer, to the line scan coils and the flyback commences. The duration of the flyback depends on the tuning of L_1, C_1, which, together with the inductance of the scan coils, form a ringing circuit. During the second half of the flyback oscillation C_1 discharges through the line output transformer, D1 and L_1. This, in effect, places a reverse bias across TH1 and the flyback thyristor also switches off.

Flyback ringing stops when the beam has reached the left-hand side of the screen. At that point the e.m.f. in the scan coils reverses and the efficiency diode, D2, is biased ON. The diode shorts out the tuned circuit L_1, C_1, the scan current decays gradually and the first half of the scan is produced.

The second half of the line scan takes place when thyristor TH2 switches on. However, it is not the line oscillator waveform which is applied to the line output thyristor. Instead a suitable pulse is developed by the input transformer, T1, when its current is interrupted at the time of TH1 switching off. This pulse is then

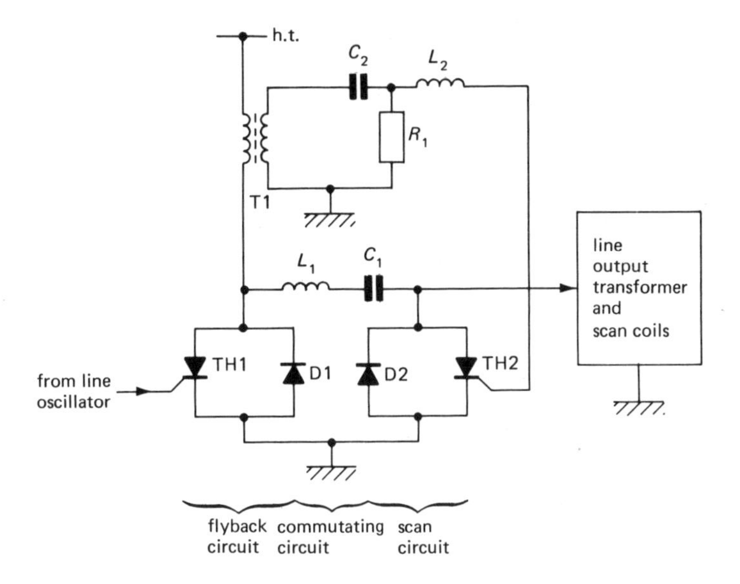

Figure 10.11 Principle of thyristor line output stage. The division of functions between the devices is due both to their inability to conduct in reverse and to the need for a thyristor 'turn-off' pulse. Commutation is the changeover of conduction from one thyristor circuit to the other

delayed by C_2, R_1 and L_2 to allow the efficiency diode to conduct before the output thyristor turns on.

The changeover of conduction from TH1 to TH2 and back again is called *commutation.* Since this action is produced by C_1 (during flyback) and T1 (during scan), the tuned circuit, C_1, L_1, is called a *commutating circuit*, and T1 is sometimes referred to as the *commutating transformer.*

A practical thyristor line output circuit might include a number of damping components connected across the thyristors. This is necessary to prevent either unwanted ringing or voltage 'spikes', which could turn one of the switching devices on and upset the timing of the entire sequence.

10.10 Faults in Line Time Base Circuits

Broadly speaking, the main fault symptoms due to a defect in the line time base may be grouped as follows: incorrect line frequency, insufficient width, horizontal non-linearity, blank screen.

In fault-finding it is important to establish that the fault is likely to be within the given stage, since identical effects could be produced by defects elsewhere in the receiver. For example, lack of line synchronisation could also be due to a weak signal, sync. separator faults, etc. A blank screen might be traced to a problem in the power supplies, the video amplifier, and so on.

Whenever valves are used, they are often the first suspect components; however, they are also easily replaced and consequently eliminated from the enquiry. It will be noted that ordinary valve testers are not always useful, because the high amplitude pulses present in the circuit cannot be simulated by the tester.

Faulty transistors are easier to discover, as they usually develop either definite short or open circuit symptoms. On the other hand, it can be both time-consuming and often undesirable to remove them from the circuit unless it is reasonably justified.

Line output transistors can be accidentally damaged by careless fault-finding. Shorting of transistor electrodes with test probes can either apply excessive drive or remove it at the wrong time interval—the resultant surge could be destructive. Many manufacturers suggest reducing the h.t. to the line output stage during fault-finding (especially in mains-operated receivers). This can be done by connecting a resistor of a suitable value *and* power rating in series with the h.t. feed or by reducing the output voltage from the relevant power supply regulator. The same technique can be adopted in the *initial* setting-up after the replacement of the output transistors. The servicing engineer must consult the makers' manual in this respect.

An oscilloscope is extremely useful in fault-finding in time base circuits, as one deals most of the time with waveforms which can be easily displayed. However, a multimeter can also indicate the presence or the absence of a drive. This is due to the fact that most pulse waveforms are not symmetrical about the horizontal axis and they have a net d.c. value which the voltmeter indicates. For example, the drive waveform maintains the output valve in a cut-off state for a considerable amount of time; hence, the d.c. readings tend to show as a high negative grid bias. Similarly, in transistor circuits a very low, or perhaps a reverse, base bias can be indicated in the presence of a pulse waveform.

Test equipment should not be connected directly to the anode of the line output valve or the cathode of the efficiency diode. Flyback pulse voltages of several kV can damage the instruments! Similarly, it is usually not recommended to measure the output voltage from the line output transistor. A low capacitance attenuator probe should be used when connecting an oscilloscope to the base of a line output transistor.

E.H.T. should be measured by means of a suitable meter with a probe connected to the final anode of the tube. In all-valve receivers it used to be common practice to draw a spark with a screwdriver from any desired part of the line output circuit. Where semiconductors are used, whether in the line output stage *or* elsewhere, this method could easily damage them. E.H.T. can often be felt as static electricity in front of the tube screen; alternatively, a neon tester, *slowly* brought towards the e.h.t. lead, might be used as an indicator.

The reader is asked to refer to the notes on I.C. fault-finding which were given in some detail in section 7.11.

The major types of line time base faults will now be discussed, together with some general suggestions regarding possible suspect areas.

Incorrect line frequency

This symptom can be classed as lack of perfect line synchronisation. Suspect areas are: the sync. pulse discriminator or the line oscillator itself. The discriminator can normally be disabled and the effect of oscillator adjustments noted. The discriminator relies on the symmetry in component values (capacitors, resistors and forward/reverse resistances of the diodes), and it also needs both waveforms (synchronising pulses and feedback) to develop the control signal.

Insufficient width and picture non-linearity

Lack of width can be accompanied by non-linearity of the picture. In general terms, defects on the left-hand side of the screen are due to the efficiency diode circuit and problems on the right-hand side are caused by the line output device (valve or transistor). In a valve line output stage inadequate drive from the oscillator can reduce the width and so can the faults in the width/e.h.t. stabilising circuit. Excessive loading of the line output transformer by an external circuit, defective windings of the transformer or flyback tuning capacitors can also be responsible.

Non-linearity towards the centre of the picture can be caused by excessive drive to the line output valve (faults in width stabilising circuit). If the picture becomes *trapezium*-shaped, it is due to a fault in one section of the line scan coils or their associated components.

Blank screen (no raster)

This fault suggests that there is no e.h.t. (apart from problems in other sections of the receiver, as mentioned earlier), since in most receivers the final anode supply is linked with horizontal scanning circuits. It is necessary to establish whether the fault is due to the e.h.t. section itself (overwind, rectifiers, etc.) or the remainder of the line time base. Some people may hear the high pitched (15 625 Hz) line whistle which is caused by the vibrations set up in the line output transformer; this could quickly confirm that the oscillator and most of the output stage are functioning.

Lack of drive to the output stage can be due to the failure of the oscillator or the coupling components. Where valves are used in the output, a loss of drive presents serious problems. The absence of the negative-going grid waveform leaves the valve with zero bias and a very high current is drawn. As a result, the line output and/or the efficiency diode can be destroyed, or, worse still, the transformer itself could be damaged. During the investigation of this type of fault the screen grid supply to the output pentode can be disconnected to cut the valve off.

A short circuit across any of the line output transformer terminals will load the output stage to such an extent that flyback oscillations can not take place; the e.h.t. will then be very low or non-existent. Since the transformer supplies a number of circuits, each one has to be disconnected in turn until the fault has been cleared. Both the boost capacitor and various tuning capacitors are a common source of short circuits, because they are subjected to relatively high pulse voltages. Failure of a number of protective components, in both the line output stage and the driver, can result in a breakdown of the line output transistor.

The line output transformer is the last item on the 'suspect' list. The cost of a replacement and the work involved must be fully justified by reasonable evidence of the fault being inside the transformer.

Other faults

(1) *'Ballooning'* (*picture 'breathing'*). The picture expands as the brightness control is advanced. This is due to the e.h.t. falling rapidly as the demand for beam current increases, and can be frequently caused by faulty e.h.t. rectifiers or the tripler capacitors, as applicable. Similarly, reduced h.t. or boost h.t. can also cause 'ballooning'. Tripler breakdown may be accompanied by its overheating, flashes on the screen and unpleasant smell!

(2) *Corona discharge and arcing.* Corona is in the form of violet 'glow', which can surround any component or lead carrying very high voltages; arcing means that flashover takes place to other parts of the circuit (normally to receiver earth). Usually corona is a warning sign of imminent arcing. High voltage discharge can be caused by dust or moisture on the suspect components, sharp points on soldered joints, defective insulation, etc. Arcing can damage semiconductor devices and produce interference on the picture.

(3) *Vertical striations.* Alternate narrow dark and bright vertical bars especially noticeable on a blank, darkened raster. These are caused by ringing in the width, in the linearity or even in the scan coils, owing to faulty damping components.

10.11 Safety Aspects of Line Time Base Operation and Servicing

The line output stage, including the e.h.t. section, is a high power and a high voltage circuit. Under fault conditions it can carry large currents which, in turn, could cause dangerous overheating. A number of components are designated as '*safety components*'; any replacements *and* the methods of mounting them in the receiver must conform to the makers' specifications. Failure to do so could cause the repair to be deemed dangerous in terms of British legislation. Some of the more obvious components in this category are: the line output transformer, the boost capacitor, the screen grid feed resistor, etc.

Valves in the line output stage, including the tube, can emit X-rays; the risk increases with the operating voltage (usually in excess of 5 kV). The danger is that such radiation is not immediately noticeable, but its effect accumulates and can be a serious health risk.

Valves which are subjected to high voltages are mounted within a metal screen, which should always be in position unless it is absolutely necessary to remove it for access. The screen not only absorbs the X-ray radiation, but also helps to prevent the spread of r.f. interference and acts as a mechanical barrier to the sections which operate at high voltages.

11 Picture Tubes and Associated Circuits

11.1 Basic Principles of Picture Tube Operation

In principle, both monochrome and colour tubes have much in common. Fundamentally, they are a form of thermionic valve in which the electrons are made to strike a special material, known as *phosphor*, which, in turn, emits light. The amount of light produced depends upon the number of electrons hitting the screen material at any time. The flow of electrons gives rise to the tube or *beam current*, which is governed by the magnitude of voltages applied to the various electrodes. In a TV tube some of the voltages are kept constant, others can be controlled by the viewer (or the service engineer) and yet others are amplified versions of the actual received signals which produce the picture.

The picture itself is formed by variations in the amount of light emitted from the screen as the electron beam scans its entire area. The scanning is under the control of the field and the line time bases. The light output persists for a fraction of a second after the beam has left any given position on the screen. This persistence, together with the chosen scanning rate, ensures that the eye retains a complete image without excessive flicker.

The colour of the emitted light depends upon the chemical composition of the phosphor. In monochrome tubes maximum beam current gives rise to white light output (in practice, it might be 'cool white'–see the discussion in chapter 1). The screen of a colour tube is coated with three types of phosphors deposited in a carefully arranged pattern. Each phosphor is capable of emitting light of different colour–namely red, green or blue. From a normal viewing distance the light outputs appear to mix and create an impression of any desired hue, as described in chapter 1. The entire process is under the control of the decoded and matrixed colour signals applied to the tube electrodes–which are usually the cathodes.

The electrons have to travel over a considerable distance between the cathode and the screen, which means that they need a lot of energy. They must also have enough energy left when they reach the screen to allow some of it to be converted into light of adequate intensity. For this reason the operating voltages are high; the voltages increase in value with the size of the screen and, for reasons given later, they tend to be higher in colour receivers than in monochrome receivers.

A thin layer of aluminium is deposited on the inner surface of the tube screen.

This allows the electrons to pass through, while acting as a barrier to the heavier and destructive ions. The coating also acts as a 'mirror' which reflects the emitted light forwards and improves the picture brightness.

11.2 The Electron Gun

The part of the tube which is responsible for the emission and acceleration of the electrons is called the *electron gun.* As shown in figure 11.1(a), the electron gun includes the heater, the cathode, the control grid, the first anode (sometimes referred to as the screen grid or the 'A1'), the second anode, the focus

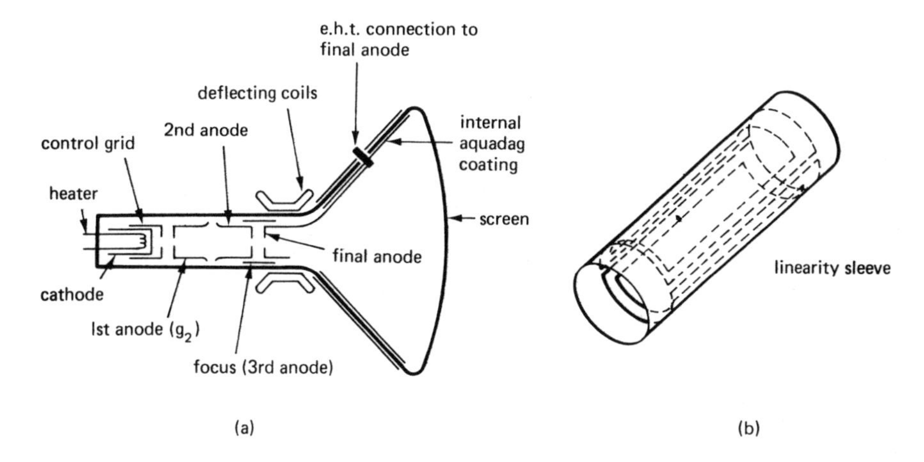

Figure 11.1 Principle of construction of monochrome tube: (a) arrangement of electrodes in the electron gun; (b) linearity sleeve is used in some receivers; the sleeve is pushed under the scan coils over the tube neck

anode and the fourth anode (commonly termed the final anode). The electrodes are in the form of hollow cylinders to allow the beam to pass through. The electrical supplies are so arranged that the electrode voltage levels increase progressively with respect to the cathode. There is one exception–the control grid is at a lower potential with respect to the cathode, the value of that voltage depends upon the setting of the brightness control and tube design. Typical electrode potentials of a monochrome tube are: heater, 6.3 V; cathode, 80 V; control grid, 20 V; A1, 300 V; focus, 350 V; final anode, 17 kV. The corresponding voltages of a colour tube are: heater, 6.3 V; cathode, 150 V; control grid, 40 V; A1, 500 V (average); focus, 4 kV; final anode, 25 kV.

Complex electrode assemblies are sometimes used, especially in colour tubes, their aim being primarily to improve the focusing of the electron beam over the entire area of the screen. There are involved reasons, beyond the scope of this discussion, why the electrons forming the beam tend to 'spread out', not only during their straight passage to the screen, but also during the deflecting process. The degree of such distortion depends on the amount of deflection from the centre; therefore, in the setting-up procedure optimum focus has to be aimed at. Because of the photographic similarities in the problems encountered, the electron gun is sometimes called the *electron lens.* Early designs had magnetic rather than electrostatic focusing, using permanent magnets or electromagnets mounted on the outside of the tube neck.

Most colour tubes have three complete guns housed side by side in the neck of the tube. Each gun is associated with a particular 'colour' beam, which, in turn, must always land on the appropriate colour phosphor. If all three do not come close enough on the screen, the impression of colour mixing is lost, and that leads to a colour fringing effect. For this reason there is an additional arrangement both on the inside and on the outside of the neck of the tube for the purpose of *converging* the three beams.

If one or more of the three colour beams land on the 'wrong' colour phosphor, then incorrect colours are produced in patches over certain parts of the screen, as though some paint had been spilled on the picture. To counteract this effect, known as lack of *colour purity*, adjustable magnets are placed on the outside of the tube neck. Both the purity and the convergence adjustments will be considered separately.

The scan coils are positioned over the neck of the tube and against the flared part; this coincides with the space immediately following the electron gun and any correction magnet assemblies.

The electrical connections to the electrodes of the gun, except that to the final anode, are brought out to the base of the tube. The final anode operates at such a high voltage that it could create serious insulation problems both outside and inside the tube. Therefore the e.h.t. supply is made via a special connector on the flared part of the tube and brought back to the final anode in the gun assembly by means of a conducting coating of graphite ('Aquadag'). This arrangement has additional advantages: firstly, together with a similar layer on the outside of the tube, it forms a reliable e.h.t. capacitor which is used for smoothing the final anode supply (the capacitance is in the order of 2000 pF); secondly, it prevents internal light reflections between the screen and the glass; finally, the coating completes the electrical circuit from the screen of the tube to the anode.

11.3 Arrangements for Correct Picture Geometry

Correct picture geometry implies a rectangular picture with reasonably straight edges, of a correct width and height, properly centred within the available screen area and, finally, of a linear, undistorted shape.

The linearity, as well as the height and the width, is governed by the shape and amplitude of each scanning waveform. This, in turn, is controlled by the time base circuits and components external to the tube. In some receivers the linearity of the horizontal scan can also be controlled by means of the so-called *linearity sleeve.* The sleeve consists of two thin copper pressings made into loops [see figure 11.1(b)] and secured onto a tube of an insulating material. The whole assembly is placed on the tube neck and is partly underneath the scan coils. The position of the linearity sleeve with respect to the scan coils can be readjusted if absolutely necessary. The loops work on the principle that the current in the scan coils induces currents in the linearity sleeve. Consequently, the electron beam is subjected to two magnetic fields—one produced by the current in the copper loops and another set up by the scan coils themselves. The resultant distribution of the deflecting magnetic field is such that a more linear raster can be achieved.

Pincushion distortion is another form of deformation of the picture. As the name suggests, the edges of the raster, instead of being straight, tend to become bowed inwards; the picture seems to be overscanned towards the corners (see figure 11.10). This is particularly pronounced in wide angle, 'flat' screen tubes. The maximum deflecting force of both the field and line scan coils—coinciding with the extremities of the raster—causes the overscan. In monochrome receivers pincushion distortion can be corrected by means of permanent magnets mounted on brackets close to the scan coils. The magnets can be adjusted in any desired direction until a satisfactory, straight raster is displayed. The use of freely adjustable magnets close to colour tubes (at either the front or the back) is not allowed, since it would alter the relative paths of the three beams and result in lack of colour purity on the screen. For this reason pincushion distortion in colour receivers becomes part of scan waveform correction. In 110° tube circuits the arrangement is particularly complicated, as will be shown in a later section of this chapter.

The position of the scan coils upon the tube neck can be altered in the majority of TV tubes. The assembly must be pushed forward so that the raster fills the screen without 'corner cutting'. Where appropriate, the coils can be rotated to ensure that the picture is not sloping. Final centering of the display in monochrome receivers is by means of adjustable ring magnets on the neck of the tube and positioned before the deflecting coils. Again, this method cannot be applied to colour tubes, so that *picture shift* takes place by injecting a suitable d.c. into the respective scan coils. To allow for an up-or-down and left-or-right movement, some form of polarity reversal of such a current is included.

In many colour receivers the adjustment to the position of the scan coils is also used to set up colour purity, as explained in chapter 13. There are some colour tubes which use precision scanning to reduce convergence problems; the scan coils are then permanently fixed to the tube, no adjustment is available and the tube must be replaced with its own deflecting coils.

11.4 The Shadowmask Tube—Delta Gun Version

The screen of a colour tube has three phosphor materials—one for each primary colour, red, green and blue. There are two principal methods of distributing the phosphors—either in the form of dots or as vertical stripes. The dot arrangement is an early but still very popular design. The dots are deposited in a triangular formation; each group of the three primary colours is known as the *triad.* Spacings between the triads are so small that from a normal viewing distance the dot structure is not evident.

The three electron guns in the neck of the tube are also arranged in a corresponding triangular fashion known as the delta gun formation. The electrons emitted from each gun are then aimed at their respective colour dots. To ensure the correct landing positions for each beam, a perforated steel plate, called the *shadowmask*, is mounted inside the tube and just in front of the screen. This acts as a masking plate –only the beam of one particular colour may strike its own phosphor dots; the other two beams cannot reach the 'wrong' phosphor, because the dot lies in the 'shadow' cast by the mask. This happens because each beam approaches the holes in the mask from a different position in the gun assembly. The diagram in figure 11.2 illustrates the principle of the shadowmask. The three beams converge upon the holes in the plate before they reach their appropriate phosphor dots.

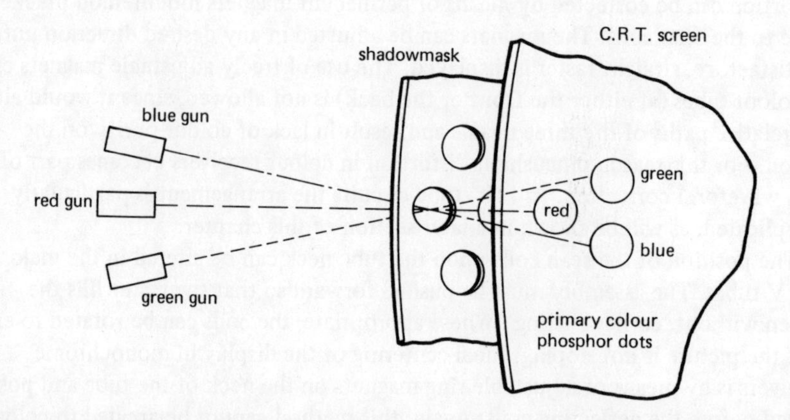

Figure 11.2 Principle of shadowmask. The three colour beams converge upon the holes in the mask, which ensures that the beams strike their own colour phosphor dots

The shadowmask is an obstruction in the path of the beam, and the majority of the emitted electrons never reach the screen. Those which do must have sufficient energy to compensate for the missing contribution from the remainder in order to produce adequate light output. As a result, the total beam current of a colour tube could reach 1.5 mA and the anode voltage is on average 25 kV. Both figures are well above their counterparts for a monochrome tube of the same screen size.

Those electrons which are captured by the mask generate heat in the plate, causing it to expand. (This power loss is in the region of 15–20 W.) This must be taken care of in the construction of the tube; otherwise the 'shadow' effect would alter, giving rise to serious purity errors.

The shadowmask, being made of steel, can become accidentally magnetised from domestic electrical appliances, toys, tools, power cables, etc., if these are placed close to the TV receiver. If it happens, the plate will deflect the electrons in a random fashion, producing irregular patches of unwanted colour over the picture. This can, however, be corrected by either automatic or manual demagnetisation (*degaussing*).

The chief disadvantage of the delta gun arrangement is the difficulty in achieving perfect convergence of the three beams over the whole area of the screen. Convergence implies that the beams land on the phosphor dots of one triad at a time. If only two beams struck the dots of one triad, while the third beam reached the phosphor of its correct colour but in a different triad, the three light outputs would not merge and colour fringing would take place. There is an obvious tendency for the three beams to diverge as they become deflected by the scan coils. This is caused by the unequal distances covered by each beam in order to reach their common destination. For example, if the arrangement of the guns is as shown in figure 11.3,

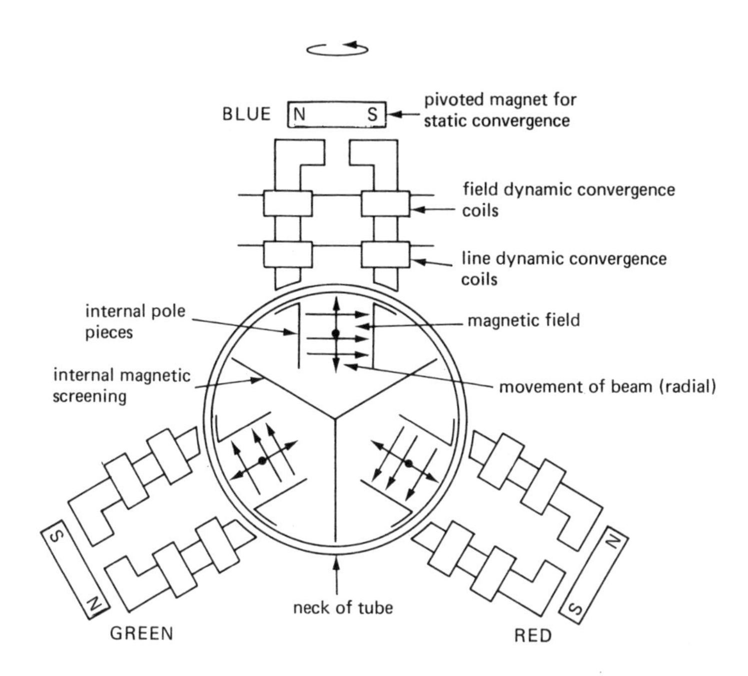

Figure 11.3 Delta gun tube-magnet assembly which provides radial convergence correction. The three beams are separately predeflected by their convergence magnets before entering the scan coils. The direction of the magnetic field is between the internal pole pieces, which results in a radial movement of the beam. A separate lateral convergence assembly (not shown here) is used for the blue beam

the 'blue' beam needs less vertical deflection to reach the top of the screen than either the 'red' or the 'green'. On the other hand, the 'red' beam requires less horizontal deflection to arrive at the extreme left edge of the picture than either of the other two. The amount of misconvergence increases with the degree of deflection from the centre of the screen. The process of correction which maintains the convergence of the beams during their line and field deflection is known as the *dynamic convergence.* However, the beams must be converged first at the centre of the screen before any deflection takes place; this is known as the *static convergence.*

The principle of convergence is based on predeflecting the beams by means of special permanent magnets and electromagnets ahead of the magnetic field of the scan coils. The combined action of the scan and the convergence fields produces a converged raster. Each 'colour' requires a different amount of correction to ensure convergence at the top or bottom, and at the left- or right-hand side of the screen.

Usually the adjustment of static convergence is by means of permanent magnets (although electromagnets operated from a d.c. source are also used), and dynamic convergence is obtained from electromagnets whose coils are fed with suitably shaped waveforms. An appropriate magnet assembly is placed on the neck of the tube and the resultant magnetic fields are guided into the path of each beam by means of internal pole pieces, as shown in figure 11.3. Since the deflection caused by each magnet is perpendicular to the lines of magnetic field, it follows that the resultant beam movement is along the radius of the neck–this is sometimes referred to as *radial convergence.* A closer study of the arrangement in figure 11.3 reveals that the red and the green beams can be moved diagonally, which in effect gives an up-or-down as well as left-or-right shift. On the other hand, the blue beam can move either up or down but not sideways. For this reason a separate magnet assembly is also placed on the tube neck; this time the magnetic field is guided into the neck of the tube at such an angle as to move the blue beam from side to side. In practice, the arrangement deflects the blue beam in one direction while shifting the green and red in the opposite direction until the three colours converge. The control magnet is known as the *blue lateral* (as opposed to the blue radial previously discussed).

11.5 The Shadowmask Tube–In-line Version

The convergence process in a delta gun tube is relatively involved. This problem has been overcome by the introduction of the in-line gun assembly, in which all three guns are in line along the horizontal diameter of the neck. It can be seen from the diagram in figure 11.4 that the amount of vertical deflection is the same for all three beams. Horizontal deflection, however, is symmetrical only for the centre colour, and it is unequal for the other two beams. In some tubes the green gun occupies the centre position, in others the red beam is centrally placed. The in-line arrangement requires less convergence correction, which leads to simpler circuits and fewer adjustments; it also reduces the overall power losses in both the line and the field time base circuits.

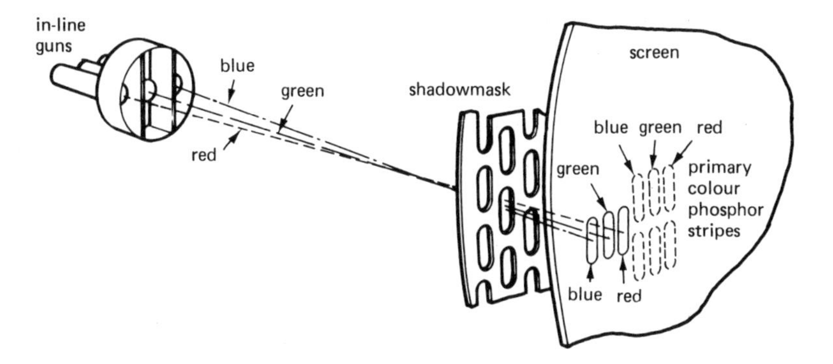

Figure 11.4 Principle of in-line tube and its mask. The guns are in line across the tube neck, the shadowmask is slotted and the phosphors form stripes on the screen

The simplicity of an in-line arrangement has been applied to the *precision-in-line* (PIL) tube. All dynamic convergence adjustments have been dispensed with, although static convergence may be corrected if absolutely necessary. This apparent simplification is achieved by very close tolerances in the manufacture of the tube, by the position of the gun and, in particular, by the use of a highly specialised design of scan coils. The latter produce a magnetic field whose distribution provides the correct beam deflection. Naturally, the scan coils then become part of the tube to which they are permanently bonded.

Because of the in-line positioning of the guns, the phosphors on the screen are in a similar configuration. Instead of dots there are now vertical stripes; the shadowmask has vertical slots which effectively make it more 'transparent' to the electron current, unlike the previous type of mask described in section 11.4. Thus, for similar values of e.h.t. and beam current, the light output is increased, giving a brighter picture. Some makers of either type of tube give the phosphor stripes, or the dots, a black surround, which improves the contrast between the illuminated and the dark areas of the screen, creating an impression of added 'brightness'.

11.6 The Trinitron Tube

The Trinitron was introduced by the Sony Co. before the in-line shadowmask version appeared on the market. The similarity between the two types of picture tube lies in the fact that they both have vertical phosphor stripes. Instead of the rather more solid shadowmask there is now an aperture grill; the three cathodes of the electron gun are in line, but beyond the cathodes there is only one set of electrodes common to all three beams. These electrodes are known as the *electron lens*, since their effect on the three beams is similar to the action of a complex optical lens arrangement. The three beams are brought to a common point before they enter the area of the deflecting coils. In effect, there is now only one apparent point of origin of the beams, which makes for good focus and easy convergence. Convergence in the smaller screen Trinitron tube is achieved by applying suitably shaped voltage waveforms to two pairs of plates which are inside the neck of the tube. External convergence magnet assemblies are used in the later versions of the tube.

11.7 Dynamic Convergence Circuits–General Requirements

The requirements of a delta gun tube will now be considered, but these can be easily adapted to the needs of an in-line tube, where the problems are similar (although considerably simplified).

Since the guns are situated away from the tube axis, the amount of both the horizontal and the vertical deflection must be somewhat different for each colour beam. As the scan coils set up symmetrical magnetic fields with respect to the axis of the tube, the convergence correction fields must be made non-symmetrical to provide the required unequal deflection.

The convergence electromagnets shown in figure 11.3 are supplied with waveforms known as *parabola* and *tilted parabola*; examples of these are shown in figure 11.5. The convergence waveforms are derived from the two time bases, because the correction process must be synchronised with the action of the main scan. Initially, the time base waveforms are available in the form of either rectangular

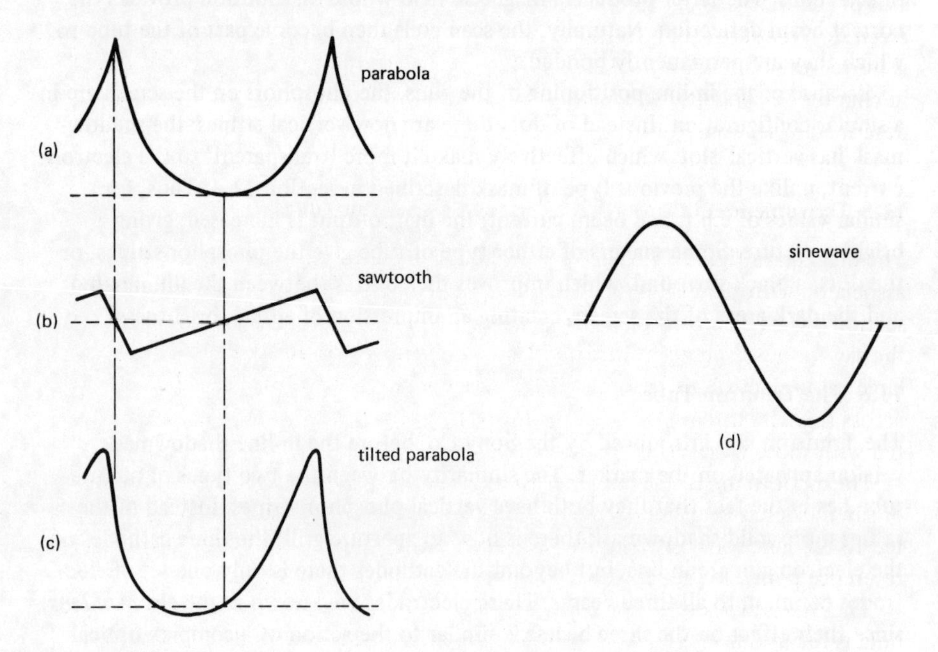

Figure 11.5 Basic waveforms used in convergence correction circuits

pulses or sawtooth. The change from these into a parabola and a tilted parabola takes place in the convergence control circuits. Various forms of pulse integrating circuits are used following the principle of integration described in chapter 9. The reader will recall how a rectangular pulse is changed into sawtooth by an integrating circuit; if, however, a sawtooth waveform is fed to an integrating circuit it becomes parabolic in shape [figure 11.5(a)]. A tilted parabola [figure 11.5(c)] is produced by adding a sawtooth waveform [figure 11.5(b)] to a parabola. The available controls vary either the *amplitude* of the parabola (these are often labelled '*amp*' controls) or the amount of *tilt* sawtooth to be added (consequently, the latter are referred to as '*tilt*' controls).

The most satisfactory method of setting up the receiver convergence is in conjunction with a test pattern display of *crosshatch.* It consists of squares formed by vertical and horizontal white lines. Any misconvergence shows up as colour fringing or, more aptly, as colour splitting in a particular area of the screen where the impression of white lines is lost. The correction circuitry is arranged and labelled according to the defects which could be noticed in the crosshatch display on the screen. Invariably a suitable, if somewhat exaggerated, diagram of the misconvergence accompanies each control to facilitate their identification. We shall return to the adjustment procedure in chapter 12.

In the delta gun tube the red and the green beams are controlled together for ease of operation. However, one set of controls—*red/green (matrixed)*—shifts both beams diagonally in the same direction—that is, either towards or away from the axis of the tube neck (see diagram in figure 11.3). The other set of controls—*red/green differential*—shifts the red beam diagonally in one direction, while the green is moved the opposite way. The reason for such an arrangement will be described in chapter 12. The need for the separate *blue lateral* and *blue radial* adjustment was described in section 11.6.

11.8 Dynamic Field Convergence Circuits—Delta Gun Tubes

The diagram in figure 11.6 shows a dynamic field convergence circuit which is associated with a transistorised field output stage (similar to the one described in section 9.6). The return path from the field scan coils is via C_1 and R_1; therefore the sawtooth scan current develops two waveforms: a parabola across C_1 (the large value electrolytic capacitor integrates the input waveform) and the sawtooth across R_1. The three sets of field convergence coils L_1/L_2, L_3/L_4 and L_5/L_6 are fed with those waveforms via their respective adjustments.

The input to the BLUE coils consists of the parabola from VR1, to which a degree of sawtooth has been added via VR2 and R_5. The return from L_5 and L_6 is connected to the junction of two equal value resistors R_2 and R_3; this establishes a centre reference point, so that the blue beam can be moved either up or down.

The other two sets of coils, RED and GREEN, are effectively in series; the parabola is fed at one end (VR3), the sawtooth at the other end (VR4). The *red/green differential* control, VR5, acts as a variable shunt; when adjusted towards its top

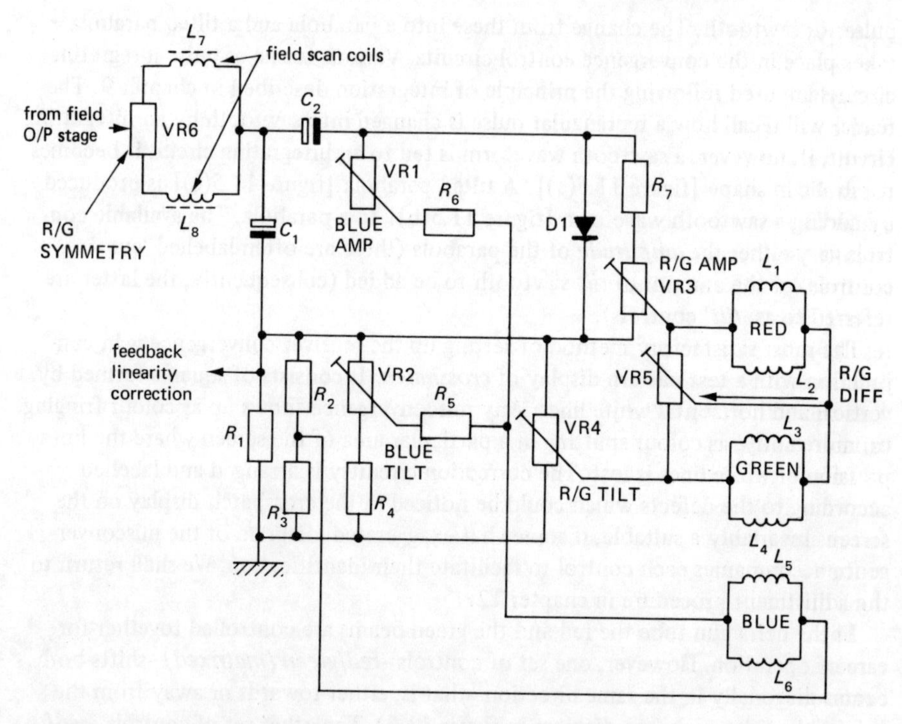

Figure 11.6 Field dynamic convergence circuit–delta gun tube. Field scan current develops a parabola across C_1 and a sawtooth waveform across R_1. Both waveforms are fed in appropriate proportions to the convergence coils (Thorn)

end position, the current through the RED coils is reduced and that through the GREEN coils is increased, which provides for an opposing diagonal movement of the two beams.

The various adjustments to the shape of the convergence waveforms tends to alter their effective d.c. levels, which, in turn, could upset *static* convergence. Diode D1, together with the coupling capacitor C_2 acts as a d.c. restoring circuit; the parabolic waveform is now clamped at a fixed level, resulting in less need for constant readjustments to the static convergence. A transistor with its base and emitter strapped to form a diode may sometimes be used in place of D1.

In valve receivers the required parabola can be derived from across the cathode bypass capacitor of the field output valve, while the sawtooth is provided by a winding on the field output transformer.

The field scan coils are in two sections which are connected in parallel and placed on both sides of the tube neck. If there is a lack of symmetry between the two sections, convergence errors are likely to occur. For this reason a balancing control–the low value variable resistor, VR6–links the coils as shown in figure 11.6.

11.9 Dynamic Line Convergence Circuits–Delta Gun Tubes

The principle of operation of the line convergence circuitry is similar to that of its field counterpart. The initial waveform in the shape of either rectangular or sawtooth voltage pulses can be obtained from a separate winding on the line output transformer, or even from across the line scan coils. A tilted parabola is needed again; therefore the necessary waveform shaping is provided by the correction circuit. Use can be made of inductance as an integrating component, because the required time constants are now relatively short compared with those of the field scan. If a rectangular voltage pulse is applied to an integrating inductor, the resultant *current* has sawtooth waveform; when a sawtooth voltage is applied to such a coil, the resultant *current* waveform becomes parabolic. The convergence coils themselves are often used as integrating components in addition to external variable inductors.

Tuned circuits can sometimes be employed for convergence correction. The sinewave they produce resembles a parabola, which is further shaped by diodes and resistor–capacitor networks. Since the line frequency is much higher than that of the field, the component values tend to be more critical and their precise functions difficult to predict without makers' information.

The reader has to appreciate that is is the *current* waveform through the convergence coils which does the actual correction. Such waveforms are not always easily obtained by means of an oscilloscope, as the instrument is usually connected to display *voltage* waveforms. Current waveforms may be viewed if a special current probe is connected between the oscilloscope and the leads to the convergence coils.

Figure 11.7 shows an arrangement for a line convergence circuit in a delta gun tube. Effectively the various convergence coils are in series with the line scan coils, L_1 and L_2. The scan current from winding L_{15} on the line output transformer flows through the parallel connected scan coils; balancing between the two sections is by means of the inductor L_3 (see the field counterpart VR6 in figure 11.6). The current returns via the linearity coil, L_4, S-correction capacitor, C_1, and another section of the output transformer–L_{16} and L_{17}. This section feeds the convergence correction circuits and their respective coils.

The *blue lateral* convergence coil, L_6, is supplied with a current whose amplitude is varied by the *blue width* adjustment, L_5 (lateral movement of the blue beam effectively increases the width of the blue raster). Since the correction may be needed to provide movement from left to right or vice versa, the connections–B_1 and B_2–to the coils can be reversed. In some receivers dynamic blue lateral action is not necessary and the coils may be disconnected. After all, the blue gun is in the central position as far as the line scan is concerned, which should not lead to line convergence errors.

The red, green and blue *radial convergence* coil circuits are effectively in series with one another. The adjustments are across the coils to modify the *voltage* waveforms applied to them; this in turn gives rise to the tilted parabola *current* waveform

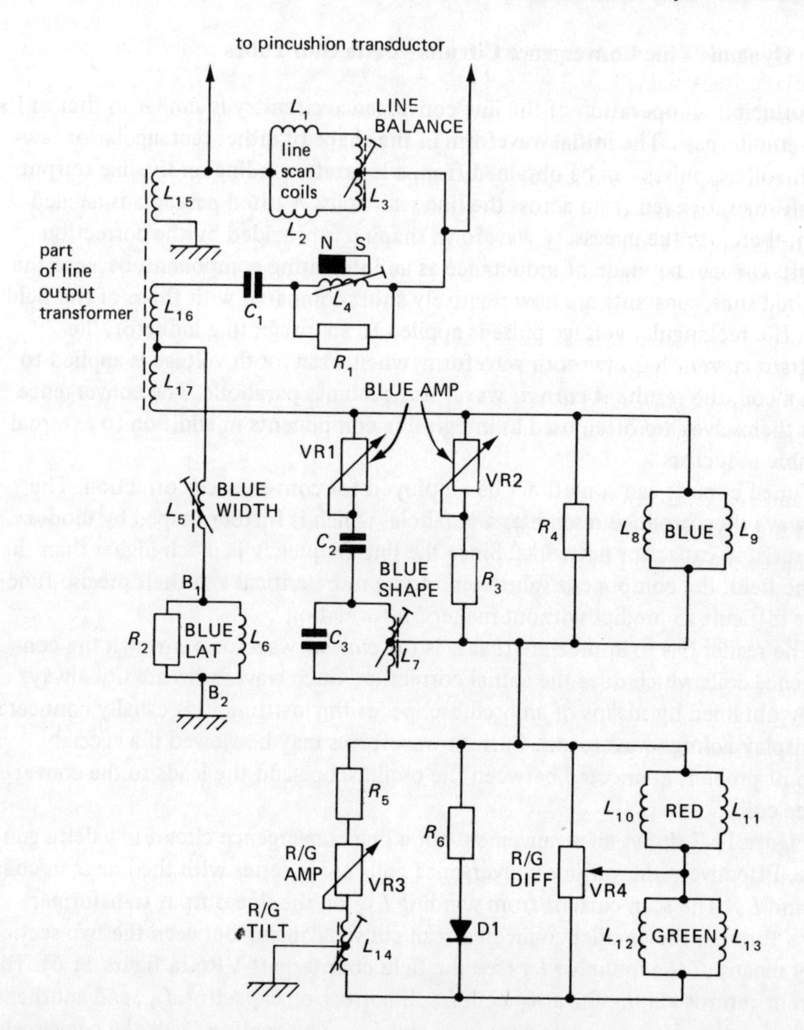

Figure 11.7 Line dynamic convergence circuit–delta gun tube. The entire assembly is fed from an auxiliary winding on line output transformer. The adjustments provide the required waveform shaping. The sense of the Blue Lateral convergence coil is reversible (at terminals B1–B2). A pincushion transductor is illustrated in figure 11.11. (Decca)

in coils L_8 to L_{13}. The blue radial adjustments in figure 11.7 include a tuned circuit L_7, C_3; here it is tuned to twice line frequency and the control is labelled *'blue shape'*.

The red and the green convergence currents are controlled together in a manner similar to that used with their field counterparts. The *red/green differential* control, VR4, produces an opposing flow of currents in the two sets of coils. The two matrixed controls are VR3 (*red/green amplitude*) and L_{14} (*red/green tilt*); these, as previously explained, affect the two beams equally. Diode D1 in series with resistor R_6 is connected across the red/green coils to clamp the d.c. level of the correction waveform. Consequently, the centre (static) convergence is relatively unaffected by the dynamic adjustments; in addition, the clamping effect also helps to make the various dynamic controls less interdependent.

The line and field convergence arrangements described so far are known as *passive circuits*—all the energy supplied to them originates from the line and the field output stages, respectively. These circuits are normally associated with 90° tubes; convergence errors in large screen, 110° delta gun tubes could be considerable, and the correction field must be made stronger. Instead of placing an additional burden on the time base circuits, separate amplifiers are used to provide the driving power for the convergence coils. Such an arrangement is called an *active convergence circuit.* The initial waveform at a low power level is derived from the line and field scan circuits and it is then fed to suitable transistor amplifiers. Apart from the amplifying stages, however, the principles of waveform shaping, etc., are similar to those used in passive circuits.

11.10 Dynamic Convergence Circuits—In-line Tubes

The in-line positioning of the guns simplifies the convergence requirements—even in 110° tubes. One of the guns, often the green gun, occupies the central position along the tube axis; consequently, only two beams need convergence correction with respect to the centre colour. The necessary circuits are very much simpler and the total number of adjustments is reduced.

The diagram in figure 11.8 shows the *line dynamic convergence* circuits associated with a 110°, in-line tube. There are two separate convergence coils—L_4 for the

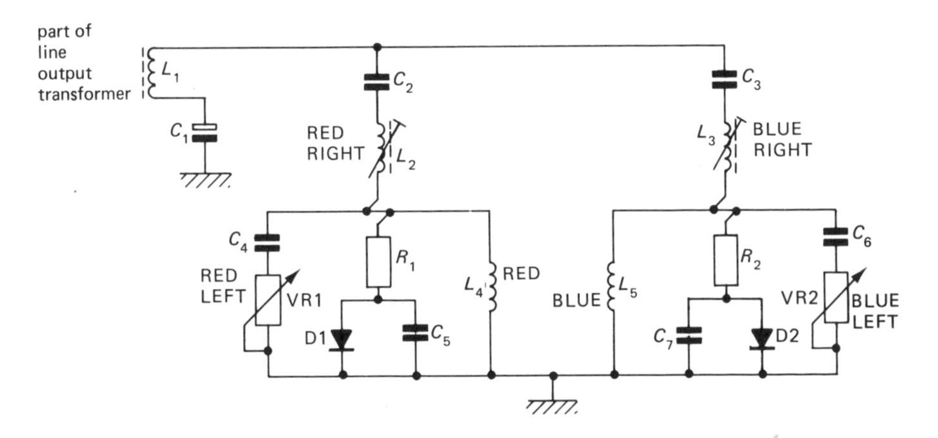

Figure 11.8 Line convergence circuit—in-line gun tube (Rank)

red and L_5 for the blue beam. Pulses from winding L_1 on the line output transformer are changed into sinewaves by the tuned circuits C_2, L_2 and C_3, L_3; the coils are the equivalent of the parabola *amplitude controls.* The shape of the initial sinewave is then altered by the presence of the diode clamp circuits R_1, D1 and R_2, D2 and the parallel networks of VR1, C_4 and VR2, C_6; the variable resistors are now the equivalents to *tilt controls.* The waveforms which are applied to the convergence coils, again, are in the form of a tilted parabola. This requirement is similar to that for a delta gun tube, since both the red and the blue beams are still off-centre as far as the line deflection is concerned.

Some manufacturers have simplified the circuitry further by connecting the red and the blue line convergence coils in series. Only one amplitude and one tilt adjustment are needed, because the resultant magnetic fields are arranged in such a way that one beam moves to the left while the other one shifts to the right and they converge.

The *field dynamic convergence* circuit of the same receiver is shown in figure 11.9. First thoughts suggest that an in-line tube should not require any field convergence correction. The three beams are in the same horizontal plane and they should be deflected in the vertical direction without parting company. Indeed, it is possible to find receivers which dispense completely with the field convergence. Any

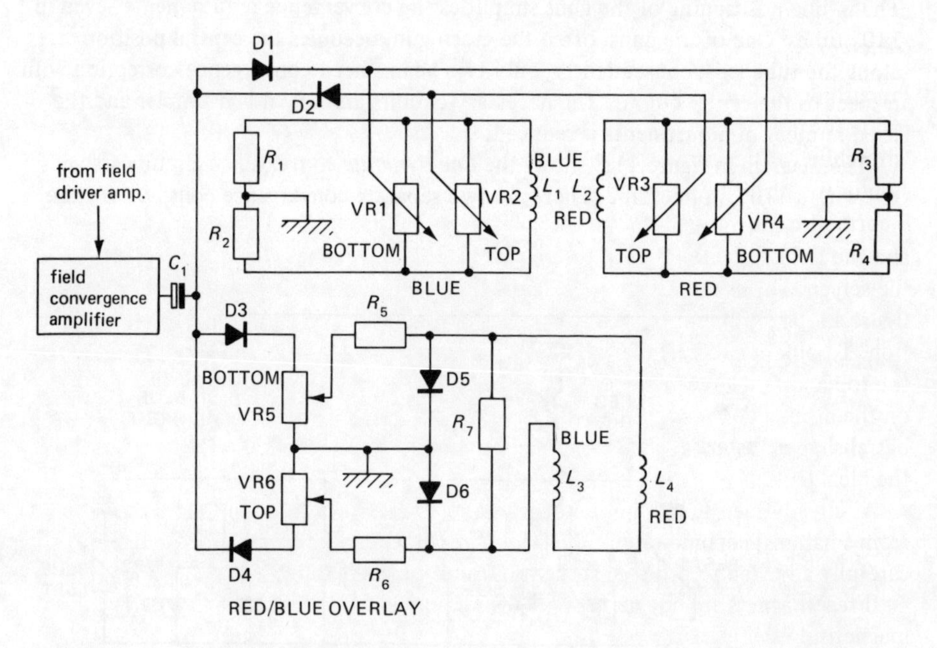

Figure 11.9 Field convergence circuit—in-line gun tube. To reduce the loading on the field output stage, the convergence correction assembly is supplied from a separate convergence waveform amplifier. (Rank)

residual errors can then be corrected by slight repositioning of the convergence coil assembly or even by readjusting the purity magnets! (This should only be undertaken where specific instructions are given in the makers' manual.)

The circuit illustrated in figure 11.9 is of interest, since it uses a technique also found in delta gun receivers. The sawtooth waveform developed in the field time base is taken from the output driver stage to be amplified separately by the field convergence amplifier (shown as a block). The a.c. coupled waveform is then split into both positive and negative half-cycles by the two pairs of diodes D1, D2 and D3, D4. Each half of the waveform is responsible for the field deflection between the centre and either the top or the bottom of the screen. The arrangement makes the convergence adjustments easier because of lack of interaction between the controls. The centre-tapped potential dividers, R_1, R_2 and R_3, R_4, form zero reference points, so that the beam may be moved either upwards or downwards as required.

The *red/blue overlay* convergence coils are connected in series, but their magnetic fields act in opposition to one another to converge the beams at the top and bottom of the screen. Practical effects of these adjustments are discussed in the next chapter.

11.11 Pincushion Correction

General requirements

Pincushion distortion gives bowed edges to what should be a rectangular picture. It is due to the effect of scanning a flat, wide angle screen, which tends to produce 'stretching' at the corners of the raster. There is a tendency for the general expansion of the raster in any case, as described in the section on S-correction. Despite S-correction, there is additional stretching due to the combined effect of the line and the field deflection which reaches its maximum towards the four corners of the screen. This is represented diagrammatically, if somewhat exaggerated, in figure 11.10; outline (a) illustrates the natural tendency towards pincushion distortion. To obtain a rectangular raster (b) the corners would have to be 'pulled-in' and the top and bottom centres slightly expanded. This is achieved by superimposing on the normal deflecting field a correction field which, by itself, would produce a barrel shaped raster (c). The combined effect of (a) and (c) should then be close to the ideal (b).

As already mentioned, in monochrome receivers the correction field comes from small magnets mounted on flexible brackets near the tube flare. These are carefully adjusted to produce straight outlines to the picture.

'Free' magnets are not permissible near colour tubes, because of the danger of magnetising the mask, or any other steel parts associated with the screen, leading to a complete loss of colour purity. Therefore pincushion correction is obtained electronically by modifying the scan waveforms, which in turn alters the deflecting fields. Correction is needed along the vertical axis of the screen, North-South, to be

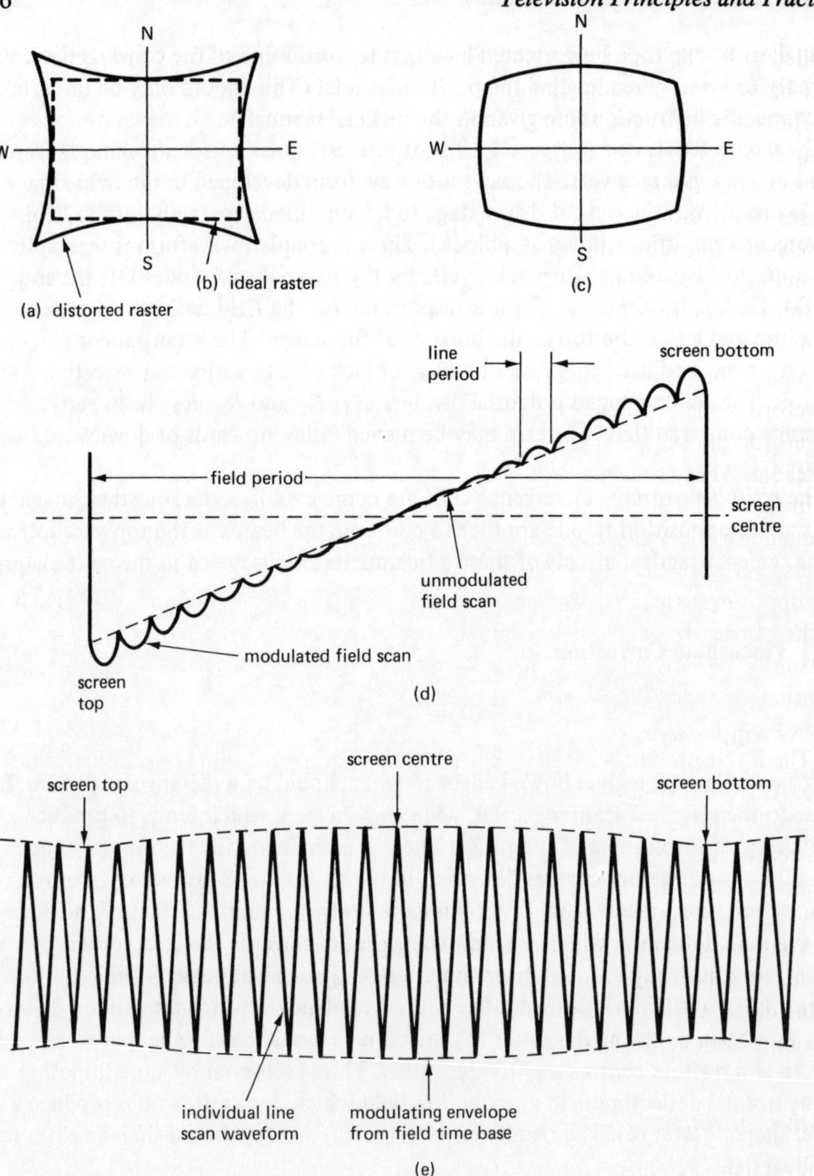

Figure 11.10 Pincushion distortion and correction waveforms: (a) distortion shows up as stretching of the raster towards screen corners and is accompanied by inward bowing of the verticals and the horizontals; (b) ideal, rectangular raster; (c) correction applied to both time bases aims to produce magnetic field which, if acting alone, would set up outward bowing raster; (d) field sawtooth waveform modulated by line parabolic waveform (N–S correction); (e) line sawtooth waveform modulated by field parabola (E–W correction)

applied to the field time base, and along the horizontal, East-West, fed to the line time base.

The outlines (a) and (c) in figure 11.10 indicate that the amount of correction along N-S and E-W depends upon the position of the beam at any particular time. For example, the *field* time base must increase its deflection as the *line* scan progresses towards the centre of the screen, and must reduce it near the edges. Conversely, the *line* deflection is at its maximum as the *field* scan nears the centre, but it is reduced in the top and bottom parts of the picture. It is, therefore, necessary to 'inform' one time base of the action of the other. This information is exchanged by modulating the field deflection current with a parabolic waveform at line frequency (*N-S correction*), and, conversely, by modulating the line scan current waveform with a parabola at field frequency (*E-W correction*). Such modulated waveforms are shown in figure 11.10(d) (for clarity only a few lines per field, instead of $312\frac{1}{2}$, are indicated).

In figure 11.10(d) the basic field sawtooth (broken line) is corrected at line frequency by a small parabolic waveform. The direction of the line parabola is reversed after the centre of the screen, because the bowing in the picture occurs in the opposite sense in the two halves of the scan. The amount of correction is nil at the centre of the screen, and it increases progressively towards the top and bottom. The individual parabolas show that maximum correction takes place at the centre of each picture line in order to produce the barrel shaped deflection from figure 11.10(c).

The E-W correction modulates the basic line scan sawtooth waveform at field rate. As a result of this, the amplitude of the current in the line coils follows a parabolic outline which is indicated by the broken line in figure 11.10(e). At the centre of the screen, along the E-W axis, the line deflection is a maximum, progressively reaching its minimum in the top and bottom parts of the picture, as suggested in figure 11.10(c).

There are two basic methods of pincushion correction used in colour TV—by means of a special device known as a *transductor*, or by employing separate modulating circuits using diodes and transistor amplifiers. Some manufacturers may even combine both methods in one receiver.

Passive correction circuits

The principle of construction of a pincushion correction transductor is shown in figure 11.11(a). At first glance it looks like a *transformer*, but the method of its operation also implies a variable *inductor*, thus giving rise to the name *transductor*. Its purpose is to modulate the two scan waveforms in the manner described above. The winding on the centre limb is known as the *control winding* and it is connected usually *in series* with the *field* scan coils. The two outermost windings are called the *load windings* and are then *in parallel* with the line scan coils. These connections apply when the transductor is used as the only method of correction, which is normally found in 90° tube circuits. As the amplitude of the *field* scan current

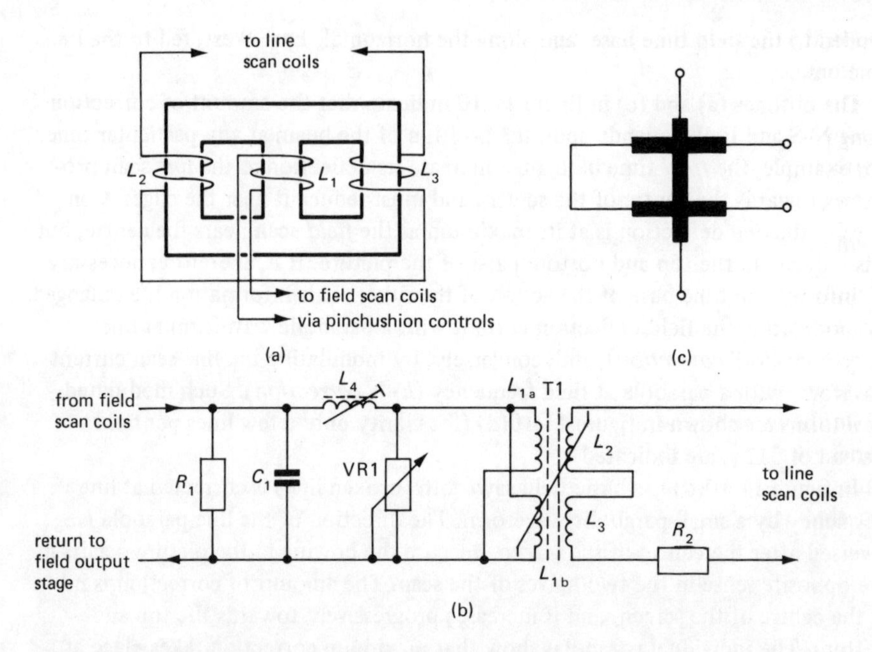

Figure 11.11 Pincushion correction by means of a transductor: (a) principle of construction of a transductor; (b) typical circuit diagram; (c) alternative circuit symbol for a transductor

through L_1 increases, the magnetic flux in the core rises towards saturation. Consequently, the inductance of L_2 is reduced during one half of the field scan and that of L_3 during the second half of the scan. (If the core is already magnetically saturated, it cannot accept any further increase in its flux and the effective inductance of a coil wound on this core is reduced. This principle was also exploited in the line linearity coil described in section 10.6.) Since the two load windings are connected across the line scan coils, they divert some of the line deflecting current away from the coils, which, in turn, reduces the line scan itself. Maximum current is diverted when the inductance of either L_2 or L_3 is a minimum–corresponding to the extremities of the field scan. This process generates the waveform shown in figure 11.10(e), which satisfies the requirements of E–W correction.

The N-S correction takes place as a result of the field scan current in L_1 being modulated by the line waveform from either L_2 or L_3. The two load windings are wound on their respective sections of the core in such a way that when saturation occurs in one coil, the other one functions normally. Modulation is achieved by transformer action between the non-saturated winding and L_1. In effect, this induces e.m.f.s of a suitable polarity and waveform which modify the field scan current as shown in figure 11.10(e).

The diagram in figure 11.11(b) shows how such a transductor can be incorporated in the correction circuit. The control winding, L_1, consists of two parallel sections, L_{1a} and L_{1b}. The available pincushion correction adjustments are: *phase* control (L_4), which governs the position along the raster where control takes place (along the N-S axis); and *amplitude* control (VR1), which governs the amount of correction actually applied.

Active correction circuits

Colour receivers which feature 110° tubes use active pincushion correction circuits, sometimes including a transductor as well. Owing to the relative complexity of the whole arrangement, partial block diagrams are used to illustrate the general principles; these are shown in figures 11.12 and 11.13.

In figure 11.12 the N-S correction is by means of the transductor connected in series with the field scan coils in the manner described above. The modulating line pulses are fed in from the line scan coils, which, again, are in parallel with the transductor windings.

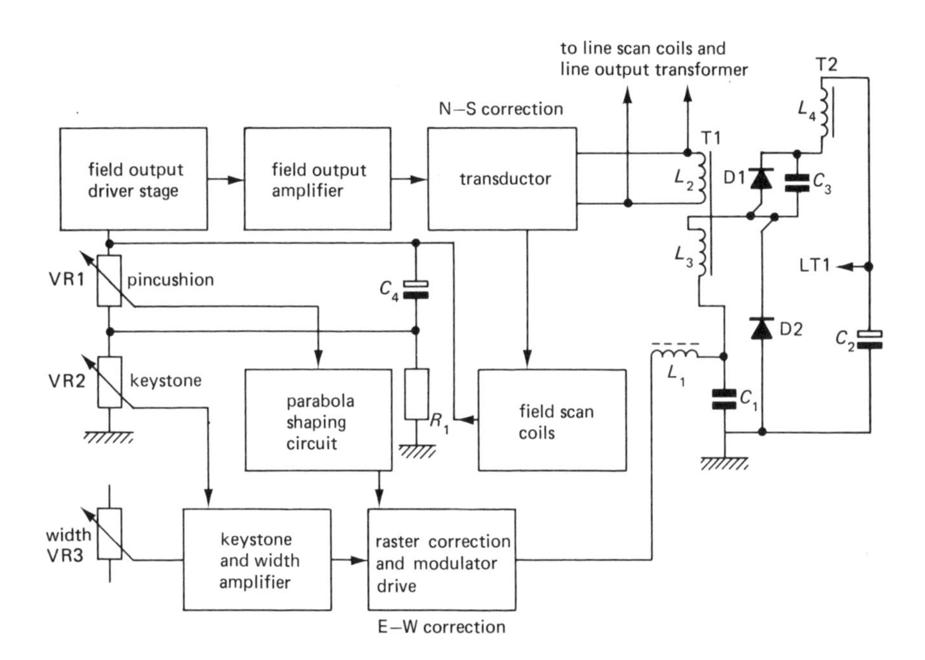

Figure 11.12 Active pincushion correction arrangement–circuit A. This type of circuit is used in conjunction with a 110° tube where large errors have to be corrected. N-S correction is by means of a transductor, but E-W correction uses a diode modulator (D1, D2). A number of additional features are provided by the arrangement (see text)

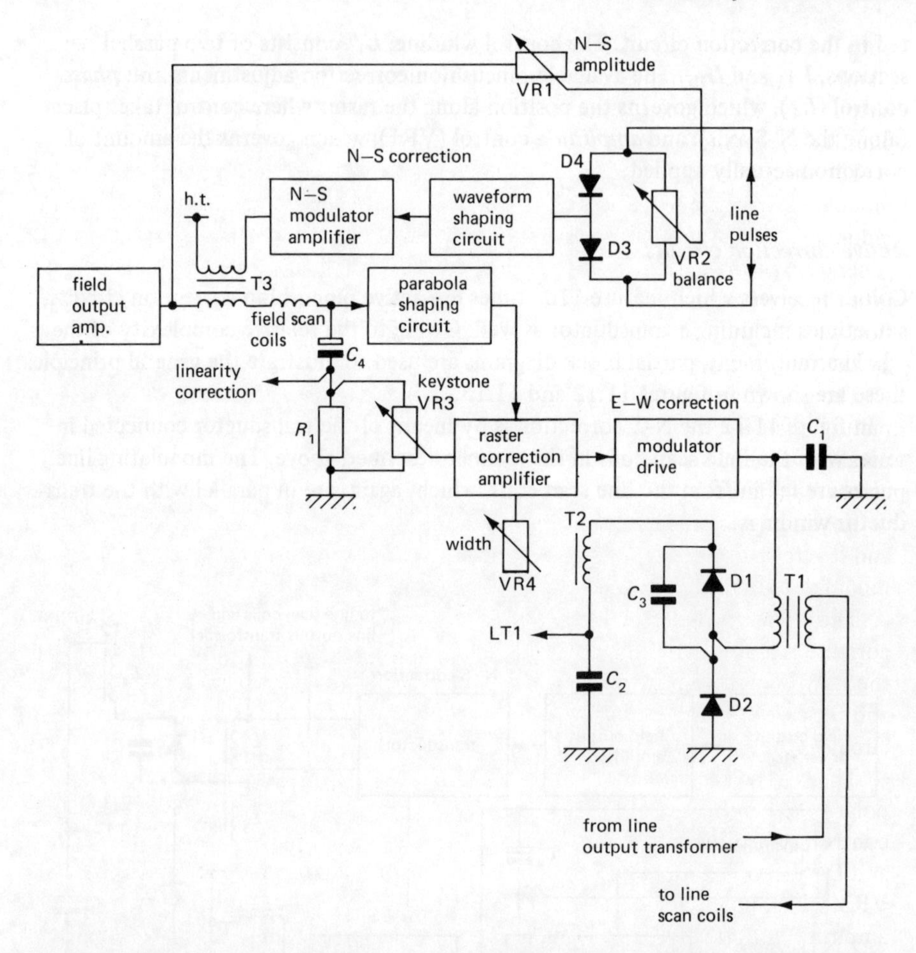

Figure 11.13 Active pincushion correction arrangement–circuit B. E-W correction circuit is similar to that in figure 11.12. N-S correction no longer uses a transductor; instead a separate modulator amplifier provides the modulation of field scan waveform by line pulses

In the E-W correction circuit the modulating field waveform is obtained from the voltages developed by the return field scan current across C_4 and R_1. The waveform across C_4 is a parabola (the capacitor integrates!), while a sawtooth is produced across R_1. The parabola is fed to a shaping circuit and then amplified before being used to modulate the line scan. This, in turn, takes place via the modulating transformer T1: the correction current is driven into winding L_3 and, because of transformer action, it modifies the line deflection current that flows in L_2. The path of the driving current from L_3 is completed via D2 to chassis. This diode, together with D1, forms the *diode modulator*, which is connected to winding L_4 on the line output transformer (T2). Both diodes conduct during the line scan period, thus clamping the top terminal of L_3 to earth. At the same time, D1 and D2 rectify the output from L_4, which is smoothed by C_2 to generate one of the receiver l.t. supplies (l.t.1). In some designs one of the diodes operates as the efficiency diode.

The variations in the current induced in L_3 (and L_1, for that matter) by the modulator drive are seen by the entire line output stage as changes in the effective inductance of the scan circuit. As this could interfere with flyback tuning and the generation of the e.h.t. (see chapter 10), an additional capacitor—C_3—is brought into the circuit during line flyback. Its value is carefully chosen so that in conjunction with L_1 the flyback tuning remains unchanged. (D1 and D2 are reverse biased by the retrace pulse, leaving only C_3 in circuit.)

The circuit also includes *keystone* distortion correction. If uncorrected, the picture could be somewhat trapezium shaped, as the sides of the raster are not parallel. The improvement is achieved by the addition of variable sawtooth from across R_1 (via VR2) to the field parabola.

Keystone and pincushion correction are forms of dynamic width adjustment, and it is logical to introduce overall width control to the picture through the modulator circuit (control VR3).

The circuit in figure 11.13 presents an alternative arrangement. E–W pincushion correction is almost identical in principle with that in the previous example. T1 is the E–W modulating transformer, T2 represents part of the line output transformer, and for ease of comparison the component numbering in the diode modulator circuit is the same as the equivalent in figure 11.12.

The N–S correction circuit in figure 11.13 does not use a transductor. Opposite polarity line pulses are fed to a bridge circuit given by D3, D4 and VR2 (balance control). When the bridge is balanced, there is no output from the junction of D3 and D4. However, a variable amount of field sawtooth is introduced to the slider of VR2, which upsets the symmetry of the bridge. The resultant output is then shaped, amplified and, finally, fed to the N–S modulating transformer, T3. Since the secondary winding of T3 is in series with the field scan coils, the line frequency correction waveform is impressed upon the field deflecting current.

In some receivers the N–S modulation waveform is fed to one of the field time base amplifiers as though it were part of linearity control arrangement.

11.12 Grey Scale Tracking

Grey scale tracking applies to the correct adjustment of colour tube voltages. These settings ensure that the operating conditions of the three guns result in a black and white (plus grey!) display when no colour signal is applied. The individual guns, although housed in the same glass envelope, are likely to have different characteristics—namely equal beam currents could be produced only by unequal voltages on the electrodes of each gun. At the same time, the sensitivity of phosphors to the beam current depends on the chemical composition of the material.

For practical reasons some of the voltages are common to all three beams, so that any balancing between the guns is achieved by trimming the potentials at the three cathodes and at the first anodes (A1). If the voltage levels are wrong, then the guns are not driven together into *cut-off*, giving some colour tinting in what should be a black, or dark grey, area of the picture. Similarly, if the *peak beam currents* are incorrectly balanced, a coloured display will appear instead of white. Correctly adjusted guns will track over the entire range of picture brightness from white to black. Methods of adjustment are described in the next chapter, since they are part of colour receiver setting-up procedure.

The circuit diagrams of video output stages given in chapter 6 showed the necessary provisions for varying the drives to the tube cathodes. The supply to the A1 electrodes can be derived from the boost voltage in a valve line output stage, or from a suitable winding on the line output transformer. The supply must be rectified, smoothed and adequately decoupled to ensure freedom from often unexplained colour tinting. The adjustments are based on preset potentiometers in the A1 supply lines; in a few receivers there is also a *tint* control (in addition to *saturation* control), which can bias the beams towards a particular colour to suit the viewers' taste.

11.13 Automatic Degaussing

One of the disadvantages of the steel shadowmask is that it can become accidentally magnetised, and this would cause serious purity errors. The irregular patches of unwanted colour which would then appear on the screen cannot be corrected by ordinary purity adjustments. In severe cases the entire picture is covered with unrelated colour patterns. Magnetisation can affect not only the mask, but also other steel parts associated with the tube—for example, the metal shield surrounding the cone of the tube, the rimband which reinforces the junction of the screen with the flared portion, etc. This effect can be produced by some domestic electrical appliances brought too close to the receiver, magnetised tools, electric toys or current carrying cables, and even by the changes in relation to the earth's magnetic field. The latter can occur if the receiver is moved from its original position in the room.

Severe cases require manual demagnetisation, or degaussing by the service engineer—a method will be described in the next chapter. In most instances the amount of accidental magnetisation is slight, and each receiver has a circuit which automatically applies a degaussing field around the tube flare and the sides whenever the set is switched on from cold.

Two circuit diagrams are shown in figure 11.14. The degaussing coils—L_1 in circuit (a) or L_2 in circuit (b)—are placed around the flared portion of the tube to include the front edge of the screen and the magnetic shield which surrounds the cone of the tube. In circuit (a), when the receiver is first switched on, thermistor R_1 is cold and has a low resistance. The voltage across the VDR is then high and its resistance is also low. A large a.c. current (about 1.5 A) flows through the coil; since this current raises both the temperature and the resistance of the thermistor, the voltage applied to the VDR reduces. The combined effect of the increasing

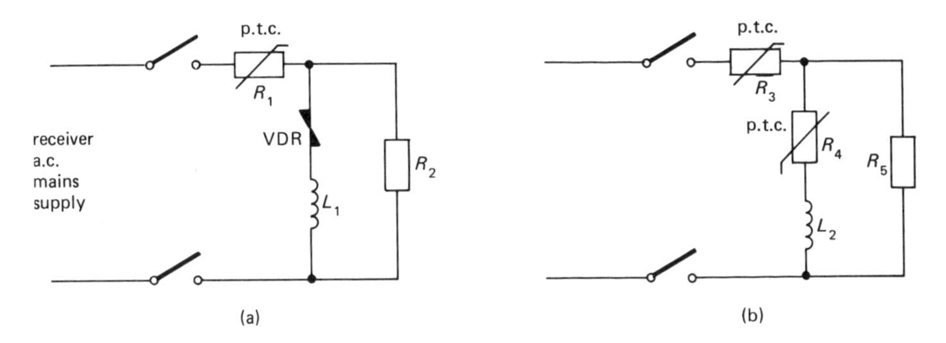

Figure 11.14 Automatic degaussing circuits

resistance of R_1 and that of the VDR reduces the amplitude of the a.c. current; the alternating magnetic field set up by the coils gradually decays to demagnetise the tube. The resistance of the VDR is finally very high, and any remaining current through R_1 is diverted into R_2. In this circuit R_2 is necessary to maintain the temperature of the thermistor and its high resistance.

In circuit 11.14(b) the VDR has been replaced by another thermistor, R_4; actually R_3 and R_4 form a single unit, as one resistor maintains the temperature of the other. The principle of operation is similar to that of the previous arrangement. Initially, there is a high demagnetising current due to a low resistance of both R_3 and R_4; as they warm up, however, the current reduces until it is finally diverted into R_5 to maintain the temperature of R_3. Since R_4 is physically close to R_3, its temperature and resistance also remain high. The entire process of automatic degaussing is like the erasing of magnetic tapes, which are similarly subjected to decaying magnetic fields.

The disadvantage of the circuits described is that if the coils were open-circuited, a large current could flow through the shunt resistors (R_2 or R_5), creating a risk of overheating. Therefore a dual thermistor can be used to dispense with the parallel resistor—an open circuit in the coils would simply result in no current!

11.14 Tube Safety Features; Precautions when Handling C.R.T.

The tube operates under vacuum, and because of its large surface area it is subjected to a large compression force. The glass must be strong enough to withstand this in normal circumstances. Many tubes have a steel band which reinforces the edge of the screen and the cone. This is to ensure that in the event of a fracture the resultant implosion will be contained, to prevent glass being scattered in all directions.

Very high voltages are applied to the tube electrodes—the focus and the final anode, in particular—and they can lead to internal flashovers between the various electrodes. Such a discharge, although momentary, is accompanied by a flow of peak current of several hundred amperes. The flashover potentials could be fed to the amplifiers supplying the particular electrodes and destroy some of the components. To prevent this happening, the electrodes are connected to *spark gaps* which usually form part of the tube base panel. The width of the gap is critical and its condition is important for correct operation and safety. The gaps are returned to the main earth of the receiver via a separate connection to prevent the discharge current flowing through any other section of the chassis and inducing destructive overvoltages.

The danger of flashovers is reduced if the e.h.t. is maintained at its correct level, which often means that the h.t. supply to the line output stage is as recommended by the makers.

The level of e.h.t. applied to the tube is such that it can be responsible for the production of X-rays. The latest colour tubes are said to be free from dangerous radiation for up to 30 kV. It would be inadvisable, however, to operate a receiver while working close to the tube for unduly long periods of time. The effect of X-rays is cumulative and is not immediately obvious!

The tube Aquadag coatings form a very efficient capacitor, which in dry room conditions can store its high voltage charge for a considerable length of time. This is not sufficient to cause any direct danger, but it might give an electrical shock, leading to an involuntary reflex which could cause an accident. For that reason the e.h.t. must be discharged from the tube connector to chassis before one works on the tube itself.

The tube rimband might also acquire a high electrostatic charge, which could interfere with the main electron beam or give shocks if the edge of the screen was touched. It is therefore connected to chassis via a parallel combination of a high value resistor and a capacitor. A direct connection to the chassis in mains-operated receivers is unacceptable, since it could bring a dangerous mains voltage to the screen. The components must be suitably rated and considered to be a safety feature replaceable only with the makers' recommended type.

The magnetic shield around the tube, where fitted, and the external Aquadag coating must be connected to the receiver chassis. A spring-loaded contact is used to afford the necessary connection–its condition must be good, to ensure freedom from interference or even flashovers.

When handling the tube, great care must be taken to prevent damage. No undue strain on the glass is allowed when either removing or refitting it in the receiver. The tube must not be carried by its neck alone, which could fracture under its own weight. The face of the tube could be easily scratched–if it is necessary to rest the screen on any surface, it must be placed on a suitable protective mat.

12 *Receiver Setting-up Procedure*

12.1 Fundamentals of Receiver Setting-up Procedure

The purpose of setting up a receiver is to ensure that it is capable of displaying a picture according to its design specifications and to the high standards of broadcasting. The TV receiver must also operate within the margins of safety prescribed by the manufacturers. The need for setting-up arises either as part of the installation checks on a new receiver or as the adjustments after certain repairs or as a result of receiver ageing. The procedure is more involved—and more critical—in colour receivers than in transistorised monochrome sets. In turn, those receivers which use a delta gun tube require a greater amount of effort to produce an acceptable picture, since their convergence is more complex than that of the in-line, or precision-in-line, tube.

One can consider both colour and monochrome receivers together, since it is initially essential to establish a good monochrome picture in either case.

Because of the much higher voltage and power levels found in colour receivers, the presets associated with power supplies (for example, h.t. and/or e.h.t., protection features, etc.) must be adjusted strictly in accordance with the makers' instructions. The colour tube itself presents different requirements from its monochrome counterpart in order to display a good black and white picture.

The list of checks in section 12.2 summarises typical setting-up procedure for a colour receiver, but this can be easily adapted to suit the requirements of a monochrome receiver. The actual order of operations must be confirmed with the service manual, since their sequence is influenced by any interaction between the adjustments. In particular, any presets which affect the operation of the time bases (linearity, amplitude, e.h.t., etc.) must be checked *before* the purity and convergence adjustments are carried out because the correction waveforms are derived from the main scan waveforms.

12.2 Installation Checks on a Colour Receiver

Some of the following checks are best carried out in the workshop, while others have to be done, or repeated, in the viewer's home.

(1) Check for best *positioning* of the set in the room. This is possible with tactful consultation with the customer, ideally no strong direct light (natural or artificial) should fall on the screen; the receiver should be away from large ferrous objects or electrical equipment, as they might introduce magnetic fields which affect the shadowmask. The position must be reasonably permanent, since moving the receiver about can cause a change in the effect of the earth magnetic field. At the same time, the set must not be close to a source of heat or dampness.
(2) Check *mains connection*–ensure correct mains voltage tapping and the polarity of the live/neutral leads where applicable.
(3) Switch on the receiver and assess the *overall picture quality*–checking for the freedom from noise, ghosting, etc., which could be caused by an inadequate aerial installation, incorrect tuning, excessive or insufficient signal strength; ensure precise line and field synchronisation, etc. Allow at least 10 min for the receiver to warm up before any circuit adjustments are to be carried out. A static display, such as a test card, is best suited for many checks.
(4) Check the receiver *h.t. level* and adjust if a suitable preset is provided in the power supply regulator; in some receivers it is the e.h.t. which must be measured.
(5) Check the *brightness* and/or *contrast* control–ideally with the aid of a test card; ensure that peak white and all shades of grey are clearly reproduced (ignoring at this point any colour casts). Often the two controls have associated presets in the beam limiter, black level clamp, video amplifier drive, a.g.c. and others which need careful setting strictly to the makers' instructions.
(6) Check the *height, width, linearity* and *centering* of the picture; the correct dimensions of the required display are given by the test card (see section 12.9). In some receivers no separate width adjustment is available; in these cases the correct e.h.t. level should automatically give an adequate line scan amplitude. On the other hand, if the width and the horizontal linearity had to be altered by their respective controls, it might be necessary to *recheck the e.h.t.*
(7) Check picture *focus* on bright details near the central part of the screen. Exact adjustment may have to be performed at the end of the entire setting-up procedure in order to arrive at an optimum which maintains good focus over the full range of brightness control and gives minimum *moiré effect*. The latter is in the form of irregular dark line pattern (somewhat reminiscent of a 'fingerprint' pattern) particularly noticeable on a blank raster.
(8) Check picture *purity*–if necessary, degauss the receiver manually and then adjust the purity; the more typical procedures will be described separately in section 12.3 and 12.4.
(9) Check *static convergence* and readjust the purity if the convergence had to be altered–see section 12.5.
(10) Check *dynamic convergence*–see section 12.6.
(11) Check for *pincushion distortion*–see section 12.7.
(12) *Recheck* picture *height* and *width.* The convergence adjustments tend to alter the time base loading and affect the scan waveforms; if large convergence errors have to be corrected, it may be necessary to repeat the sequence from that point. In a

really badly misadjusted receiver the procedure might be repeated three times; on each occasion the errors are reduced and finally optimum settings are obtained.
(13) Check *grey scale tracking*—see section 12.8.
(14) Check *saturation* (*colour*) control for correct flesh tones, and trim the brightness and contrast controls if required.

12.3 Manual Degaussing

The auto-degaussing circuit of a colour receiver (section 11.13) can remove all but the most severe magnetisation of the shadowmask and the steel parts associated with the tube. If random purity errors persist on the screen, it may be necessary to demagnetise the tube using a portable coil of a suitable design.

The coil is connected to the a.c. mains and, with its push button ON-OFF switch depressed, it is moved very slowly over the top, bottom and sides of the receiver at the front of the cabinet. Next, in a circular motion, the coil is moved over the face of the tube and then slowly brought away from the screen as if following a spiral path until the coil is about 2.5 m (8 ft) from the receiver. The whole procedure should take approximately one minute; at the end, turn the degaussing coil face downwards and release the ON-OFF switch.

The following precautions must be observed to ensure complete and safe demagnetisation:

(1) The coil must not be brought close to the back of the cabinet, as it would demagnetise the various permanent magnets mounted on the neck of the tube.
(2) The coil must be moved away from the receiver to simulate a decaying magnetic field; the specified distance ensures that the field reaching the receiver at the moment of switch-off is practically nil; as an added precaution the coil is first turned face downwards. In a workshop it is essential to check that the coil is at least 2.5 m from *any* colour receiver prior to switching off; otherwise, the field from the coil could induce magnetisation.
(3) The coil has to pass a relatively heavy current; therefore it must not be switched on for longer than about two minutes at a time, to prevent it overheating.
NOTE: The receiver may be left switched either on or off during the operation.

12.4 Purity Adjustments

Bad colour purity produces irregular shaped random patches of colour which are most noticeable on a blank, white raster and they persist despite prior degaussing.

Purity adjustments ensure that the given colour beam can only strike its own phosphor dots, or stripes, over the entire screen area. This is achieved by the shadowmask, provided that the angle of approach between each beam and the mask is correct. The approach angle depends upon the position of the scan coils along the neck of the tube and on the paths of the beams in relation to the tube axis before they even reach the deflecting coils. The scan coils of all colour tubes except the precision-in-line type (*PIL*) can be moved along the neck axis. The paths of the electron beams can be modified by means of ring magnets placed on the neck of the tube.

The more typical procedure is now summarised:

(1) The receiver displays an unmodulated (no picture) raster.
(2) Switch off the blue and the green guns of the tube, leaving only a plain red raster—a single colour display helps to observe even slight purity errors. (Most receivers have suitable gun switches which usually disconnect the supplies to the first anodes; there are also signal generators whose outputs include plain, single colour rasters.)
(3) Slacken the wing nuts which secure the scan coils to their housing and slide the assembly as far back as possible—initially, this will result in a very impure raster, especially around the edges, and possibly with 'shaded' corners.
(4) Adjust the purity magnets until a pure red area is displayed approximately in the centre of the screen only, ignoring the raster impurity elsewhere. The ring magnets have tabs which, apart from facilitating the adjustments, act as position indicators. It may be necessary to rotate the rings to bring the tabs closer together or spread them further apart; alternatively the magnets are turned as a pair in either direction on the tube neck.
(5) Slide the scan coils forward, noting that the centre red area will now expand; the correct position of the deflecting coils is reached when a pure red raster completely fills the screen.
(6) Switch off the red gun and switch on the green gun, and check the purity of the raster; repeat for the blue raster. If the purity of either of these two is not acceptable, a slight readjustment of the scan coils should be sufficient. Failing that, slide the scan coils back again and readjust the ring magnets. In some rare cases a compromise setting will have to be adopted for the best overall purity of all three rasters.
(7) Lock the scan coil securing nuts.

NOTES: The action of the purity magnets shifts the three beams from their original positions; hence, the picture centering controls may have to be readjusted. The purity magnets also affect static convergence (see section 12.5); conversely, the static convergence adjustments may alter the colour purity, because of somewhat similar effects on the beams in each case. Consequently, it is often recommended to check and adjust the static convergence before and after 'purifying' the display. It has been reported in a few rare instances that damage to the tube resulted if the scan coils were left for too long in their rearmost position.

PIL tubes have fixed deflecting coils, and overall purity can be obtained by adjusting the ring magnets. In this type of tube the purity and the convergence magnets form a group of rings on the neck which are clamped in position by a locking ring in front of the assembly. Often there is an identification line marked on the magnets—when all of them are in their optimum positions, the individual markers should form a continuous line across the assembly. Normally, the purity magnets occupy the rearmost position in the arrangement.

12.5 Static Convergence Adjustments

Errors in static convergence give coloured outline ('fringing') to picture detail in the *central area* of the screen. This is especially noticeable on a monochrome display. The term 'static convergence' suggests that the beams converge at the centre of the screen—the landing area if the beams remained *static* or undeflected.

The ideal display for this adjustment is either a crosshatch or a dot pattern; alternatively, test cards can be used. Converged crosshatch consists of intersecting *white* vertical and horizontal lines, and is produced by a specially designed signal generator. Such an instrument can also have a dot pattern facility which is often recommended as being more suitable for static convergence adjustments; a converged display consists, then, of rows of *white* dots across the screen. The impression of white in either case can only be obtained if the beams land on their respective phosphors within one triad (or group of stripes). Misconverged beams hit the correct colour phosphors, but they are too far apart to create satisfactory colour mixing from a normal viewing distance.

In most receivers static adjustments are carried out by means of permanent magnets which are associated with the convergence assembly mounted on the neck of the tube. In a few receivers a suitable d.c. supply is used for this purpose.

Delta gun tubes use a *radial* convergence arrangement for each beam, together with a *lateral* adjustment for the *blue.* It will be recalled from chapter 11 that in the delta gun tube the *radial convergence* magnets move the red and the green beams *diagonally* across the screen, but the blue beam is shifted *vertically* up or down; the *lateral* magnet moves the blue beam 'sideways'. The procedure for static convergence adjustments is as follows:

(1) Apply crosshatch or another suitable signal to the receiver.
(2) Switch off the *blue* gun.
(3) Converge the *red* and *green*, horizontal and vertical lines in the centre of the screen, which should give *yellow* crosshatch; ignore any misconvergence away from the middle.
(4) Switch the blue gun on and converge the *blue* lines with the *yellow* to obtain *white* at the centre; both the radial and lateral controls are used as required.

Ordinary **in-line** tubes often have the static convergence magnets so arranged that the beams are moved diagonally across the screen in a manner similar to that with the delta gun system. The reason for this is illustrated in plate 1 (opposite p. 292), where part of a crosshatch display is shown. The solid lines indicate the beam positions; the green and the blue are close together, and the red is not converged. To converge the red lines with the other colours, the horizontals would have to be shifted upwards and the verticals moved to the left until their correct locations, as indicated by the broken lines, were reached. Only one diagonal movement of the beam is now needed to achieve this. The same comment applies to the blue beam should any correction be required. The green gun usually occupies the central position, and it is not affected by the adjustments. Some tubes have the red gun in the middle, so that the other colours are converged with respect to it.

Precision-in-line tubes have four ring magnets to adjust static convergence. The rings form part of the assembly with the purity magnets already described. The front pair, next to the locking ring, is adjusted first, followed by the second pair. Whichever beam is in the centre, it will not be affected by the magnets. The following instructions are given for a tube with the red gun in the middle.

(1) Apply a crosshatch signal to the receiver.
(2) Switch off the red gun.
(3) Slacken the magnet locking ring.
(4) Observing the centre of the display, adjust the front pair of magnets: separate the identifying tabs slightly by rotating the magnets in opposite directions, and then rotate them *together*, to reduce the amount of separation between the blue and the green lines of the display.
(5) Adjust one magnet with respect to the other, to converge the blue and green into cyan. If necessary, repeat step (4) for the best results.
(6) Switch the red gun on and, using the method outlined in steps (4) and (5), adjust the second pair of ring magnets to converge the cyan with the red, to obtain white.
(7) Recheck the blue and green convergence and adjust the front pair of rings again, because of possible interaction between the magnets.
(8) Carefully tighten the locking ring so as not to disturb the adjustments.

The somewhat different method of convergence adjustment of the PIL tube is due to the fact that four-pole and six-pole ring magnets are used. The arrangement gives a greater freedom of movement to the beams (and the magnets!) than is provided by the two-pole magnets found elsewhere. Identifying markers can be provided on the rings to denote their correct position—see section 12.4.

12.6 Dynamic Convergence Adjustments

Errors in dynamic convergence cause colour fringing away from the centre of the screen which usually tends to get worse towards the edges of the screen. The diagram in figure 12.1 indicates the areas on the screen where maximum misconvergence may occur. The centre is affected by static convergence as already explained. The top and bottom of the display are influenced by the *field dynamic convergence* since it corresponds to maximum field scan. Similarly, the left- and right-hand sides are governed by the *line dynamic convergence* as the errors coincide with the extremities of the line scan.

To correct dynamic convergence errors, a cross-hatch display is again used, although test cards may be acceptable. The receiver controls are labelled according to the effect they have on the lines of the crosshatch pattern. Some manufacturers provide 'flow-charts' to show the sequence of adjustments together with colour diagrams which indicate errors to be corrected.

A delta gun tube needs a large number of adjustments; the various combinations are presented diagrammatically in figure 12.2 and their effects on a crosshatch are

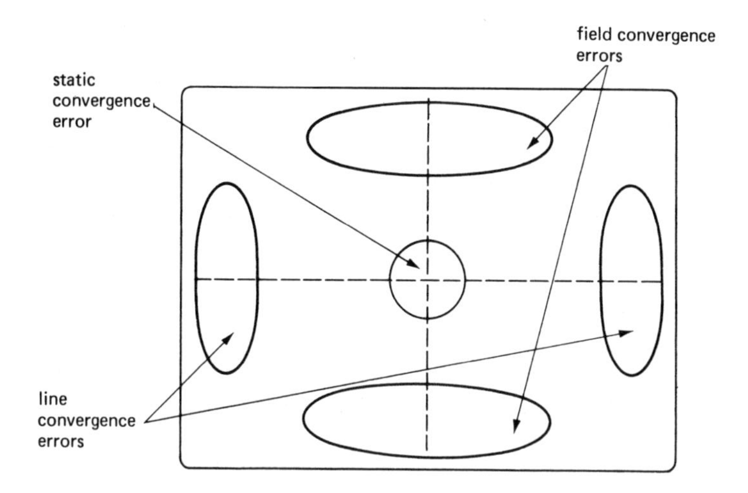

Figure 12.1 The areas on a TV screen identify source of convergence errors. Correction is applied while concentrating on the errors along the vertical (field) and the horizontal (line) axes of the picture. Details of adjustments are given in plates 2 and 4

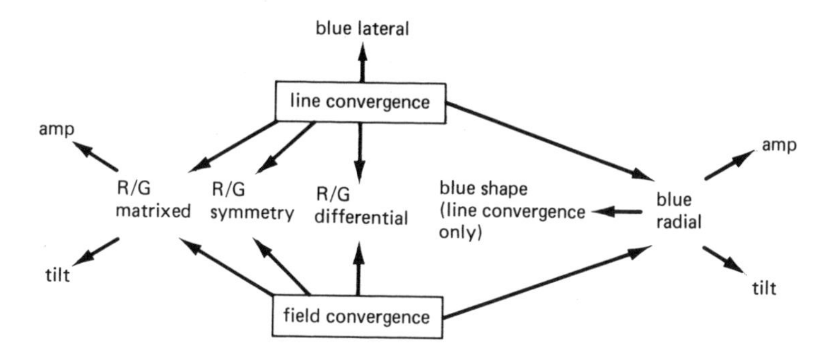

Figure 12.2 Typical dynamic convergence arrangements for a delta gun tube

listed in plate 2. They are not difficult to memorise if one studies this list in conjunction with figure 12.1. *Differential controls* are used to converge the red and green *horizontal* lines, while the *matrixed* adjustments bring together the corresponding *vertical* lines of a crosshatch. The diagram in plate 3 illustrates the need for the two types of controls. Part of a misconverged crosshatch display is represented here by the lines of individual phosphor dots. The red/green *matrixed* controls move the two beams *diagonally* in the *same direction*, and the *differential* adjustments cause a shift in *opposite directions.* For example, if the matrixed controls were to be adjusted to move the beams upwards, the red and green vertical lines of the crosshatch would be brought together, while the horizontal lines would also move upwards, but their mutual relationship would not be affected. Conversely, a differential adjustment which takes the red horizontal line downwards shifts the green line upwards and they converge; at the same time, the verticals will effectively move to the left, with their relative position otherwise unchanged.

If the complete sequence of dynamic convergence adjustments is to be carried out, then:

(1) Apply crosshatch signal to the receiver.
(2) Switch off the blue gun.
(3) Carry out the *red–green* adjustments in the sequence indicated by the numbers in the table in plate 2–field convergence first, followed by the line controls.
(4) Switch the blue gun on and converge with the yellow to obtain white.
NOTES: (a) There may be some loss of static convergence during the above procedure; it is therefore recommended to check and adjust this as necessary between the dynamic adjustments.
(b) The respective symmetry and differential controls should be adjusted in conjunction with each other.
(c) Large initial errors can only be corrected by repeating the entire sequence.
(d) The '*tilt*' and '*amp*' adjustments tend to affect each other; the degree of interaction can be reduced by operating both controls together.
(e) In some cases it is very difficult to obtain perfect convergence over the entire screen area; therefore it may be necessary to settle for compromise settings. As much of the central area of the display as possible should then be converged satisfactorily, since the eye tends to concentrate on this part of the picture.

In-line tubes require very little dynamic convergence; the adjustments are usually straightforward, as listed in the table in plate 4. In fact, this list is comparatively long, since many makers may omit certain controls altogether. The gun which occupies the centre position will not be affected by the adjustments (usually the green, sometimes the red). The following sequence is recommended by one manufacturer (Rank):

(1) Apply crosshatch signal to the receiver.
(2) Adjust the line symmetry (balance) control to correct red/blue crossover of central horizontal lines; if difficulties are experienced, the convergence coils may be rotated slightly or the individual coil assemblies can be tilted.
(3) Carry out line convergence adjustments (see plate 4).
(4) Carry out field convergence adjustments (see plate 4).

12.7 Pincushion Distortion Adjustments

Pincushion errors show up as bowed verticals and horizontals, especially near the edges of the screen. This could be particularly important in large screen receivers, or if 110° tubes are used. Because of the effects produced by these errors, they are best observed on a crosshatch display.

Where a simple transductor is used, the adjustments aim at straightening the crosshatch pattern. In a more involved arrangement the *E-W keystone* and the *E-W pincushion* controls are adjusted for straight verticals at the sides. These are followed by the *N-S pincushion amplitude* and *N-S phase*, to achieve straight horizontal lines at the top and bottom of the picture. Effects of various pincushion correction adjustments on the shape of the picture are shown in figure 12.3.

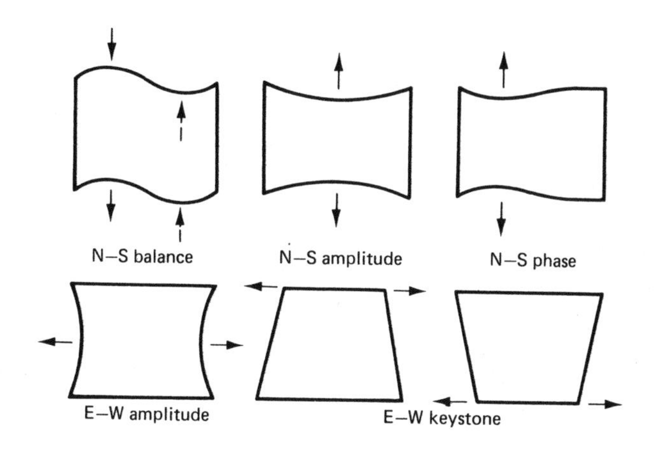

Figure 12.3 Effects of various pincushion correction adjustments on the shape of the picture. Arrows indicate the direction and the position of the applied correction

12.8 Grey Scale Adjustments

Incorrect grey scale is indicated by colour 'tint' or 'cast', especially noticeable on a monochrome display. Such unwanted colour may appear over the whole area of the screen, or it could be associated with either the brighter (*highlights*) or the darker (*lowlights*) parts of the picture. Correct grey scale is necessary, not only because of its effect on a black and white programme, but also because an impression of wrong colours could easily be produced if the balance between the three guns were upset. It has already been stressed that before investigating a 'wrong colour' type of fault, the servicing engineer must ensure correct grey scale tracking.

The adjustments are in two groups:

(1) Highlights–to control the amount of video drive (whether R, G, B or luminance) to each of the three cathodes of the tube.
(2) Lowlights–to control the beam cut-off voltage for each gun; this is governed by the first anode (A1), sometimes referred to as the screen grid (G2), potential.

The procedure is best carried out with the aid of the standard colour bar display which has the colour information completely removed; it is then known as the *grey scale pattern.* Some signal generators have the facility to produce the luminance waveform only; otherwise the colour control has to be turned down or the receiver detuned *slightly* until colour drops out. Test cards are a suitable alternative. The adjustments should be carried out in subdued lighting, as the changes in the tube output are then easily noticed.

Correctly adjusted video drives should produce white light output which corresponds to the standard known as *Illuminant D 6500.* The figure refers to the equivalent temperature to which a black object would have to be heated to reach the required 'shade' of white. Ideally, a reference source of D 6500 should be used, since the broadcasting standards are also based on it. The reference source consists of a fluorescent tube which provides the desired light output, e.g. Tropical Daylight (Atlas Lamps). Sometimes a suitable grey scale filter is fitted over the tube to give not only white, but also a few shades of grey and black. The necessary adjustments are made while holding the tube against the receiver screen.

If the grey scale is grossly out of adjustment, then it is advisable to consult makers' instructions before any setting up is carried out. In some receivers a number of conditions have to be fulfilled prior to such adjustments. For example, it may be necessary to set the grid-cathode voltages to a prescribed level, or a switch needs to be operated which turns off the field time base and applies certain fixed voltages to the tube electrodes etc.

For **delta gun** and ordinary **in-line** tubes the recommended procedure of readjusting the grey scale (thus disregarding any special conditions mentioned above) is summarised as follows:

(1) Apply a grey scale pattern to the receiver.
(2) Adjust the contrast and brightness control to ensure equal brightness change between each step of the grey scale.
(3) Observe the black and the dark grey portion of the display ('lowlights') for any sign of coloration; the appropriate A1 control(s) have to be turned down to reduce the unwanted tint.
(4) Observe the white and the light grey portion of the display ('highlights') for any sign of coloration and adjust the video drive controls accordingly. Some receivers have only two adjustments, with the cathode of the third gun (usually the red) being at a fixed potential.

Precision-in-line tubes have a common A1 potential applied to all three guns. Therefore the cut-off point and the maximum video drive are set at the video output stages for each gun in turn. The lowlights are controlled by altering the d.c. bias of the individual output transistors; the highlights are adjusted by varying the gain of the respective video output amplifiers.

12.9 Test Cards

The purpose of test cards is to provide a stationary TV display which can be used for the assessment of the receiver performance (both colour and monochrome) and for setting-up. The latter application has its limitations; more specialist display patterns may be preferred in some cases, as outlined in the preceding sections.

Test cards are designed jointly by the broadcasting companies and the manufacturing industries. The display is revised periodically to include new techniques, extend its usefulness, etc. The chronological order of the appearance of test cards is indicated by the appropriate letter. Currently in the UK two types are broadcast: Test Card F and Test Card G.

Test card F

Test card F is shown in plate 5; it is televised from a picture seen by the studio camera except for the colour bars, which come from a suitable generator. The purpose of each section of the display will now be explained.

(1) *Standard colour bars* at the top—the first 24 lines of each frame are modulated with the colour bar signal to facilitate decoder fault finding (chapter 7), both by viewing the bars on the screen and by waveform examination with the aid of an oscilloscope. In the latter case it may be necessary to trigger the oscilloscope time base with the field flyback pulse. The visibility of the colour bars depends upon the receiver height adjustment—see item (5) below.
(2) Coloured castellations at the bottom and on the left- and right-hand side of the picture, together with the colour bars at the top, indicate the performance of the 4.43 MHz *subcarrier oscillator* and the associated control circuit (a.p.c. loop). If the top bars are correct, then the decoder is not affected by the field synchronising sequence during which the colour burst is not transmitted. A temporary lack of the burst signal should not upset the a.c.c. (variations in saturation) or the oscillator (variations in saturation or 'unlocked' colour). The castellations at the bottom show whether the oscillator regained full synchronisation at the end of the picture. The castellations on the left (red in top half, blue at the bottom) introduce colour at the beginning of the picture line, which could upset the oscillator if the burst gating was incorrect (allowing picture content through the gate). Such a fault would affect the part of the centre colour picture level with the coloured castellations. By comparison, the alternate black and white pattern has no chrominance content and the oscillator should remain unaffected. The right-hand side yellow castellations indicate the oscillator stability at the end of the picture line.

The end-of-line castellations—both the black and white and the colour—test the performance of the *sync. separator*, which must not allow the video (and the chroma) information to affect its action. A fault in this respect would show as 'wavy' verticals level with the bright sections of the pattern.
(3) The background grid pattern—this forms a crosshatch display; some of the white lines have a black outline, to make the assessment of receiver *dynamic convergence* easier. *Static convergence* is indicated by the white cross in the centre

('noughts and crosses' on the blackboard). Adjustments, however, ought to be carried out with the aid of a proper crosshatch generator.
(4) The colour picture in the centre–includes areas of flesh tones and bright colours to assess the overall quality of colour reproduction. Any defects in the *decoder* will clearly reappear in the centre picture. Sound/chroma beat pattern, Hanover blinds, subcarrier dot interference, ghosting and other effects can be easily observed. The surrounding circle, together with the background crosshatch can be used to check picture *linearity*.
(5) White 'arrowheads' in the centre of the edge castellations (top castellations and the top 'arrowhead' are almost completely obscured by the electronically inserted colour bars) indicate the limits of the picture. This allows the receiver *height, width* and *centering* controls to be set up where applicable. Modern TV receivers use picture tubes of the same aspect ratio (width/height) as that of the camera tube at the transmitter (4:3). Therefore the display should be set to within the limits of the points of the 'arrowheads' (the top colour bars to be inside the picture). Older receivers employ an aspect ratio of 5:4; the height could then be set as above, while the width is adjusted until a linear picture is obtained.
(6) Grey scale rectangles in the column to the left of the centre circle–the six small rectangles of varying brightness can be used to check the *grey scale, contrast* and *brightness.* The difference in brightness between adjacent rectangles should be approximately constant. Inside the bottom and the top rectangle there is a brighter spot. Too much brightness causes 'white crushing'; not enough, 'black crushing' (it can also be caused by faulty video amplifier biasing). This appears when the top or the bottom spot, respectively, merges with its surrounding. The top rectangle in the grey scale does not correspond to peak white; therefore the pattern may not be suitable for the adjustment of the 'highlights'.
(7) Frequency gratings in the column to the right of the centre circle–six rectangles with vertical black and white stripes of varying thickness. These are produced by a square wave signal which turns the beam alternately on and off. The size of the stripe depends on the frequency of the signal (see the chequerboard concept in chapter 2); hence, the display shows the *frequency response* of the video amplifier and indirectly the bandwidth of the i.f. strip. The order of the video frequencies represented from the top to the bottom rectangle is: 1.5 MHz, 2.5 MHz, 3.5 MHz, 4 MHz, 4.5 MHz, 5.25 MHz. In a colour receiver the 4.5 MHz pattern will be very blurred because of the chroma notch filter in the luminance amplifier. The 4 MHz and 5.25 MHz gratings might also be difficult to resolve, owing to a bluish colouring that could be displayed. This is known as *cross colour* and it is caused by the high frequency content of the luminance signal which lies within the passband of the chrominance amplifiers (the action of the demodulators gives rise to spurious colour).
(8) Black rectangle within the white rectangle above the centre circle gives indication of the *low frequency response* of the video amplifier (and to some extent the alignment of the i.f. vision carrier). Poor response produces streaking at the right-hand edges of the rectangles.

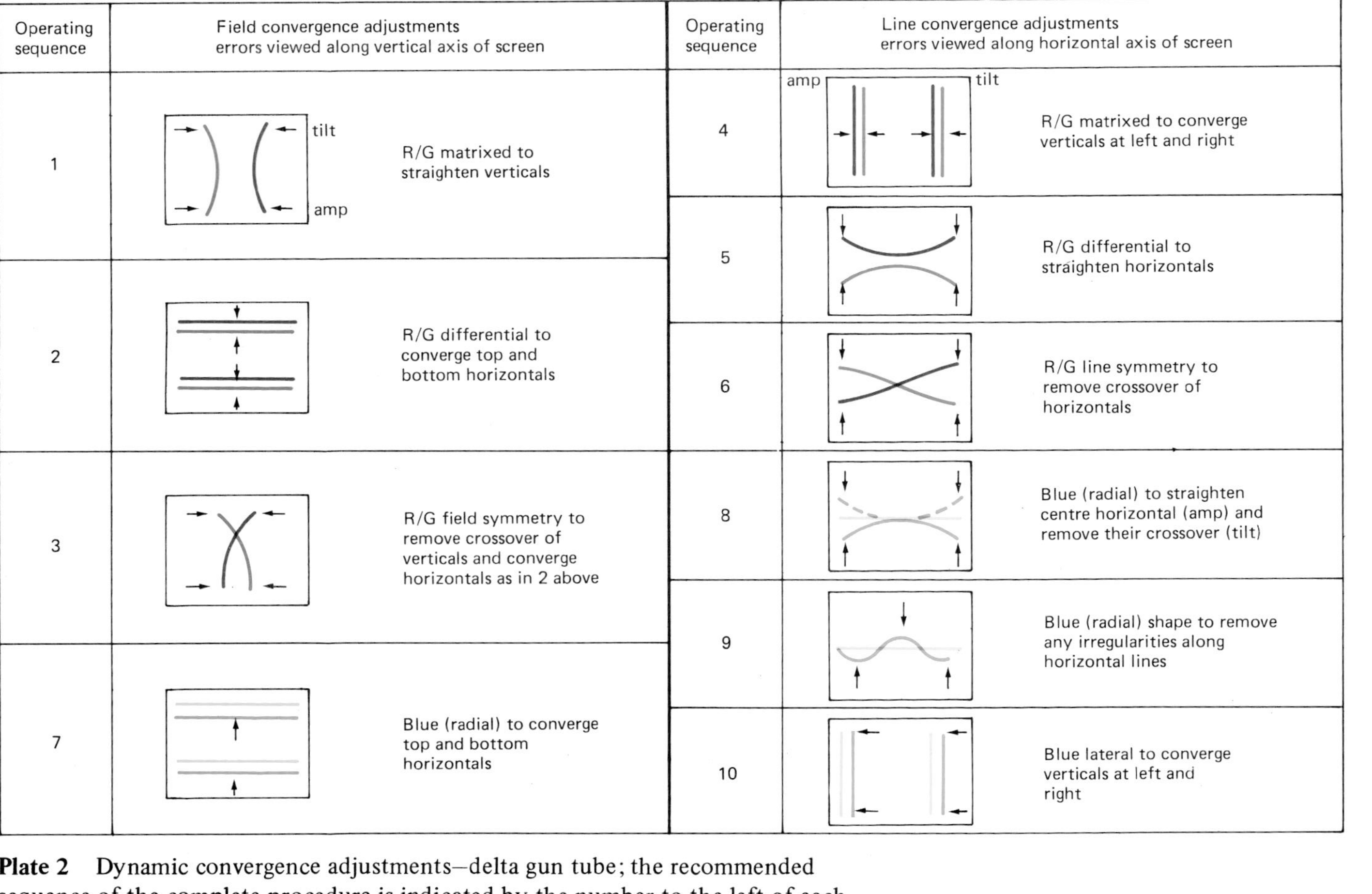

Operating sequence	Field convergence adjustments errors viewed along vertical axis of screen	Operating sequence	Line convergence adjustments errors viewed along horizontal axis of screen
1	tilt amp R/G matrixed to straighten verticals	4	amp tilt R/G matrixed to converge verticals at left and right
		5	R/G differential to straighten horizontals
2	R/G differential to converge top and bottom horizontals	6	R/G line symmetry to remove crossover of horizontals
3	R/G field symmetry to remove crossover of verticals and converge horizontals as in 2 above	8	Blue (radial) to straighten centre horizontal (amp) and remove their crossover (tilt)
		9	Blue (radial) shape to remove any irregularities along horizontal lines
7	Blue (radial) to converge top and bottom horizontals	10	Blue lateral to converge verticals at left and right

Plate 2 Dynamic convergence adjustments–delta gun tube; the recommended sequence of the complete procedure is indicated by the number to the left of each diagram.

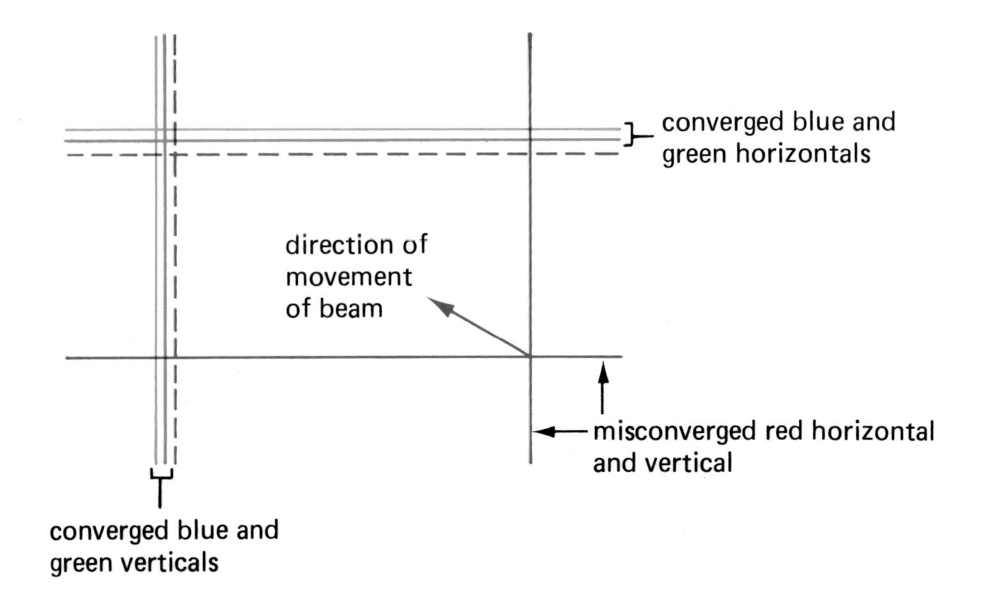

Plate 1 Effect of static convergence adjustment–in-line gun tube. Static correction fields act diagonally–as in delta gun tube. This ensures convergence of both vertical and horizontal lines of crosshatch display. Solid red lines represent misconvergence; these must be moved to take up the converged position–broken red lines.

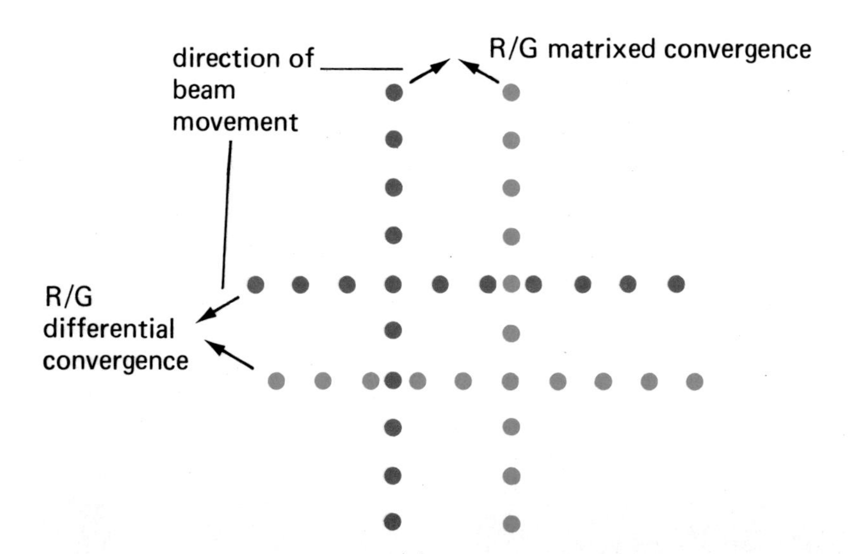

Plate 3 Dynamic convergence in a delta gun tube. The 'lines' of crosshatch consist of rows of illuminated phosphor dots; the movement of the beams aims at striking rows of dots which are closer together. Owing to the radial (diagonal) shift of the R/G beams (see figure 11.3), matrixed adjustments converge the 'verticals', and differential adjustments converge the 'horizontals' of the display.

Plate 5 Test Card F.

Plate 6 Test Card G.

	Line convergence adjustments		Field convergence adjustments (if included in receiver)
1	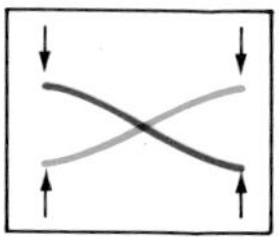 Line symmetry (balance) to remove crossover of red and blue horizontals	4	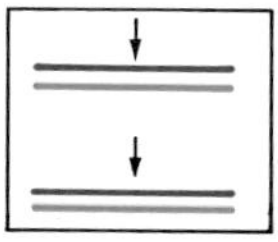 Red top and bottom to converge horizontals to yellow
2	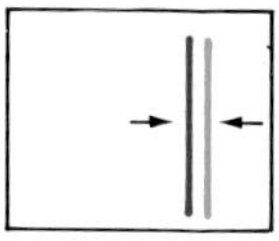 Red–Blue right (amplitude inductor) to converge right-hand side verticals	5	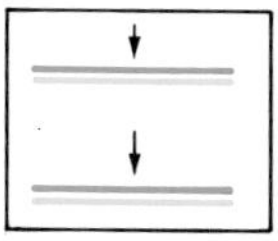 Blue top and bottom to converge horizontals to white
3	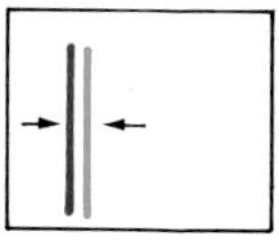 Red–Blue left (tilt potentiometer) to converge left-hand side verticals	6	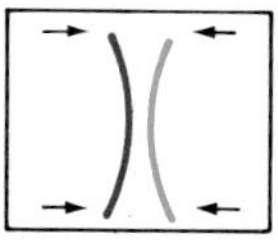 Top and bottom overlay to straighten centre verticals

Plate 4 Dymanic convergence adjustments–in-line gun tube. The recommended sequence is indicated by the number to the left of each diagram. Apart from the electrical controls shown, some movement of the convergence assembly may be needed to straighten red and blue 'horizontals'. If required, the purity magnets are *carefully* adjusted to eliminate crossover of red and blue 'verticals'; ensure the rings shift the raster up or down–*not* sideways, or purity will be affected. *Note:* The green beam is central in this example, and is not influenced by the adjustments.

Plates 5 and 6 (photographed from TV screens) are reproduced by courtesy of the BBC.

(9) Diagonally striped areas originating from each corner of the display—used to check the uniformity of picture *focusing*. Good focus is most important in the centre of the picture, but it should also extend towards the edges. Most of the striped area should be well focused, although in wide angle colour tubes some form of compromise may be necessary.

Test card G

This pattern is transmitted from a special signal generator. It is based on a design by Philips, but certain modifications have been introduced by the BBC. Other broadcasting authorities may use either the original design (type PM 5544) or their own version of it. Many facilities are similar to those offered by test card F; therefore in the following description cross-reference will be made as appropriate. A photograph of test card G is given in plate 6.

(1) Black and white castellation at the sides—check on the effect of picture content on the performance of *line synchronisation* and the synchronising pulse separator. A fault here produces wavy verticals level with the white blocks.

Picture *width* and *centering* are adjusted to ensure that the side castellations are just visible with the aspect ratio 4:3, and excluded if the ratio is 5:4 (see item 5, test card F).

The stability of the receiver *black level clamp* is judged by comparing the black (and/or the white) at the beginning of the line and at the end.

(2) Top and bottom castellations—check the performance of the *field synchronising* circuit by introducing a pulsed video waveform. These are also used to *centre* the picture and adjust the *height* to include the two borders.

Sometimes the top castellations may be replaced by standard colour bars, as item (1), test card F.

(3) Background crosshatch pattern—check the *dynamic convergence; static convergence* is judged by the intersecting lines at the centre of the picture.

The horizontal lines are used to check receiver *interlace*—if faulty, the centre line would be of a different thickness from that of the others.

The second and third squares of the crosshatch down from the top on the left-hand side, together with the neighbouring pair of castellations, are made up of unswitched (R – Y). The combined action of the PAL delay line and the PAL switch should cause no chroma output—the pattern appears grey ('colourless' information). Faults in the *delay line adjustments* would show up as Hanover blinds in that small area or in the subsequent coloured pattern (see item 4 below). Defects in burst gating (at the beginning of the line) would cause either variable saturation or unsynchronised colour later on the same level as the test squares.

(4) Coloured 'brackets' on the left- and right-hand sides of the centre circle—these coloured areas are used to further check the performance of the *decoder.* The left-hand side vertical stripe of the 'bracket' is produced by a chrominance signal for which (B – Y) = 0 and (G – Y) = 0; however, the top half corresponds to negative

(R – Y), which gives turquoise colour, and in the lower half is the positive (R – Y) –purple. The smaller sections at the top and bottom next to the (R – Y) signals consist of (B – Y) and (R – Y) of such polarities and proportions that the matrixed (G – Y) = 0. Two colours are then possible, since the constituent (R – Y) and (B – Y) can change their polarities (bluish at the top and reddish brown at the bottom).

The vertical columns of the right-hand side 'brackets' consist of the negative (B – Y), top half; positive (B – Y), lower half. The small (G – Y) = 0 areas are as before. This part of the test card can be used to check the operation of the synchronous demodulators and the (G – Y) matrix. When viewed on an oscilloscope (triggered at line rate), the decoder waveform will consist of sections corresponding to the colour information displayed across the TV screen. For example, the (B – Y) demodulator should have no output waveform in the two left-hand side sections immediately after the line pulse [no (B – Y) in the 'colourless' information on the left, nor in the ± (R – Y) which follows]. There should be an ouput in the last section of the waveform, though, which is due to the ± (B – Y) towards the end of the picture. Following the above reasoning, both the (R – Y) demodulator and the (G – Y) matrix can be checked.

(5) The centre circle–for picture *linearity* checks.

(6) Large black rectangle at the top–*low frequency response* (see item 8, test card F).

(7) Thin, short vertical line below and to the left of the low frequency rectangle–check for '*ghosting*'.

(8) Row of alternate dark grey-white rectangles–checks for *transient response*: freedom from ringing or other distortion primarily in the video amplifiers.

(9) Standard colour bars–check overall colour quality of the receiver.

(10) Frequency gratings–row of six rectangles with vertical stripes: the frequency range is the same as in test card F.

(11) Grey scale rectangles below the frequency gratings–the full range from black to peak white is available; hence, complete adjustments can be carried out if required.

(12) Coloured area at the bottom of the circle, red on a yellow background–check the luminance delay line circuitry; if correct, the red block should line up exactly with the centre square of the crosshatch immediately below.

13 Television Aerials and Systems

13.1 Properties of V.H.F. and U.H.F. Signals

As we have seen, the bandspread of one TV channel can be 8 MHz or more, depending on national standards. In order to accommodate a number of channels of such large bandwidth, it is necessary to use high carrier frequencies, in both the v.h.f. and u.h.f. regions; microwave transmissions may be adopted in the future. The propagation of electromagnetic waves is very severely affected by the order of their frequency. The higher the frequency the more pronounced is the 'line of sight' distribution of the signal. That is, at the higher frequencies the receiver installation ought to 'see' the transmitting aerial. The process of reflecting the signals at frequencies above 40 MHz from the upper layers of the atmosphere is no longer reliable for normal communication purposes. It will also be appreciated that the signal strength will be considerably reduced as the reception area becomes more distant. In countries where the population density is low it is necessary for the transmitter to cover a large land area; in these cases v.h.f. transmission is particularly useful, because it provides increased coverage. In the UK, with its large population centres, often only 40 miles apart, u.h.f. transmissions offer better facilities in allocating the available channels without adversely affecting the reception in the neighbouring areas. Additionally, since very little man-made interference is generated at u.h.f. frequencies, transmissions in this frequency band result in better quality reception.

The reception area in the true 'line-of-sight' region is governed not only by the earth's curvature, but also by the terrain. In fact, TV signals can be received beyond the horizon, owing to a small amount of 'bending' (*refraction* and *diffraction*) in the line of the radiated signal. This effect is specially noticeable at frequencies in Band I. The usable reception area can be further extended by raising the transmitting aerial. Whenever possible, the primary transmitters are on high ground and have very tall masts. The positioning of relay or 'fill-in' stations is governed by other factors; in fact, it may be desirable to restrict their range, to prevent interference with neighbouring transmitters. The signal coverage is also influenced by the radiated power of the transmitter, which may be expressed in watts or kilowatts.

Although the signals at such high frequencies are not normally reflected from the ionosphere, they can be reflected from the surface of the earth, from hills, tall buildings and other obstacles. When this happens, the receiving aerial will get at least two signals—the *direct* and the *indirect wave.* The indirect, or reflected, wave has to cover a longer distance and arrives at the receiver after the direct signal. The resulting time delay produces a second image on the screen—known as a *ghost image*; it is possible to receive a number of reflected signals in this way, which result in several ghost images. The latter must not be confused with the symptoms of instability in the receiver amplifiers. This manifests itself as a series of equally spaced outlines, while 'ghosting' produces irregularly spaced images.

Aircraft passing in the vicinity of the receiver can give rise to a fluttering picture (aircraft flutter), which is due to the reflected signal from the aeroplane. As the aircraft moves, so the reflected and direct signals go in and out of phase with each other, because the length of the indirect path alters. The rate of picture flutter is about two or three times per second. Some improvement is obtained by using a highly directional aerial, and receivers with fast-acting a.g.c. are less prone to this effect.

The signal strength at any particular location is expressed in mV/m or μV/m; it gives the voltage induced in an aerial of effective length of 1 m. Broadcasting authorities issue maps of the transmitter service areas showing signal strength 'contour lines'. From such information it is possible to anticipate the effectiveness of a particular aerial design before its installation. In some cases it is necessary to use a portable field strength meter in order to help choose and position the aerial for best reception.

13.2 Fundamentals of the Receiving Aerial; the Dipole

The receiving aerial must capture some of the transmitted energy and feed it to the tuner via the cable. In this respect, the design of the aerial must match the remainder of the installation to achieve maximum signal transfer—that is, the various impedances involved ought to be equal: the impedance of the aerial should have the same value as that of the feeder cable, which, in turn, should be equal to the input impedance of the first r.f. amplifier in the tuner.

The aerial should be directional in order to discriminate against any unwanted signals (e.g. reflections) and it must have adequate gain to produce sufficient noise-free output from the available signal strength.

V.H.F. and u.h.f. receiving aerials are based on a *half-wave dipole*, which in its simplest form consists of two metal rods mounted on a suitable support. The length of each rod is equal to one-quarter (approximately) of the wavelength to be received; thus the overall dimensions of the dipole approximate half the wavelength of the signal [see figure 13.2(a)]. Electrically, such an arrangement behaves like a resonant circuit, which makes it frequency-selective. The rods themselves offer an inductance (similarly to a length of any conductor); the necessary capacitance is distributed along these elements—in this respect, it resembles the action of stray capacitance.

The transmitting aerial is also based on the dipole and is fed with energy from the transmitter amplifier. The flow of current in the elements gives rise to an electromagnetic field and an electric field which simultaneously radiate outwards from the aerial. The two fields are radiated at right angles to each other. The electric field can be arranged to spread in the required direction in either the vertical or the horizontal plane, depending upon the orientation of the transmitter dipole. Therefore it is said that the transmitted signal is either vertically or horizontally *polarised.*

When the two fields reach the receiving aerial, the position of its dipole must be such that, for a maximum signal pick-up, the elements are in the plane of the arriving *electric field.* This principle gives rise to either vertically or horizontally polarised aerials. Normally, in the UK the main u.h.f. transmitters employ horizontal polarisation; signals from relay stations are usually vertically polarised, to prevent mutual interaction. The effect of horizontal polarisation makes the dipole somewhat directional. The receiving aerial is capable of picking up equal amounts of signal from the 'front' or from 'behind' (that is, the 'broadside' facing the transmitter); radiation approaching from the 'sides' will not be readily accepted. In a vertically polarised system the dipole receives signals from all points and the aerial is no longer directional.

The ability to develop different voltage outputs, depending upon the direction of the incoming signal, is often illustrated by the *polar diagram* of the aerial. Generally, it consists of a loop, or a number of loops, drawn in relation to the common point, known as the *pole*, and placed upon the sketch of the aerial itself. In practice, the junction between the cable and the aerial marks the pole of the diagram; the distance from there to any point on the curve(s) represents the output developed by the signal arriving from that particular direction. Therefore the curve becomes a form of directional pattern of the aerial. This principle is illustrated in figure 13.1; a polar diagram for a dipole is shown when vertical polarisation is used in (a) and horizontal polarisation in (b). The direction of each arrow represents the possible line of signal pick-up; the length corresponds to the output which would be fed to the cable. Often the manufacturer quotes the outputs in relation to the possible maximum, so that they are expressed as ratios or in dB. As the strongest signal is always received from the 'straight ahead' direction, those arriving from elsewhere are weaker and the figures given are negative dB.

Vertical polarisation gives a circular polar diagram, which confirms that the dipole has no directional properties. As a result of horizontal polarisation there are two loops in the form of a figure of eight. It indicates that the signals arriving from the front, or from the back of the dipole, can develop equal outputs. Those from the sides are relatively ineffective.

The polar diagram shows the *angle of acceptance* of the aerial—that is, the angle between the directions of the received signals when the output is reduced by 3 dB (that is, 0.707 of maximum sensitivity). As shown in figure 13.1(b), this angle (α) is approximately 80° for a simple horizontal dipole (±40° from the maximum).

The aerial must have a certain bandwidth to ensure relatively constant response

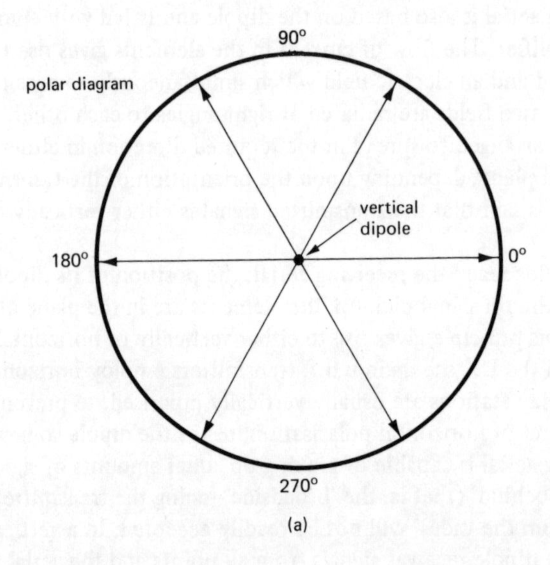

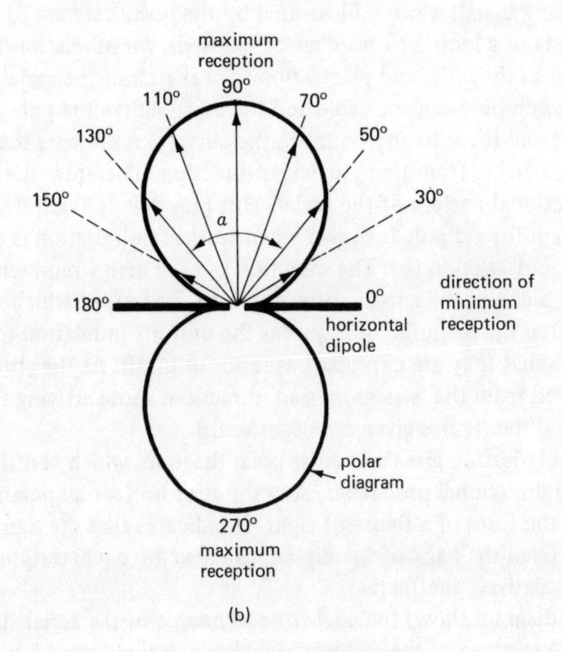

Figure 13.1 Polar diagrams of a simple dipole aerial: (a) vertical polarisation–equal pickup from all directions; (b) horizontal polarisation–the dipole now has some directional properties; angle α is the acceptance angle

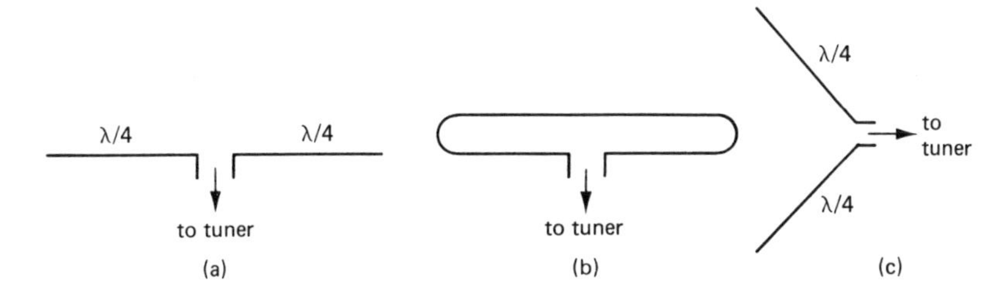

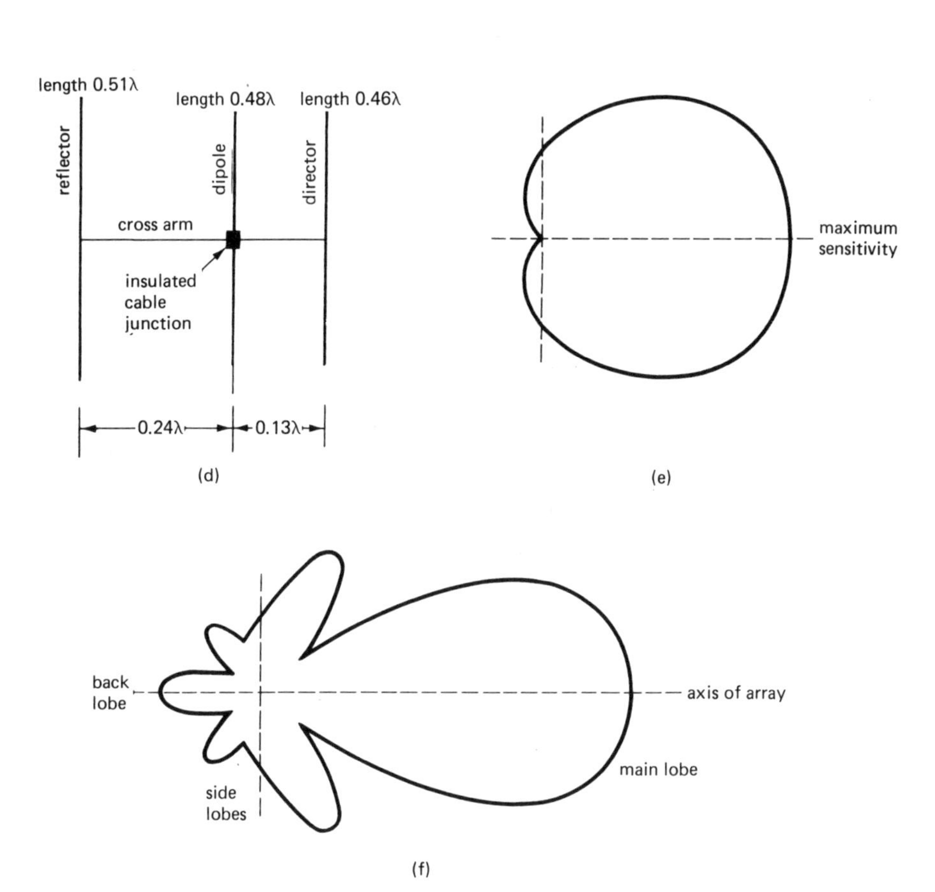

Figure 13.2 Examples of receiving aerials: (a) dipole; (b) folded dipole; (c) V-shaped dipole; (d) three-element Yagi array; (e) polar diagram of a three-element array; (f) polar diagram of a nine-element array. *Note*: λ (lambda) is the wavelength of the received signal:

$$\lambda = \frac{300}{\text{frequency (MHz)}} \text{ m}$$

to the frequencies associated with one channel, or even a group of channels. The bandwidth of each u.h.f. channel in the UK is 8 MHz, but a single aerial installation must be able to receive at least four programmes, which in turn are separated from one another by gaps of unused channels. Therefore the effective frequency range corresponds to a group of 11 channels, which require a bandwidth of $11 \times 8 = 88$ MHz. It appears that a simple dipole cannot fulfil this, especially at the lower frequency end. V.H.F. aerials are usually designed for a particular channel and a dipole may then be useful.

An important property of an aerial is its *impedance*, which is the ratio of the induced voltage to the resultant current. The need for impedance matching has already been stressed if maximum signal transfer to the receiver is to be achieved. Serious mismatch in the complete installation can lead to a loss of signal and signal reflections which would appear as 'ghosting'. The impedance of a simple dipole is 73 Ω at the frequency of resonance; this figure is often referred to as being nominally 70 Ω or 75 Ω. The dipole is thus easily matched to a coaxial cable downlead whose impedance is also approximately 75 Ω.

Another form of dipole is the *folded dipole*, which could be constructed by folding back a metal rod on itself so as to form a very tight loop, as shown in figure 13.2(b); the overall dimensions are still based on the half-wave principle. Some of its properties are similar to those of the simple dipole, except that the impedance is now 300 Ω and the bandwidth is also increased. The impedance may be modified further by making the 'top' and the 'bottom' parts of the assembly from rods of different cross-sectional area or shape; such an arrangement is useful in the design of multielement aerials. Because of the impedance mismatch, the folded dipole alone is not suitable for connecting to the coaxial cable, but it can be used directly with a twin lead parallel line, which is favoured in some countries.

13.3 Multielement Aerials

The simple dipole aerial has a relatively low signal output, except in areas of high signal strength. Even there, its lack of directional properties—that is, its inability to discriminate against unwanted transmissions—can make it unsuitable. Finally, the available bandwidth of a dipole is restricted. Some improvement can be made by bending the arms of the dipole to produce a V-shaped arrangement. This modification improves the directional properties and widens the frequency response.

A second element can be mounted behind the dipole to form a *reflector.* If the distance between the two is carefully chosen, the aerial will produce a greater output than the dipole and have better directional properties by rejecting signals from the back. The *gain* of an aerial is expressed in dB and it is based on the signal output of a particular array with respect to the output of a single dipole. Thus the gain of an H-type arrangement can be about 5 dB—that is, approximately 1.8 times more voltage than from a dipole alone. There is also a reduction in the signal pick-up from the unwanted direction. The ratio of the maximum signal developed from the front to that from the backward direction is known as the *front-to-back* ratio, and is normally expressed in dB. The corresponding figure for the H-type aerial is up to 10 dB.

The reflector is of a slightly greater length than the dipole, but it also forms a tuned circuit at the required frequency. The signal induced in the reflector is then reradiated towards the dipole, to increase the resultant output. At the same time, the somewhat involved phase relationships between the currents in the two elements will cause partial cancellation of the unwanted pick-up.

Additional elements can be placed in front of the dipole, which are then called the *directors.* Their dimensions and mutual spacings are critical and depend upon the frequency of the channels to be received. The directors also pick up the signal energy, only to reradiate it ultimately towards the dipole. The gain increases with the number of elements in the array. There is a practical limit, however, due either to the physical size, especially in Bands I and III, or due to electrical factors which cause reduced impedance and bandwidth. The advantages of a multielement aerial of the type described, also known as a *Yagi array*, are high gain and good directional properties. A diagram showing the typical dimensions of a three-element array and its polar diagram are in figure 13.2(d) and (e), respectively.

The polar diagram of an aerial with many directors consists of a number of lobes; the main one is along the axis of the array which indicates maximum sensitivity. An example of this is shown in figure 13.2(f). The angle of acceptance tends to become narrower as the number of elements increases. The resultant reduction in the width of the main lobe makes the positioning of the aerial more critical, and a few degrees 'off beam' might produce an unusable signal. The existence of the side lobes in the polar diagrams must be taken into consideration. Either unwanted signals may be received from those directions, or the required signal, but of inadequate strength, could be produced. The front-to-back ratio of a Yagi array is good, being approximately 30 dB; the gain of a u.h.f. 10-element aerial can be typically 12 dB.

13.4 Special Aerial Designs

The disadvantage of a simple Yagi array (consisting usually of a folded dipole and straight rods as both the reflector and the directors) are: insufficient gain in extreme fringe areas and relatively narrow bandwidth to cover reception on different bands or different channel groups. It has not been practicable to eliminate both disadvantages with one design. Aerial design is based both on theoretical considerations and on the practical experience of manufacturers. The number of elements in an array does not in itself guarantee good performance.

To improve the front-to-back ratio, the reflector may be in the form of a grid, which may either be flat or parabolic.

The improvement in gain would require many additional elements. At u.h.f. the economical optimum for a Yagi array is 18 elements; to increase the gain by another 3 dB, it would be necessary to double the size of the aerial. In fact, stacked arrangements are sometimes used consisting of two arrays side by side (or stacked in the vertical direction), their dipoles interconnected and joined to the common down-lead. The additional array increases the overall gain by 3 dB, and also helps to remove some of the signal reflections from the ground.

V.H.F. aerials, because of their greater physical dimensions, produce a greater output at a given signal strength. The design of some u.h.f. aerials also attempts to present a greater effective length of the elements to the incoming signals. This is done without increasing the overall size of the aerial and still maintaining the critical dimensions of the rods related to the wavelength. Such designs use X-shaped dipoles and directors; the dipole may be cut for full instead of half wavelength. One manufacturer quotes a gain of 21 dB for an array based on this principle, with an aerial consisting of 21 directors, dipole and a reflector, the angle of acceptance being only about 12°.

A wide bandwidth aerial is usually designed at the expense of gain; hence, it is suitable where the signal strength is high. A *log-periodic* array is very suitable to cover a very wide bandwidth, which can extend from v.h.f. to u.h.f. An ordinary, Yagi-type, array has a limited frequency range since its electrical properties change drastically at frequencies considerably different from the design figure. As the diagram of a log-periodic array in figure 13.3 shows, the length of the individual rods

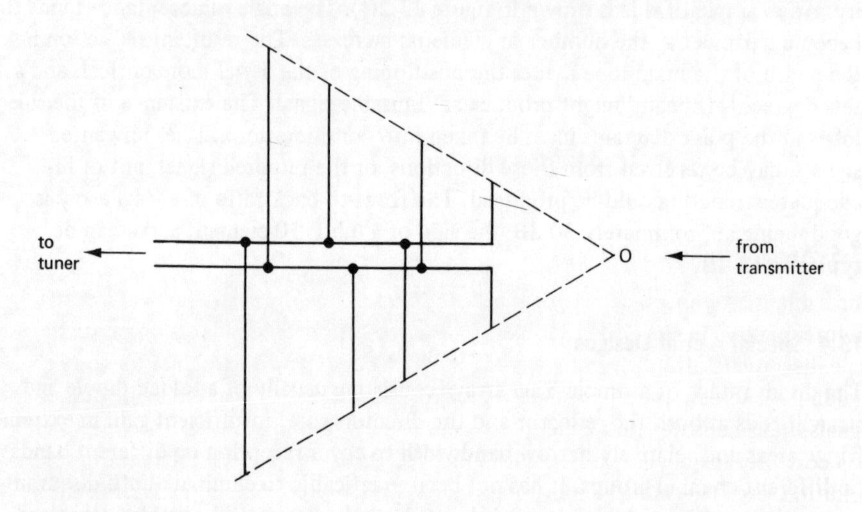

Figure 13.3 Arrangement of elements in a simple log-periodic aerial array

increases with the distance from the apparent apex (point O in the diagram) of the aerial. The opposite elements along the beam are electrically connected. The properties of a log-periodic aerial repeat themselves periodically as the frequency increases along a logarithmic scale. For example, if the array was designed to have an impedance of 70 Ω and a gain of 5 dB at 50 MHz, it will also possess the same specification at 100 MHz, 200 MHz, 400 MHz and 800 MHz. In appearance, the array has small elements towards the front and they become larger towards the back. The low gain comes from the fact that at any particular frequency only a few of the elements are actively resonating, giving a relatively low contribution to the overall output.

For the reception of Band I and III a combined aerial can be used with perhaps only one element for the lower frequency; the remainder are designed for the higher frequency range. If a large gain is needed, however, two separate aerials are necessary, but in order to simplify the installation, only one cable is provided. The two arrays feed via a *diplexer*, which is a splitting network to ensure that the signals produced by one aerial are not affected by the presence of the other one. An example of a diplexer circuit is shown in figure 13.4. It consists of two T network filters: L_1, C_1, L_2 pass signals of Band I while offering a high impedance to Band III, and C_2, L_3, C_3 pass the frequencies of Band III and block Band I signals.

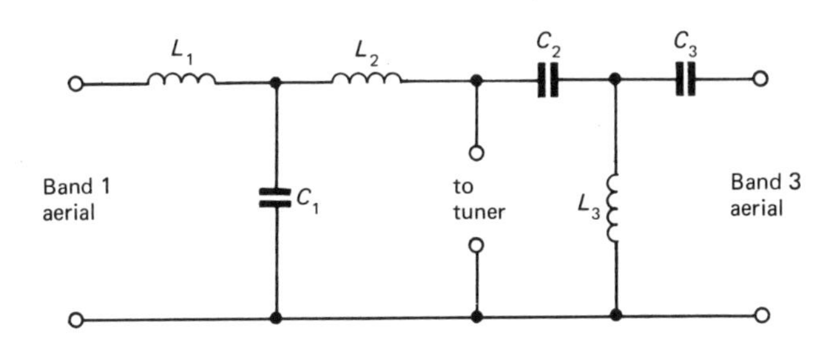

Figure 13.4 Diplexer circuit

13.5 Aerial Feeder Cables

The cable is an important part of the installation, as its properties may affect the picture quality. In the UK and in many other countries the feeder is of a coaxial type, because it offers several advantages over a parallel two-wire line. The outer braiding, or screen, prevents the cable radiating its signal to interfere with other signals, it also ensures that little interfering signal will be picked up. The impedance of a correctly installed coaxial cable is unaffected by adverse weather conditions, pollution, etc. The value of this impedance is 75 Ω, which should make it suitable for direct connection to a simple dipole. Unfortunately, at the lower v.h.f. range there can be problems, because of the current distribution in the cable, which does not produce a balanced electric field. A *balun* (*bal*anced to *un*balanced) *transformer* can be used to achieve suitable matching. Such a device may not necessarily resemble a more conventional type of transformer, as it can be in the form of a short sleeve surrounding the cable.

The presence of the cable introduces some reduction in the magnitude of the signal reaching the tuner, which can become significant at high frequencies. To keep the losses down, the conductors can be of a relatively large cross-sectional area; very low loss cables have a solid centre conductor instead of a stranded type. Signal attenuation can be also due to the losses in the insulation between the core and the sheath; at frequencies up to 300 MHz the insulation could be made of either solid or foam polythene. For u.h.f. application the insulation is 'semi-air'–namely the polythene is extruded so as to form a large air cell structure. The plastic is needed only to support the inner conductor.

When installing such cables, it is necessary to avoid large physical stresses, because either the inner conductor or the polythene dielectric can be easily damaged–often without any obvious external signs. Therefore one must not apply forces likely to cause stretching, leave a long piece of cable hanging without any support, etc. Insulation must be removed very carefully, to avoid nicking the centre core, which could cause a fracture at some later time. A suitable clamp should be used at the aerial end, to prevent water entering the cellular structure of the polythene–any moisture here will cause very severe loss in the signal strength. Kinks in the cable can compress the inside so much as to cause a short circuit between the outer and the inner conductors; also, such localised irregularities can lead to signal reflections which cause ringing (short length of cable) or ghosting (long cable run). When the cable has to be bent, the radius must be at least five times the cable diameter. Where the cable is stapled to the walls of the building, it must be done carefully so as not to compress the cable; again, irrespective of any obvious mechanical damage, the irregularities produced in the aerial downlead can cause signal problems, as already described. An interesting point to note is that signal reduction could occur when the cable run is longer than 5 m and the staples are fixed at regular intervals which happen to be equal to half the wavelength of the received channel frequency (see the wavelength formula in figure 13.2).

Signal loss in cables can be expressed in dB per 100 m run at a given frequency (e.g. for the cable when it is used at 50 MHz, 200 MHz or 800 MHz); these figures are available from the supplier and they must be considered to ensure adequate signal feed to the tuner.

13.6 Choice of Aerial Installation

The choice of the type of aerial installation is governed by the following factors: signal strength in the given reception area, possible source of signal reflection and obstruction, presence of interference and the performance of the receiver itself.

A simple dipole should only be considered where the signal strength is high *and* there is freedom from interference, reflections, etc. Directional aerials which involve multielement arrays must be used in all other circumstances. V.H.F. arrays have relatively large dimensions, particularly on Band I; hence, the number of elements must be restricted (say three on Band I and six on Band III). Special attention would have to be paid to the mechanical strength of the supporting mast and its attachment to the building. The effect of the additional stress on the chimney stack or any other structural part must not be overlooked.

Mechanical problems are reduced when installing u.h.f. aerials. There are usually

6-, 10-, 13- or 18-element arrays apart from the special types previously mentioned. A high gain aerial may be necessary in an area troubled by signal 'ghosting' or interference; the narrow angle of acceptance of this type of aerial prevents unwanted signals developing any significant output. An aerial with fewer elements has a wider acceptance angle, in which case it can be so positioned that the wanted signal arrives at an angle to the line of maximum sensitivity. It may then be possible to have the 'gap' between the lobes of the polar diagram to face the direction of the unwanted signal. Such a method of installation may be useful when there is only one major source of interference or reflection. In practice, the array is rotated until the 'ghost' disappears and a satisfactory picture remains.

The principal source of signal reflection could often be judged from the measurement of the distance between the main display and the 'ghost' on the TV screen. The electron beam takes 52 μs to scan one line of picture. If the screen is 40 cm wide (approximately 19 in tube), then the beam travels $40/52 = 0.77$ cm per μs. At the same time, the speed of propagation of the TV signal from the transmitter is 300 000 km/s (speed of light!) or 0.3 km/μs. Equating these two velocities, it can be seen that for approximately every 8 mm of 'ghost distance' the radio waves must have travelled 300 m longer than the direct path. (Thus 1 in. or 25 mm, spacing gives approximately 1 km, or 0.6 mile, extra distance.)

A high gain aerial used in an area of good signal strength can develop an output which might overload the receiver tuner and i.f. stages. The a.g.c. has only a limited ability to reduce the receiver gain. In such circumstances an *attenuator* may have to be fitted between the tuner aerial socket and the cable. The design principles of such a circuit are discussed in section 13.7.

The positioning of the aerial array is important. Only in exceptionally favourable circumstances can the so-called 'set-top' dipole be at all satisfactory. The picture quality is likely to be affected by interference signals, or even by movement of people in the room. This type of aerial is normally provided with small portable, battery receivers. A loft-mounted aerial will be an improvement on this type, and it may be a necessity if circumstances do not allow external installation. Signal reflection, or even signal absorption, can be caused by metal water tanks or pipework in the vicinity of the aerial. Because of the obstruction introduced by the walls of the building, the wanted signal may be attenuated, allowing interference signals to predominate. A highly directional array might be needed to overcome those problems.

An external aerial is the most satisfactory arrangement. The height of the mast might have to be adjusted, especially in fringe areas, to minimise the effect of obstructions in the signal path.

A very low loss cable has to be used where long runs are involved, especially for the reception of the upper number channels (Band V).

Generally, colour receivers require a higher signal strength than their monochrome counterparts. An installation which is only just satisfactory for black and white reception must be improved before an even comparable picture quality in colour can be obtained.

Satisfactory reception of Teletext requires a very efficient aerial installation. Any tendency to ghosting can cause either a loss of characters or incorrect characters in the displayed 'page'.

Aerial signal preamplifiers may sometimes be recommended to improve the picture quality. It should be considered as the last measure in the attempt to boost the signal. Modern TV tuners are of optimum design, and an indifferent preamplifier may not warrant the expense. The best design array of maximum gain, followed by a low loss cable, are the first considerations, since any signal gain is not accompanied by additional noise. It must be remembered that electronic noise will be produced by both the preamplifier and the tuner. For this reason an amplifier, if it has to be used, ought to be installed close to the aerial on top of the mast. It then handles a relatively noise-free signal and boosts the output prior to the inevitable attenuation in the cable.

U.H.F. aerial arrays are made to receive a group of channels; these can be identified by group letters and sometimes colours: Group A (channels 21-34, coding red), Group B (39-51, yellow), Group C/D (48-68, green), Group E (39-68, brown).

13.7 Aerial Attenuators

Attenuators may be necessary in exceptional circumstances; these reduce the signal strength fed to the tuner to prevent overloading of the receiver circuits. Overloading would manifest itself as an overcontrasted picture, possibly with *grey crushing* or *white crushing* (see test cards). Often synchronisation is affected, giving rise to 'wavy' verticals; sometimes instability can be induced in certain parts of the receiver. Also, severe interference patterns might be displayed (*cross-modulation*).

Attenuators are basically resistive potential dividers, but their most important property is that their effective impedance matches the impedance of the remainder of the installation. Where a coaxial cable is used, this must be nominally 75 Ω.

There are two alternative circuits which can be used. Diagram 13.5(a) shows a T network, and 13.6(b) shows a π network. The values of the resistors required are given by the following formulae

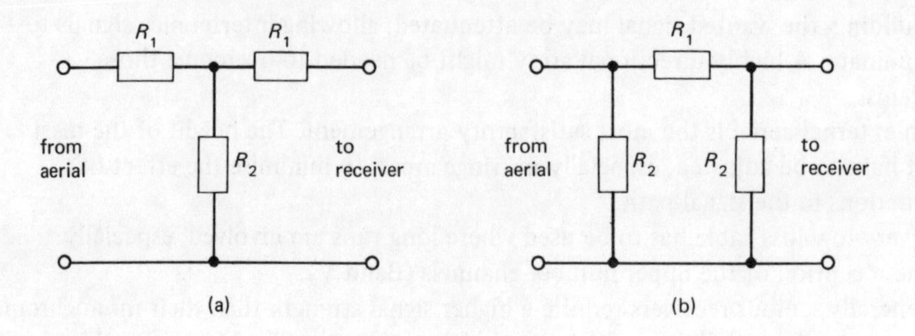

Figure 13.5 Aerial signal attenuators: (a) T network; (b) π network

T network $$R_1 = R\frac{A-1}{A+1} \text{ (2 required)}$$

$$R_2 = \frac{2A}{A^2-1}$$

π network $$R_1 = R\frac{A^2-1}{2A}$$

$$R_2 = R\frac{A+1}{A-1} \text{ (2 required)}$$

In the above expressions R is the impedance of the cable (typically 75 Ω) and A is the numerical attenuation *ratio* (V_{IN}/V_{OUT}).

Since the attenuation needed is often expressed in dB, it must be converted into a numerical ratio. For example, when attenuation is 6 dB, $A = 2$; for 20 dB, $A = 10$; etc. Of course, the nearest preferred values of resistors are chosen after the calculations.

13.8 TV Signal Distribution Systems

It may be necessary to feed several receivers from a common aerial installation. The signal is split into a number of outlets in such a way as to present to the cable a matching impedance of, say, 75 Ω.

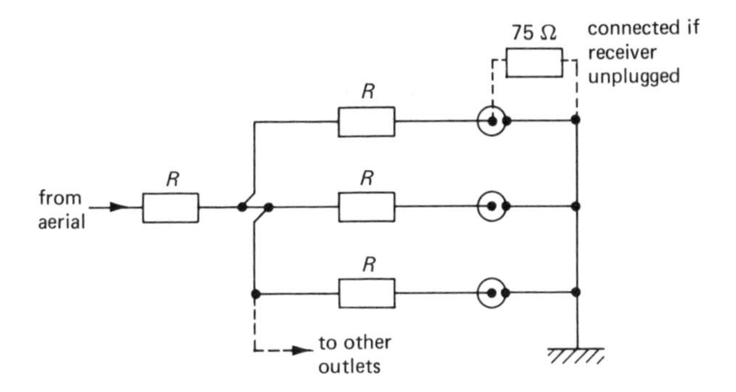

Figure 13.6 Resistive signal splitter

The method of distribution shown in figure 13.6 ensures the necessary impedance match, provided the equal value resistors, R, have been calculated as follows

$$R = \frac{n-1}{n+1} \times Z$$

where n is the number of outlets, Z is the impedance of the cable (75 Ω) and R is chosen to the nearest preferred value. Should one of the receivers be disconnected from the socket outlet, a 75 Ω resistor ought to be connected in its place, to preserve the original value of impedance; an open-ended outlet can give rise to signal reflections. The resistors in the arrangement will dissipate some of the signal energy which an aerial array may not be able to supply; therefore an amplifier will be necessary to feed the system, if a satisfactory picture is to be maintained.

Extensive distribution systems have proportionally higher signal losses, including large attenuation in the cable network itself. It can be more economical to convert the received u.h.f. signal from the aerial into v.h.f. for distribution purposes, as the effect of signal loss in the cable is considerably reduced at the lower frequencies. The receivers in this scheme can be equipped with either v.h.f. or integrated (v.h.f. and u.h.f.) tuners. A block diagram of an arrangement based on this system is shown in figure 13.7.

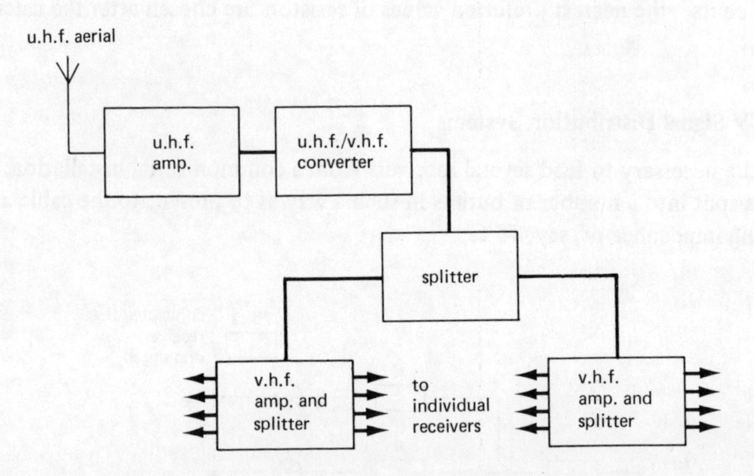

Figure 13.7 Block diagram of a distribution network with v.h.f. conversion

Alternatively, if the more popular u.h.f.-only tuners are used, the signal has to be converted back from v.h.f. to the higher frequency by converter/amplifiers situated ahead of final subsections of the system. The 'up-' or 'down-frequency' converters are basically mixer/oscillator circuits which also include suitable amplifiers.

The systems described in this section can be expanded relatively easily, provided the amount of amplification is sufficient at the 'strategic' points in the network.

14 *Receiver Power Supplies*

14.1 Power Supply Requirements

The power requirements of the various sections of TV receivers are at differing voltage levels, depending on the function of the circuit to be supplied. In fully transistorised as well as hybrid (valve and transistors) receivers relatively low voltages are supplied to the circuits which handle small amplitude signals. Large amplitude waveforms can only be accommodated if the amplifier is connected to a suitable high voltage feed. Valves, by virtue of their principle of operation, need relatively high anode (and screen where applicable) potentials, irrespective of their circuit functions. The voltage levels used on the picture tube electrodes vary from low voltage on the heater to the extra-high tension on the final anode.

The various supply lines are often distinguished from one another by abbreviations: l.t. (low tension), h.t. (high tension) or e.h.t. (extra-high tension). Since several feeds at slightly different voltages may have to be provided within a given category, an identifying number has to be included, say l.t.1 or l.t.2 etc.

The following list presents the various requirements of either a hybrid or a fully transistorised TV receiver as applicable.

a.c. supply: valve heaters, including the picture tube (0.3 A and at least 6.3 V per valve), degaussing circuit in colour receiver.
d.c. supply: *l.t.* (between 10 V and 40 V approximately) to the tuner, i.f. strip, colour decoder, sound, video drive, transistorised time base oscillators, often 'solid state' field output;
h.t. (between 50 V and 300 V approximately), all valve anode supplies, line output, video output.
boosted h.t. (between 350 V and 1000 V approximately), field charging circuit in valve oscillators, A1 supplies to the picture tube, focus voltage to monochrome tube.
e.h.t. (between 6 kV and 25 kV), final anode of the picture tube, focus voltage to colour tube.

Some of the supply lines have to be stabilised to ensure a constant voltage feed to a particular circuit.

Total consumption can be from approximately 15 W for a small screen battery portable to, say, 150 W for a large screen, hybrid, colour TV.

The design of the power supply circuitry must ensure safe operation of the receiver under normal conditions as well as under fault conditions. To this end, a number of protective features are incorporated which will be discussed later in this chapter.

14.2 Heater Supplies

Television receivers employing valves are designed so that the heaters are all connected in series with one another, each carrying a current of 0.3 A r.m.s. The basic circuit is shown in figure 14.1; the power consumed by the circuit is constant and, for a mains voltage of 240 V, is 240 × 0.3 = 72 W. The reasons for diodes D1 and D2 shown in insets (i) and (ii), respectively, are given later. The voltage requirement of the heater chain is dependent on the number of heaters in series and also on the p.d. across each heater. Consequently, a dropper resistor must be included in series with the heater chain to absorb the remaining voltage. In practice, the dropper resistor R_1 is in series with the *thermistor* R_2 shown in figure 14.1. The name 'thermistor' is a contraction of *therm*ally dependent res*istor*, and is a device, in the case considered, whose resistance reduces with temperature. This fact is indicated by the $-t^\circ$ in association with the circuit symbol; alternatively, the abbreviation n.t.c. (*n*egative *t*emperature *c*oefficient or resistance) may be written against the resistor. When the heater chain is first switched on, the temperature and resistance of the heaters are low; in the absence of any method of limiting, the heater current flow would initially be very large, introducing a possible risk of a filament failure. At switch-on the high resistance of the thermistor restricts the heater current to a small value. As the device warms up, its resistance falls and the current increases to its normal value of 0.3 A.

Capacitors C_1 and C_2 in figure 14.1 are for decoupling purposes. This prevents high frequency signals developed in one valve, or in a group of valves, from being transmitted to another section of the receiver via the heater chain. The c.r.t. heater is placed towards the chassis end of the arrangement so as to reduce the heater-to-cathode potential; this minimises the risk of a heater-to-cathode breakdown of the expensive tube.

In fully transistorised receivers the tube heater is usually supplied from a suitable winding on the mains transformer or the line output transformer.

Hybrid TV receivers employ a relatively small number of valves, and the power dissipated by the dropper resistor in figure 14.1 would be much greater than in an all-valve receiver. The reason for this is that the smaller number of valves drop a smaller voltage, while still taking a current of 0.3 A; consequently, the p.d. across the dropper resistor would have to be increased. To reduce the power consumed by the dropper resistor, diode D1 shown in inset (i) in figure 14.1 can be connected in series with the heaters. In some cases this diode is shunted by a capacitor of a few nanofarads for voltage surge limiting purposes. Owing to the rectifying action of D1, the power consumed by the heater circuit is halved from the previously quoted

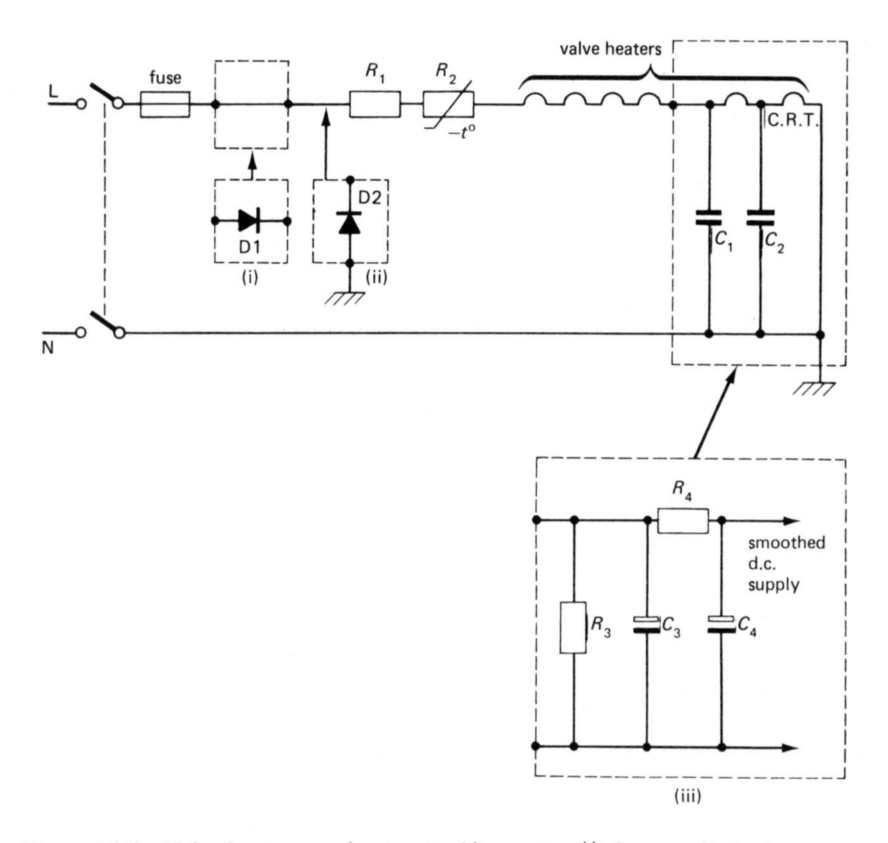

Figure 14.1 Valve heater supply circuit. Alternative (i) shows a diode dropper (D1). In (ii) diode D2 protects the heaters against effects of D1 short circuiting. Arrangement (iii) can be used to extend the heater chain to provide a low voltage d.c. supply

72 W to 36 W. Readers should note that there is no other advantage to be gained by using a d.c. to supply the heater chain. However, the available d.c. can sometimes be used to provide an l.t. feed to another section of the receiver, as shown in inset (iii), figure 14.1. The voltage developed across R_3 is smoothed by the circuit consisting of C_3, R_4 and C_4; the resultant d.c. is then directed to the appropriate circuits.

In diode-fed heater chains there is a risk that the series diode may develop a short circuit, with no apparent ill-effects until one or more heaters burn out. One method used to protect against this type of fault is the use of an additional diode, D2, connected across the supply as shown in inset (ii) in figure 14.1. Under normal operation D2 is reverse biased and does not carry current; if D1 develops a short circuit, diode D2 is forward biased in alternate half-cycles of the a.c. supply and applies a short circuit to the heater supply. This either causes the cut-out to operate (when fitted) or blows the main fuse. Alternatively, the d.c. supply shown in figure 14.1 may be used to power either the sync. separator or the field oscillator; shorted D1 will then make the picture unviewable. A similar arrangement, although tapped at a higher voltage level in the heater chain, can supply the screen of the field output valve. Again, a diode fault will cause field roll which could not be corrected by the viewer.

To be strictly accurate, voltage and current measurements taken in diode-fed valve heater chains should be made by means of either a moving iron or a hot wire instrument, since these give a true r.m.s. indication. A mean reading instrument, such as an Avometer, gives an indication which is lower than the true reading; an r.m.s. current of 0.3 A gives an indication of about 0.19 A on the d.c. range of a mean value reading instrument, and an r.m.s. voltage of 6.3 V gives a reading of about 4 V on the d.c. voltage range of this type of instrument.

14.3 Unregulated D.C. Supplies

A method of obtaining a low voltage d.c. supply was shown in figure 14.1. The polarity of the output can be positive with respect to the chassis in the circuit as shown, or it can be reversed simply by reversing the connections of diode D1.

In the case of mains-only receivers, the a.c. mains can be rectified directly and smoothed to give a d.c. supply of about 260–290 V. A typical half-wave circuit is shown in figure 14.2. The input is taken from the mains via filter network C_1, L_1, C_2, which is employed to minimise the possibility of mains-borne interference from affecting the operation of the receiver. In some circuits only a single capacitor connected across the supply is used. Series resistor R_1 and thermistor R_2 are included in figure 14.2 to reduce the effects of surges; rectifier diode D1 is protected from the effects of mains voltage transients by capacitor C_3. Capacitor C_4 acts as a reservoir capacitor, the voltage across it being smoothed by filter L_2, C_5 to provide the main h.t. supply, HT1. Additional h.t. supplies, HT2, HT3, HT4, are derived from HT1 via RC smoothing and decoupling circuits.

Electrolytic capacitors in power supplies, as in other parts of TV circuits, do not provide satisfactory decoupling at radio frequencies. Adequate decoupling is obtained by shunting them with polystyrene or polycarbonate capacitors of relatively low capacitance—for example C_{6a} in figure 14.2.

Low voltage supplies for transistor stages are frequently obtained from full-wave bridge rectifier circuits of the type in figure 14.3. Resistor R_1 is included to reduce the peak value of the charging current drawn by capacitor C_1, and typically has a value of a few ohms. The value of the smoothing capacitor C_1 lies typically in the range 1000–3000 μF. The circuit shown can be designed to provide any desired value of current or voltage. A negative output potential is obtained if the chassis connection is transferred to point X and the output is taken from point Y. Alternatively, if the chassis connection is transferred to the centre point, S, of the secondary winding, then a positive potential relative to the chassis is developed at point X, and a negative potential relative to the chassis simultaneously appears at point Y.

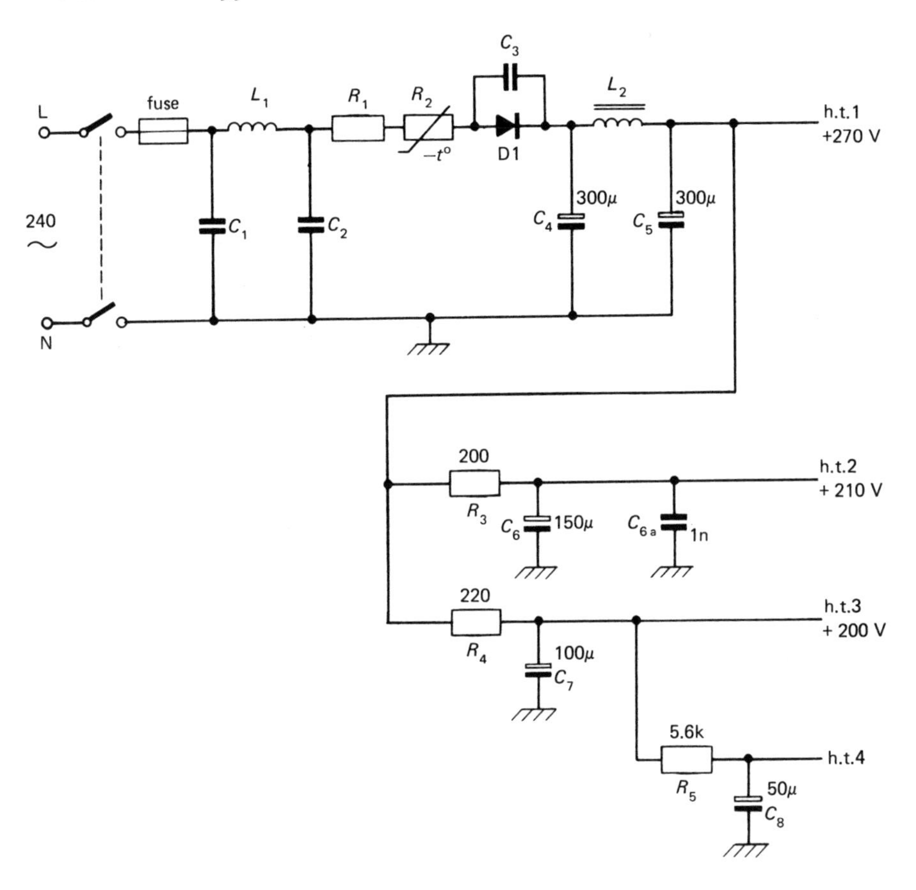

Figure 14.2 Half-wave rectifier circuit which provides a number of h.t. levels in the receiver

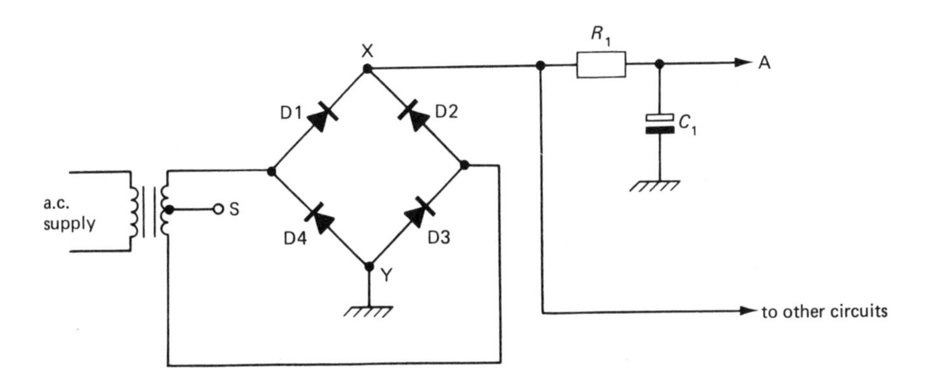

Figure 14.3 Full-wave bridge rectifier circuit

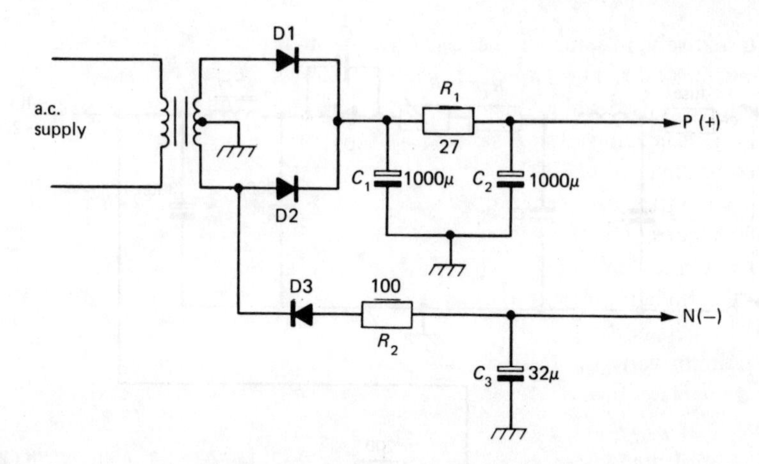

Figure 14.4 Biphase rectifier circuit (D1, D2) for the main positive l.t. rail and an auxiliary half-wave circuit (D3) to provide a negative supply

A bridge rectifier can also be connected directly across the mains to produce a full-wave output which is easier to smooth than a half-wave supply. The individual diodes are shunted by suitable surge diverting capacitors. Either the positive or the negative d.c. terminal of the bridge is connected to chassis; consequently, the neutral of the a.c. supply is isolated from the receiver earth. This introduces a safety problem when servicing such a receiver. The reader is urged to study the safety aspects of TV maintenance discussed in section 14.11.

Other types of mains-derived l.t. power supplies are shown in figure 14.4. Diodes D1 and D2 form a *biphase* rectifier circuit giving a positive supply, P, relative to the chassis. A centre tap transformer is now necessary, but the rectified output is equivalent to a full-wave arrangement. Diode D3 is a half-wave rectifier circuit which provides a negative supply, N, relative to the chassis.

14.4 Simple Stabilised Supply Circuits

The simplest form of circuit used to provide a regulated supply voltage is shown in figure 14.5. It consists of a resistor, R_1, in series with a Zener diode, the output voltage being taken across the Zener diode. Capacitors C_1 (electrolytic) and C_2 (low value, say ceramic type) are connected across the Zener diode to provide additional smoothing and high frequency decoupling. The supply voltage, V_1, is always greater than the value of the Zener diode breakdown voltage, V_2; the excess, $(V_1 - V_2)$, is dropped across the current limiting resistor, R_1. The diode works in its *reverse breakdown mode* with the cathode connected to the positive terminal of the supply. The output voltage remains relatively constant; it is equal to the breakdown potential of the Zener diode, and has a value which can be selected out of a wide range from a few volts upwards.

Alternatively, in some circuits the Zener diode is replaced by the *voltage dependent resistor* (v.d.r.; also known as a *varistor*) shown in inset (i) in figure 14.5. This arrangement is used where high values of voltage are involved, as, for example, in valve circuits needing a stabilised anode voltage supply (e.g. the feed from the boosted h.t. line to the field charging capacitor).

A disadvantage of a Zener diode stabiliser is that the output voltage fluctuates if the load current is subject to considerable variations or if the diode temperature changes. At the same time, the power losses in the series resistor (R_1 in figure 14.5) and in the diode itself tend to be considerable when relatively high current and/or high voltage loads are supplied.

The tuning voltage for varicap diodes must be particularly stable, to prevent tuner drift. Many manufacturers use an integrated circuit regulator which has only two external connections, so that it replaces a Zener diode. The I.C. also requires a limiting resistor and a suitable decoupling capacitor; this alternative is shown in inset (ii), figure 14.5.

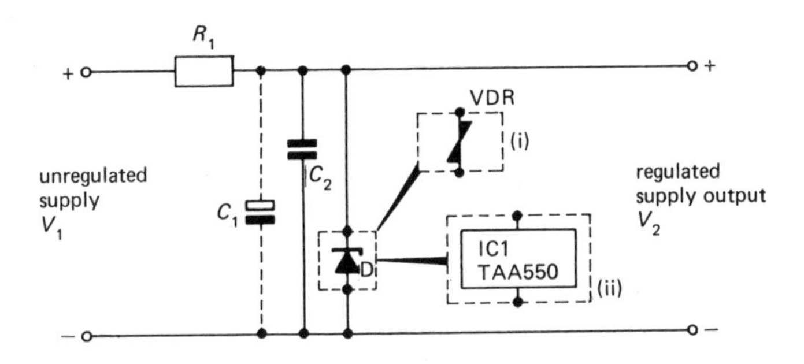

Figure 14.5 Simple voltage stabiliser circuit. The main circuit shows a Zener diode regulator; alternative (i) uses a VDR, and in (ii) a stabiliser I.C. replaces the diode

A type of arrangement frequently used to provide a stabilised low voltage supply is based on the *series regulator circuit*. In these circuits a transistor of a suitable power rating is connected in series with the load; its base voltage is controlled by a Zener diode. In more complex regulators an error amplifier is used (for example, see figure 14.8). A simple series regulator is shown in figure 14.6. The bias chain for the transistor is formed by the Zener diode stabiliser circuit, R_1, D1 (described previously). The load current is supplied by the series regulating transistor TR1; in this circuit the diode current is reasonably constant, so that the base voltage remains constant. The output voltage is approximately (V_Z – 0.6 V) with a silicon transistor, where V_Z is the breakdown voltage of the Zener diode. High frequency decoupling and low frequency smoothing is again provided by capacitors C_1 and C_2, respectively. The circuit in figure 14.6 is more economical than the one shown in figure 14.5, especially when relatively high and fluctuating load currents are supplied; the

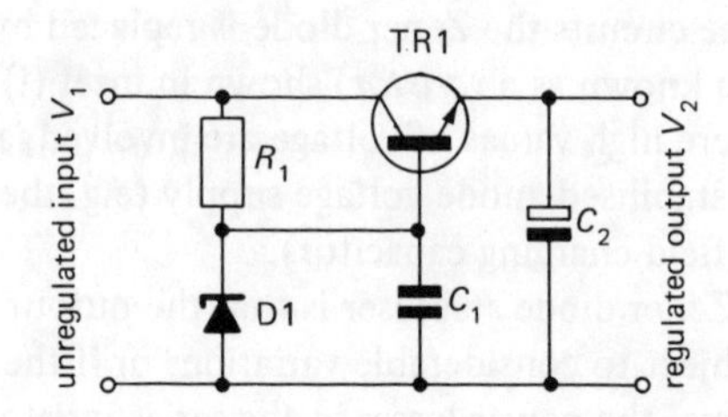

Figure 14.6 Simple series regulator circuit

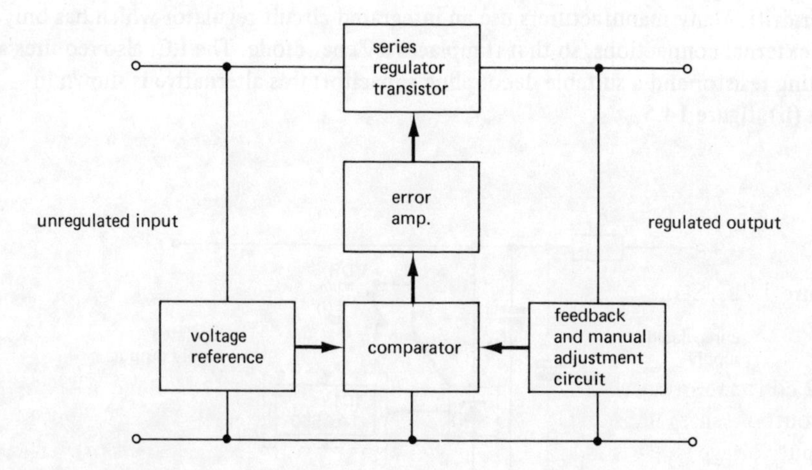

Figure 14.7 Block diagram of a series regulator with feedback

only major power dissipation is in the transistor across which the excess voltage, $(V_2 - V_1)$, must be dropped.

The circuit in figure 14.6 is an 'open-loop' type, because there is no provision to ensure that the output voltage stays constant with load current fluctuations, temperature variations and component tolerances. The block diagram in figure 14.7 shows the arrangement for a closed-loop regulator, in which the base current of the series transistor is altered to compensate for any changes in the output voltage. A sample of the regulator output voltage is fed to a *comparator* circuit which compares it with a reference voltage from a Zener diode. The resulting difference in potential is amplified and is applied to the base of the series regulator transistor to adjust its conductivity. In the circuits discussed so far the output voltage was rigidly tied to the breakdown voltage of the Zener diode. In the present arrangement the level of the regulated potential can be manually preset within a fairly wide range if desired.

A practical circuit is shown in figure 14.8. The familiar Zener diode arrangement of R_2 and D1 supplies the voltage reference which maintains the emitter voltage of

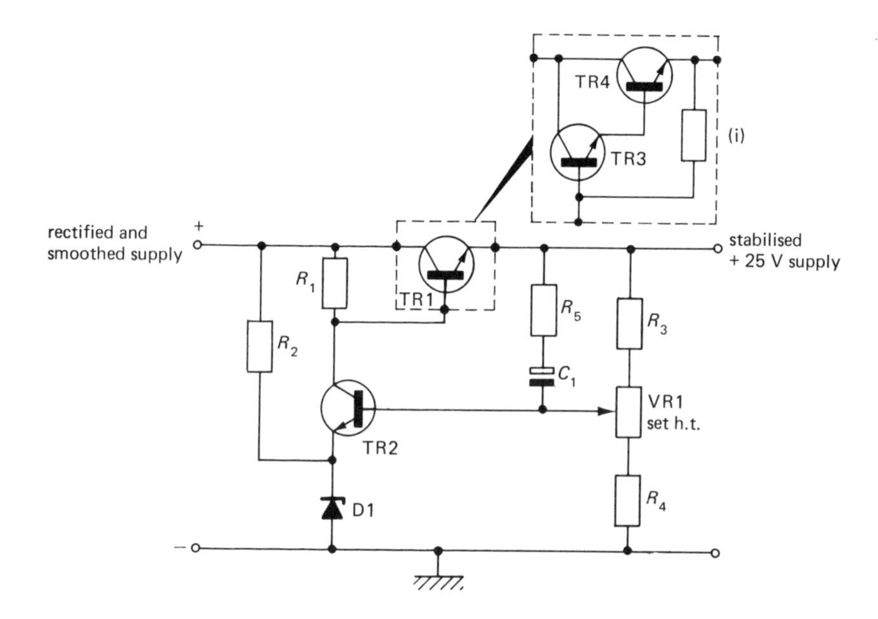

Figure 14.8 Series transistor regulator circuit with feedback

TR2 constant at, say, 6 V. The potential divider formed by R_3, R_4 and VR1 across the output senses the value of the regulated voltage, and a suitable proportion of the output voltage is fed to the base of TR2. This transistor compares its base voltage with the emitter reference potential and produces an output to correct the error. The output from TR2 is fed to the series regulator transistor, TR1, to adjust its conductivity accordingly. For example, if the output voltage from the stabiliser rises for some reason, the increased potential at the slider of VR1 is communicated to the base of TR2, which increases its conductivity; consequently, the collector voltage of TR2 falls and reduces the base voltage of the regulator TR1. The reduced base potential of TR1 causes the emitter voltage to fall (emitter follower action), thus restoring the output to its normal value. Variable resistor VR1 is used to preset the required level of the output voltage.

The ability of the regulator to oppose any changes in its output voltage can be used to perform '*electronic smoothing*'. Residual mains ripple on the stabilised supply line is fed via R_5, C_1 to the base of TR2, where it appears as a fluctuating output. The circuit takes a corrective action to improve considerably the overall power supply filtering. In effect, it looks as though the capacitance of C_1 has been multiplied by the regulator.

In some versions of the circuit in figure 14.8 transistor TR1 is replaced by the *Darlington connected pair* of transistors shown in inset (i) in the diagram. This configuration causes the current gain of the combination to be approximately equal to the product of the gains of TR3 and TR4. Such an arrangement is used in high current regulators, since a relatively small current in the control circuit can cope with large power levels in the output. Occasionally, even a third transistor may be connected in a similar manner.

There are many variations on the basic design of the circuit discussed above. For example, the series transistor can supply an output from the collector instead of the emitter. In this case a low value resistor is connected between the emitter and the collector, to provide a supply voltage to the control circuit which is placed on the output side of the regulator. The resistor also shares the load current with the transistor, reducing the power rating of the device. The comparator and error amplifier circuits can also be more complex than the basic version.

It is possible to connect the regulating transistor across the output instead of being in series with the load. A diagram of one form of *shunt voltage stabiliser* is given in figure 14.9. The output voltage is equal to the sum of the potential across the Zener diode D1 and the base-emitter voltage of TR2. Since both voltages remain relatively constant, the output is stabilised. Should the supply voltage tend to rise, the emitter potential of TR2 would increase, which, in turn, would also raise the base voltage of TR1. The increased conduction of TR1 causes more current to be taken from the unregulated source, and the additional p.d. across R_1 restores the output to its stabilised level.

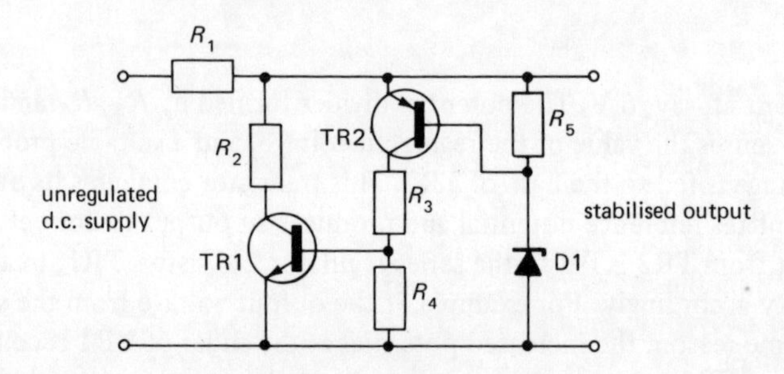

Figure 14.9 Shunt voltage stabiliser circuit

14.5 Switched-mode Power Supplies

A feature of the regulating circuits described in section 14.4 is that a large amount of power is dissipated in the principal components. Greater economy and reduced heating effects can be achieved if an electronic switch is used instead of a continuously conducting transistor. The switch conducts intermittently; it can either be a transistor or a thyristor, but when the device is ON the *voltage* across it is nearly zero and in its OFF state the *current* through it is zero. Therefore the power dissipation in the electronic switch is very low. The output from the circuit is in the form of pulses which can be fed to a transformer to be stepped either up or down and give various voltage levels. Alternatively, the pulses pass through a conventional smoothing filter, to produce a steady d.c. The mean value of the output voltage is controlled by the ON/OFF ratio, or *mark/space* ratio of the electronic switch. The switching rate depends upon the actual type of circuit used. A thyristor-fed supply operates at 50 Hz or at 100 Hz, which is governed by the half-wave or full-wave rectified a.c. mains pulses. Other switched-mode power supplies are controlled by the line time base oscillator, which runs at 15 625 Hz. Finally, a separate power supply oscillator may be used whose normal operating frequency is around 30 kHz. The higher frequencies make subsequent smoothing easier, and should a transformer be required at all, its physical size is very much less compared with one operating at 50 Hz. The reader will already have noticed how the line output transformer is used to provide a number of supply lines to the receiver. This trend is justified by the fact that the line output stage behaves like a switched-mode circuit except that in its simplest form it does not provide stabilised outputs. However, a few receivers utilise a relatively complex circuit where the line output transistor is also used as a power supply regulator.

14.6 Thyristor Regulator Circuits

The thyristor, also known as an SCR (*s*ilicon *c*ontrolled *r*ectifier), is a semiconductor device having three electrodes—namely the *anode*, the *cathode* and the *gate* [see figure 14.10(a)]. When the anode of the thyristor is negative with respect to its cathode, the device is non-conducting and cannot be triggered into conduction by a gate signal. In this respect it is similar to a reverse biased diode. It differs from a conventional diode in that, when the gate voltage is zero, the thyristor remains non-conducting even if the anode potential is positive with respect to the cathode. It is triggered into conduction, however, when a positive potential (which may be either a steady voltage or a pulse of a few microseconds duration) is applied to the gate electrode at an instant of time when the anode is positive with respect to the cathode. Once conducting, the thyristor remains in this state until its anode current is reduced to zero (or to a very low value). When conducting, the forward voltage drop across the device is about 1 V.

A somewhat similar device, known as a *silicon controlled switch* (SCS), is also used in power supply circuits [see figure 14.10(b)]. The reader will already have come across an SCS in the field time base oscillator in chapter 9. In addition to the *cathode gate*, this device has an *anode gate*; the latter is normally used for triggering the SCS into conduction. A negative potential (steady d.c. or a pulse) with respect to the anode must be applied to the anode gate to switch the device on. Once switched on, an SCS behaves in a similar manner to a thyristor. An SCS is sometimes used to provide trigger pulses for the regulator thyristor in full-wave circuits; it is also found in power supply protection schemes (to be discussed in a later section).

Both the thyristor and the SCS have four layers of semiconductor materials in a p-n-p-n arrangement. This structure may be considered to be equivalent to two transistors—see figure 14.10(c). TR1 is a p-n-p transistor whose base and collector are connected to the n-p-n transistor, TR2. The interconnections arise from the fact that the two middle p- and n-regions are shared between TR1 and TR2. The

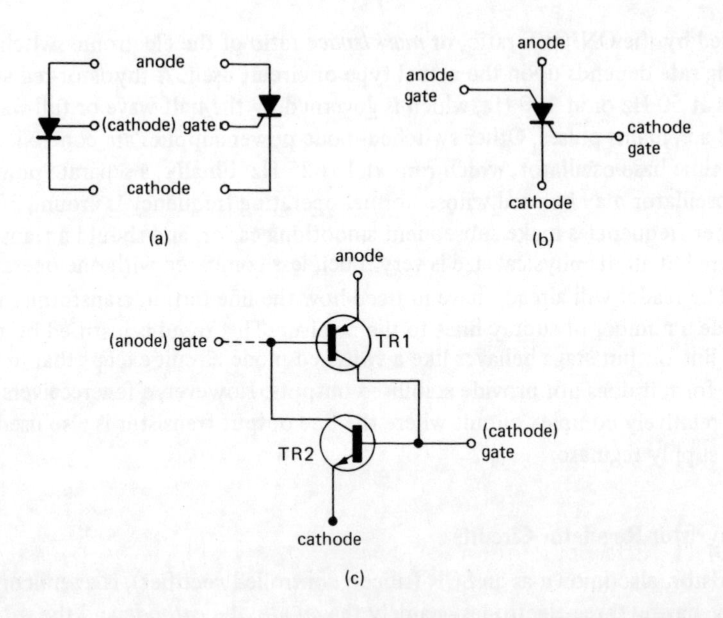

Figure 14.10 Thyristor circuit symbols: (a) alternative thyristor (SCR) symbols; (b) silicon-controlled switch (SCS); (c) two-transistor analogy of a p–n–p–n switching device

behaviour of the thyristor can now be studied by considering the operation of the two-transistor analogy. With zero signal on either gate, the device does not conduct, because of lack of base bias. Applying either a positive voltage to the cathode gate or a negative one to the anode gate causes the appropriate transistor to conduct, which, in turn, biases the other transistor through the interconnections between them. The pair rapidly drive each other into saturation, and the voltage across the thyristor falls to a very low value. The two-transistor arrangement is used in many applications instead of a thyristor or an SCS. This was shown in the various field time base oscillator circuits, but it can also be found in power supply regulators to provide thyristor trigger pulses.

The voltage regulating effect of a thyristor circuit is shown in figure 14.11. The input to the regulator may be a.c. from the mains, since the device can provide half-wave rectification; alternatively, the thyristor may be supplied from a half-wave or a full-wave rectifier. At time t_1 a trigger pulse is fed to the gate of the thyristor, which causes the device to switch on to recharge a smoothing capacitor connected across the output. The thyristor automatically turns off whenever the supply voltage falls below the capacitor voltage. The energy fed into the smoothing circuit is proportional to the shaded area in figure 14.11; this, in turn, is an indication of the value of the output voltage. The smoothing capacitor maintains the d.c. supply between the pulses. If the trigger pulse is applied later during the input sinewave–

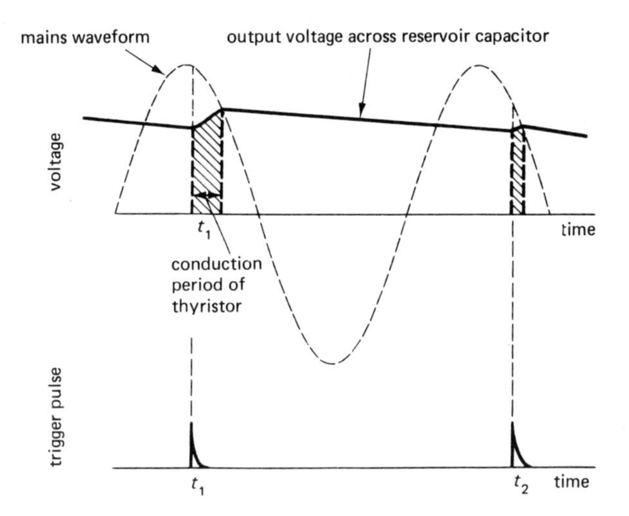

Figure 14.11 Effect of thyristor switching on power supply output voltage. At time t_1, thyristor turns on early in the mains cycle—high output voltage; at t_2 thyristor switches later—reduced output voltage

for example, at t_2—less energy will be transferred to the output and the regulated voltage will fall correspondingly. Therefore a trigger pulse timing circuit is necessary to fire the thyristor at a predetermined moment dependent upon the level of the stabilised potential.

A typical circuit is shown in figure 14.12. The regulator thyristor is TH1, which supplies the smoothing circuit comprising R_{10}, C_4 and C_5. Thyristor firing pulses are obtained by discharging C_2 through the *diac* DA1 (capacitor C_2 is charged via R_4 from the a.c. line); R_7, R_8, C_3 and R_9 are pulse limiting and shaping components. The diac—a *bidirectional breakdown diode*—is a semiconductor device which is non-conducting until the voltage across it reaches a certain value known as the breakdown voltage. When this is reached, the diac is rapidly driven into conduction and the potential across it falls to a low value. In this respect, the diac is another form of electronic switch; the type used in TV has a breakdown voltage of about 30 V.

The voltage across C_2 is thus allowed to build up to 30 V when DA1 fires, and the energy from the capacitor triggers the thyristor. The charging rate of C_2 is controlled by TR1, which shunts the capacitor. If the transistor conducts heavily, it takes longer to reach the diac breakdown voltage and the thyristor trigger pulse occurs later in the mains cycle, with consequent reduction in the output voltage. Variable resistor VR1 alters the bias on the base of TR1, which affects the stabilised output in the manner described above. The feedback resistor, R_5, senses the regulated d.c. level and influences the conduction of TR1 accordingly.

A 'slow-start' circuit is often used in the thyristor regulator. A typical arrange-

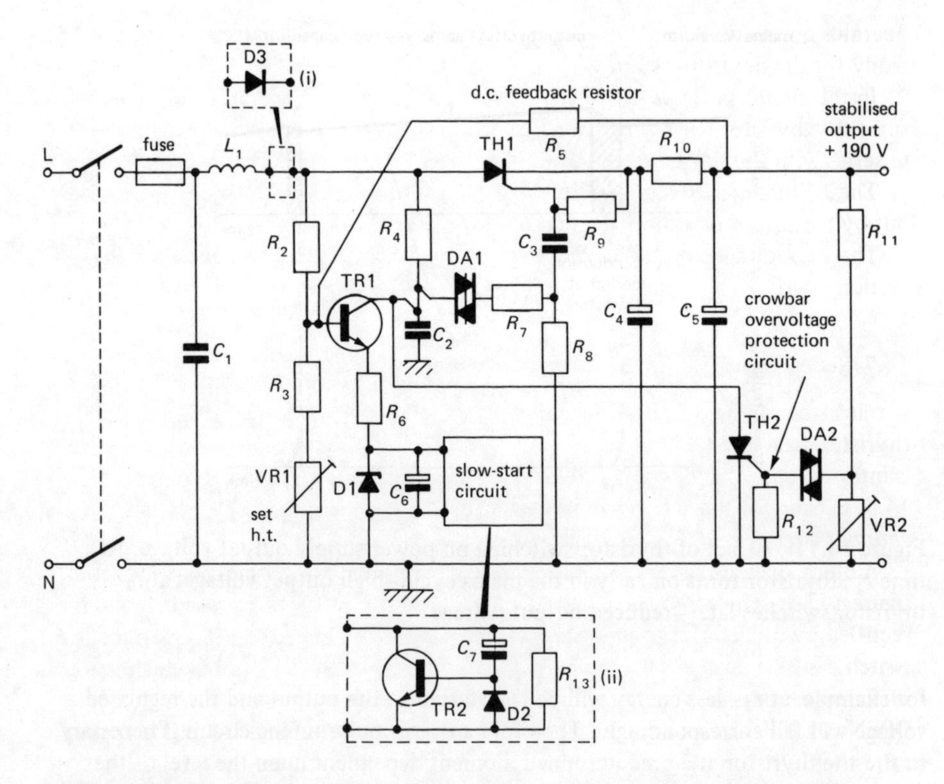

Figure 14.12 Thyristor voltage regulator circuit. The thyristor and its trigger pulse generator can be fed either directly from the mains or via a rectifier [D3—alternative (i)] . A slow start circuit—arrangement (ii)—is often used to reduce the initial current taken from the supply

ment is shown in inset (ii) in figure 14.12. When the receiver is first switched on, the large value filter capacitors have to be charged, which causes the output voltage to be low. The regulator would try to trigger the thyristor early in the mains cycle to increase the voltage; this, in turn, could overload the device, or blow the fuse or even damage the capacitor itself. The slow-start circuit connected in the emitter of TR1 makes the transistor conduct heavily to retard the firing pulses and allows the h.t. voltage to rise gradually. Transistor TR2 is connected across Zener diode D1 in the emitter circuit of TR1. Normally TR2 is non-conducting, because of the lack of base bias, but when the receiver is first switched on, C_7 begins to charge. The charging current flows through the base–emitter junction of TR2 and the transistor conducts. This causes TR2 to bypass the Zener diode, D1, causing the emitter voltage of TR1 to fall, so that this transistor conducts heavily; the net result is that the voltage across C_2 takes longer to reach the trigger voltage of DA1. Once C_7 is fully charged, the voltage across D1 is restored to its steady value and the entire regulator operates normally. When the receiver is switched off, C_7 discharges via R_{13} and D2 ready for the next slow start.

In the circuit in figure 14.12 the trigger pulse generator operates from the mains, since the thyristor itself acts as a rectifier. In some designs a diode [D3 in inset (i)] is in series with the thyristor, to share the reverse voltage when TH2 is turned off.

The switching action of the thyristor can cause severe interference; this is filtered out by the action of a low value inductor, L_1, and capacitor, C_1.

The overvoltage protection circuit associated with TH2 will be discussed in section 14.10.

14.7 Series Transistor Switched-mode Regulator Circuits

A transistor can be connected in series with the load in a manner similar to the thyristor. The input to the regulator is a rectified and partly smoothed d.c. which is interrupted, or 'chopped', by the switching transistor, as explained in section 14.5. Because of the action of the device, the arrangement is sometimes called a *chopper circuit.*

A block diagram of a power supply based on the chopper principle is shown in figure 14.13. The supply is obtained via a tap on the autotransformer, T1, and is rectified by diode D1 and smoothed by capacitor C_2. Transistor TR1 operates as a switch, its base current being switched between zero and a high value which is sufficient to saturate the transistor. When the base current is zero, the 'switch' is OPEN or OFF, and no energy is supplied to the load. When base current is applied,

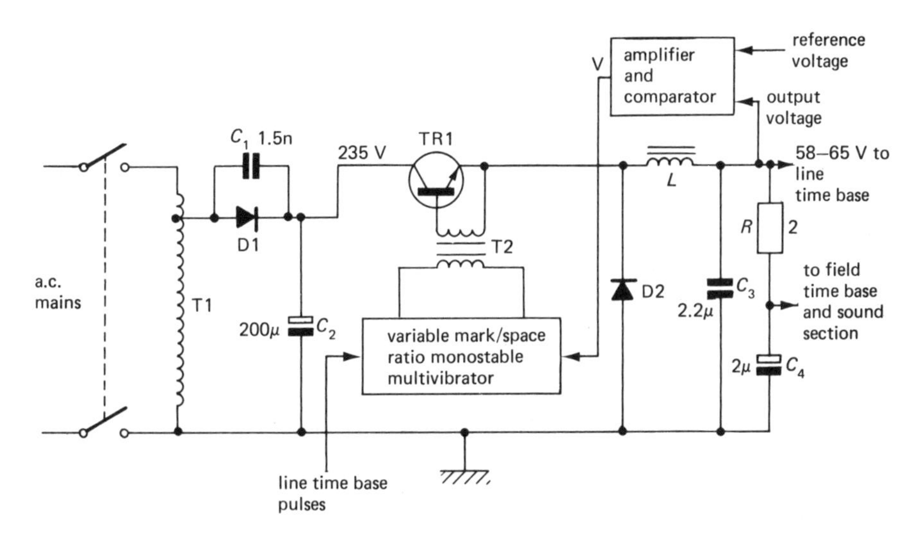

Figure 14.13 Simplified diagram of a switched-mode power supply regulator–using a series transistor (so called chopper circuit). Note the values of smoothing capacitors: C_2 (ripple 50 Hz), C_3 and C_4 (ripple at line frequency)

it drives the transistor into saturation so that the collector–emitter voltage falls to a very low value (typically 200 mV), and the switch is CLOSED or ON. The chopper operates at line time base frequency; thus its output is in the form of pulses at 15 625 Hz which are fed to the smoothing circuit consisting of inductor L_1 and capacitor C_3. The reader will notice the difference in capacitance value between smoothing capacitor C_2 (at 50 Hz) and C_3 (at line frequency). When the chopper is OFF, energy stored in the inductor discharges into the load to maintain the load voltage. Diode D2 provides a return path for the inductive current during this period of operation.

The base switching current for the chopper is provided by the *monostable multivibrator* coupled to the transistor via transformer T2. To regulate the output voltage, the ratio between the ON time and the OFF time of transistor TR1 is automatically adjusted; that is, the *mark/space ratio* of the output of the multivibrator is made variable. A monostable multivibrator is a type of switching circuit which has to be triggered by an external pulse for the output to change its state from OFF to ON. The ON state is then maintained for a period of time governed by the circuit internal time constant, after which the output returns to OFF again (in which it remains until retriggered).

In the circuit in figure 14.13 the monostable ON/OFF ratio is controlled by altering the discharge rate of a capacitor. In turn, this discharge rate is regulated by voltage V at the output of an amplifier/comparator circuit. The latter compares a reference voltage derived from a Zener diode with the chopper stabilised output. The resultant effect is that the ON time of TR1 is increased by the multivibrator in order to raise the regulator output voltage.

The circuit also includes a slow-start arrangement plus overcurrent and overvoltage protection; the latter features are discussed later in this chapter.

14.8 Shunt Switched-mode Power Supplies

An alternative arrangement to the series regulator described in section 14.7 is a switched-mode transistor which, together with a suitable transformer, is connected *across* the rectified mains supply. The transformer presents an inductive load where energy is stored when the switch is ON. The energy is transferred to some of the receiver circuits during the ON periods and to others during the OFF periods; the latter is due to the storage effect of the transformer. Rectifiers are connected to the transformer windings, their polarities being arranged so that conduction takes place during either, the positive or the negative half-cycles of the transformer e.m.f. The system resembles the action of a line output stage; indeed, many designers combine the functions of a switched-mode power supply with the line time base of their receivers. The reader will recall from section 10.5 that the switching waveform of a conventional line output transistor has to be carefully timed, not only to prevent overloading the device, but also to ensure a linear picture. The transistor used in the switched mode regulator is fed with a line frequency waveform, but the mark/space ratio is variable according to the generated voltage levels. Satisfactory operation of

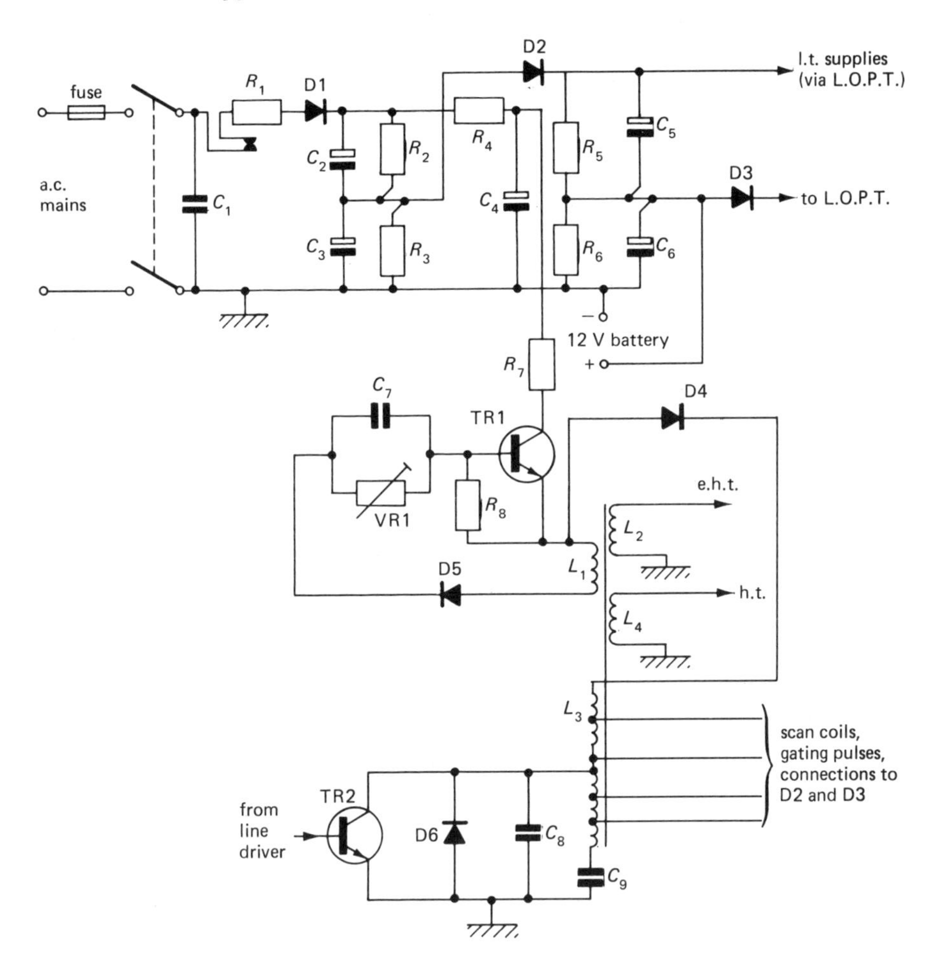

Figure 14.14 Shunt switched-mode power supply with the associated line output stage (GEC). The receiver is mains/battery-operated, but the regulator circuit (TR1) is not used during battery operation

a combined arrangement can be achieved by a simple *isolating diode* which isolates the regulator from the rest of the line output stage during the conduction of the efficiency diode.

The circuit in figure 14.14 shows the principle of operation of one type of shunt switched-mode regulator. The regulator transistor is TR1, whose collector is fed from the rectified and smoothed mains supply via current limiting resistor R_7. The transistor is switched on by line flyback pulses fed to its base from winding L_1 on the line output transformer. Therefore, during the regulator ON state, current is fed to the line output stage via TR1 and D4. The energy thus stored is supplied into

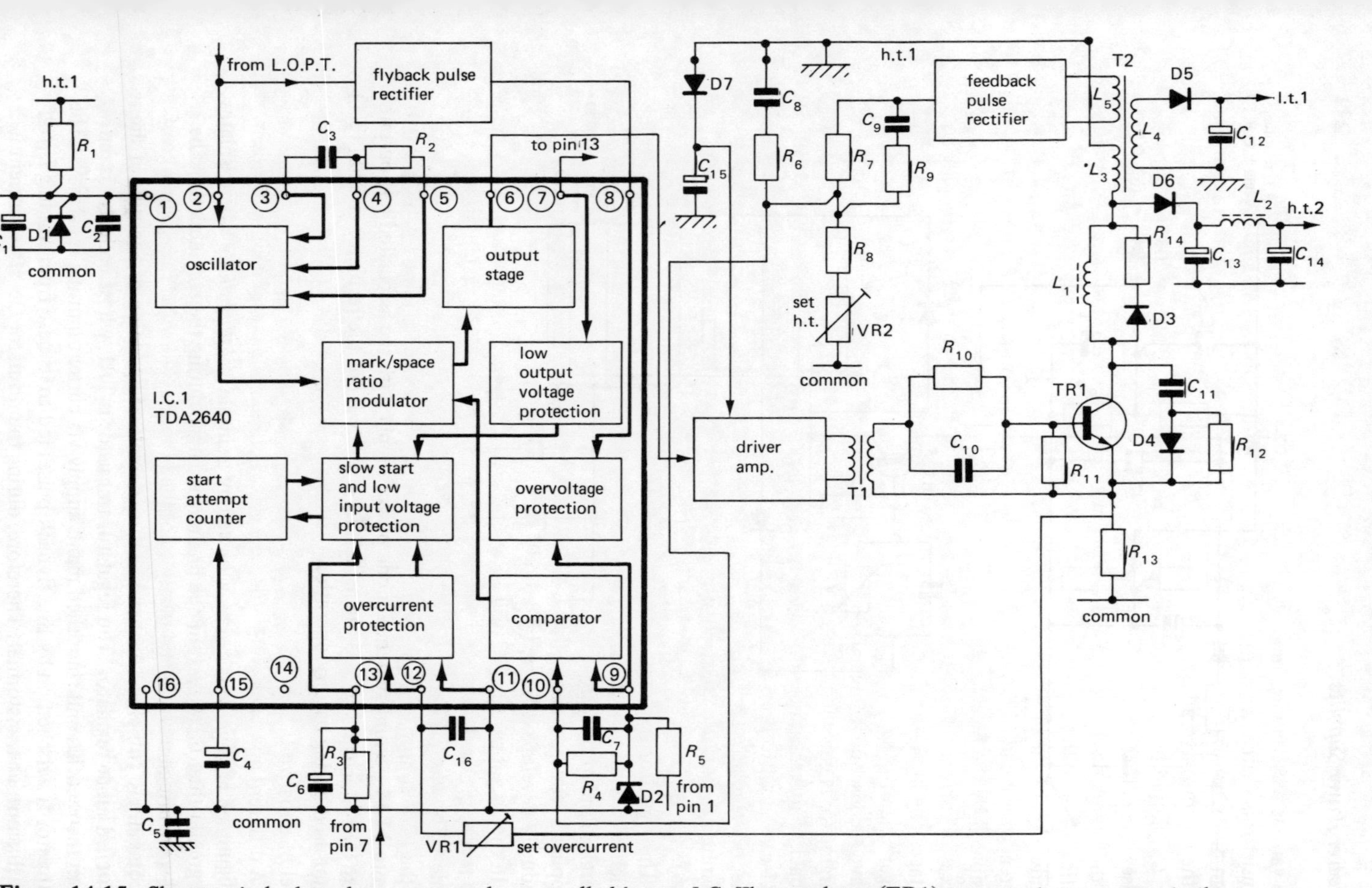

Figure 14.15 Shunt switched-mode power supply controlled by an I.C. The regulator (TR1) operates in a manner similar to a line output transistor. *Note*: Regulator 'COMMON' is not connected directly to chassis (it is decoupled via C_5). Instead regulator h.t.1 is earthed. Output rails (l.t.1, h.t.2) are conventionally arranged. Therefore servicing this type of circuit must be done with care. (Mullard)

the various receiver circuits, including l.t. rails, during line scan. When the regulator transistor is switched off, isolating diode D4 is reverse biased by the back e.m.f. from the transformer winding, L_3. (The typical line output arrangement consisting of the line output transistor, TR2, efficiency diode, D6, and flyback tuning capacitor, C_8, should be readily recognisable.) Although TR1 is switched on by flyback pulses, the actual period of conduction during the 12 μs time interval is governed by the charge/discharge cycle associated with C_7 and VR1. This adjusts itself automatically, since the magnitude of the flyback pulse is a measure of the state of conduction of the line output transistor just prior to the flyback. In turn, the current at the end of line scan depends on how much of the initial energy has been dissipated by the receiver power requirements. A large amplitude pulse caused by excessive current flow charges capacitor C_7 in the base circuit of TR1 to a higher voltage. The increased capacitor potential maintains D5 reverse biased for slightly longer when the next flyback pulse arrives, and the point of switching TR1 on is delayed; consequently less energy is fed into the line output stage. The rate of discharging C_7 is governed by the setting of resistor VR1. Since the amplitude of the flyback is also an indication of the receiver e.h.t. and the resultant picture width, VR1 can be set to provide correct picture geometry (see chapter 13). At the same time, the regulator controls the e.h.t., h.t. and l.t. supplies. The circuit action resembles that of the width stabiliser found in a valve line output stage (section 10.6).

The regulator transistor TR1 can operate only when there are flyback pulses, which means that the line oscillator, the driver and the output transistor must be energised independently. This is achieved by means of the starting diode D2, which supplies a reduced voltage to the circuits concerned until the line transformer gradually develops its correct output.

In the circuit shown in figure 14.14 the regulator circuit is only operative when the receiver is used on the a.c. mains. During battery operation stabilisation is achieved by various Zener diode stabilisers (not shown in this diagram). The level of the 12 V supply is raised by the *boost* circuit formed by D3 and C_5. This circuit is connected across one winding on the line output transformer; boost diode D3 rectifies during the scan period and the resultant voltage developed across C_5 is added to the battery potential. It will be noted that in transistorised receivers the level of boost voltage is far less than in their valve counterparts.

A different approach to the design of a switched mode power supply is shown in figure 14.15; this circuit is used in a colour receiver whose power requirements are greater than those of a monochrome set. The a.c. mains supply is rectified by a full-wave bridge rectifier and partly smoothed (neither shown in the diagram) to become h.t.1. This is fed to the regulator transistor TR1 via transformer T2. Transistor TR1 is switched ON and OFF by the signal applied to its base from the driver stage, whose action is, in turn, controlled by the integrated circuit, I.C.1. Alternate switching of TR1 converts the d.c. input from h.t.1 into a.c. by the action of the transformer. The two outputs are rectified and smoothed to provide l.t.1 and h.t.2, which between them supply all the low voltage and the high voltage requirements of the receiver.

The circuitry associated with the regulator transistor closely resembles the line output stage. The reader is asked to turn to section 10.5 to recall the operational problems connected with high voltage power transistors used for switching an inductive load. The driver amplifier stage, represented by a block in figure 14.15, is very similar to the corresponding arrangement in the transistorised line time base (see section 10.3). The driver output is coupled to the base of TR1 by transformer T1 and the pulse speed-up network R_{10}, C_{10}. When the regulator transistor is driven ON, the collector current is limited by the choke, L_1, in its collector circuit. When TR1 is switched OFF by the base waveform, the entire output circuit would tend to burst into self-oscillations (ringing) as well as to induce a large back e.m.f. at the collector. The rise in collector voltage is prevented by the protection network C_{11}, D4 and R_{12} connected across the transistor; diode D3 and choke L_1 also assist with voltage limiting. Ringing at TR1 collector is damped by the respective resistors R_{12} and R_{14} in the protection circuit.

The control circuit of figure 14.15 is now discussed. The regulator switching waveform is generated by the oscillator inside I.C.1 in figure 14.15. The free-running frequency is governed by the values of C_3 and R_2 connected between pins 3, 4, 5. This design ensures that the system is self-starting, unlike in other circuits, where the line oscillator must first be made operative. However, once the line time base becomes fully operational, the power supply oscillator is synchronised with the line frequency by means of the flyback pulses applied to pin 2. The mark/space ratio of the oscillator waveform is modified in the modulator block according to the receiver requirements, as explained in section 14.5. The suitably 'modulated' square wave is amplified by the output stage before it is applied to the driver amplifier at pin 6.

The mark/space ratio is affected principally by two features: the power demand placed on the regulator circuit of TR1, and the various protection arrangements incorporated in the overall design.

The control of power requirements is now considered. The amplitude of the pulses induced in the windings of the regulator transformer depends upon the magnitude of the current flow in the windings. Therefore a separate coil, L_5, is wound on T2 to feed back a signal proportional to the regulator output. The initial pulses are rectified and smoothed, and a fraction of the resultant voltage (given by the potential divider R_7, R_8, VR2) is then fed back to I.C.1 at pin 10. The comparator circuit compares the actual output with a reference voltage developed by Zener diode D2 (applied at pin 9). An error signal is developed by the comparator and is communicated to the mark/space ratio modulator block. Residual mains hum can also be reduced by the action of the comparator, which now affords electronic smoothing. The level of a.c. ripple on h.t.1 rail is sampled by C_8, R_6 and fed back to pin 10, where it would appear as an unwanted supply voltage variation to be corrected accordingly. The two resistor–capacitor networks, R_9, C_9 in the sampling circuit and R_4, C_7 between pins 9 and 10, govern the speed with which the regulator responds to any changes in demand while preventing the whole system becoming unstable.

The arrangement in figure 14.15 includes a slow-start feature, the need for which was explained in section 14.6. Capacitor C_6 at pin 13 charges slowly, to affect the mark/space ratio via the slow-start block inside the I.C. After the receiver has been switched off, C_6 discharges via R_3 ready for the next slow-start.

A stabilised power supply to I.C.1 (at pin 1) is obtained from h.t.1 via the Zener regulator R_1, D1 with its associated smoothing/decoupling capacitors C_1, C_2. The comparator reference voltage is also derived from this point and fed to pin 9 via R_5.

The integrated circuit includes a very comprehensive power supply protection scheme, which will be discussed in the next section.

The power supply regulator COMMON connection is isolated from the receiver chassis by capacitor C_5, since it is the positive terminal of h.t.1 supply line (which is at chassis potential). If a bridge rectifier is used, then the chassis is at half mains voltage. Care must be exercised in servicing circuits of this type to prevent a short circuit between the regulator COMMON and the receiver CHASSIS.

14.9 Overcurrent Protection

The TV receiver must be protected against any harmful effects caused by internal faults. These could render the receiver unsafe (risk of fire, X-rays, electric shock) or cause the original breakdown to induce more damage, which would be costly to repair. Various methods of electrical protection are considered below. The general aspects of safety are discussed in section 14.10.

Overcurrent protection can be provided simply by means of correctly rated *anti-surge* fuses, or by means of *thermal cut-outs, drop-off resistors* and *fusible resistors*, or by means of *magnetically operated circuit breakers. Fuses* provide protection for normal operating conditions, and also permit switch-on surges to pass without blowing. In a *thermal cut-out* the heat produced by the current causes a bimetallic strip to bend, and if an excessive current is flowing, it opens a set of contacts which must be reset manually. In the case of a *drop-off resistor* (see R_1 in figure 14.14) the heat developed in the resistor is transmitted to a soldered joint; when the current rating of the device is exceeded, the solder melts and the circuit is broken. *Electromagnetic circuit breakers* are a form of relay which opens a set of contacts when the current through the coil rises above a certain value; the circuit breaker has to be reset manually.

The above-mentioned protective devices can be installed in the main a.c. supply line, or in the various l.t. and h.t. rails. They can be used even in individual sub-circuits—for example, in the screen feed to the line output valve or in the line output transistor, etc. These methods of protection have one major disadvantage—their slow speed of operation and lack of sensitivity may not be acceptable in many semiconductor circuits. Semiconductor p–n junctions can be easily damaged before protection starts to operate; on the other hand, very sensitive devices may not allow normal supply surges to occur without tripping out. A number of receivers use *electronic overcurrent protection*, which can be incorporated into a power supply regulator circuit.

The diagram in figure 14.16 shows a constant current regulator. Zener diode D1

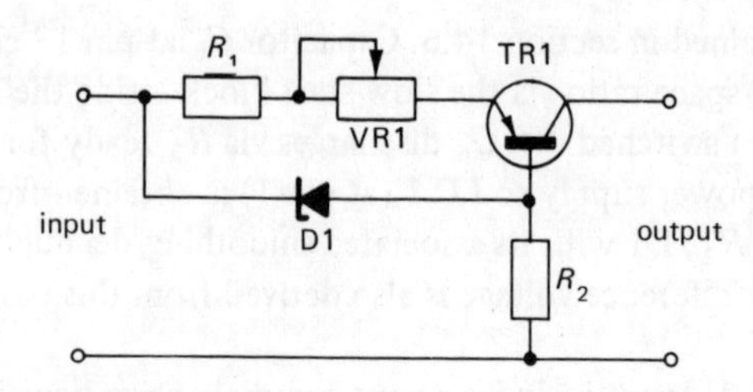

Figure 14.16 Constant current regulator

is connected across the series combination of the base–emitter junction of the regulator transistor TR1 and the two resistors R_1, VR1 in the emitter feed. The emitter current produces a voltage drop across R_1 and VR1. However, the emitter current, I_E, is limited to the value which causes the potential difference across the two resistors to be equal to the Zener diode breakdown voltage, V_Z (neglecting the effect of V_{BE}). That is

$$I_E = V_Z/(R_1 + \mathrm{VR1})$$

This type of regulator can be used to develop a constant output voltage across a fixed load resistor. The arrangement may be found in a stabiliser supplying the A1 controls of a colour receiver.

In other circuits the overcurrent is sensed by measuring the p.d. produced by the current when it flows in a low value resistor. The resultant voltage biases on a protection transistor, which, in turn, switches off, or at least reduces the conduction of the regulator device directly or indirectly. An example of this arrangement is given in figure 14.15; the I.C. includes overcurrent protection by sensing the p.d. developed across R_{13} (1 Ω) in the emitter of the regulator transistor, TR1. The control voltage is fed via preset VR1 to I.C.1 at pin 12. The internal protection circuit reduces the mark/space ratio to zero, and the drive waveform is removed for a period given by the discharge of C_6 at pin 13. The regulator automatically starts up again via the slow-start circuit; the procedure may be repeated a number of times, as governed by the restart counter and its external capacitor, C_4, at pin 15. If the overload persists, the regulator locks out. A low value capacitor, C_{16}, is connected between pins 11 and 12 to prevent protection operating in the event of very short duration surges, which are thus diverted from the control circuit.

Another form of overcurrent protection was shown in figure 10.8, section 10.5.

14.10 Overvoltage Protection

Overvoltage protection is necessary, since an excessive voltage may cause damage, particularly to the line output stage or to integrated circuits and other semiconductor devices. In addition, excessive voltages will be reflected in an increased e.h.t.; for example, a rise of 5 V in the h.t. supply can 'boost' the e.h.t. by 600 V. E.H.T.

levels above the design figure for a given receiver cause frequent tube flashovers leading to a possible, and often unaccountable, breakdown of various transistors. Excessive e.h.t. also increases the risk of X-ray radiation from the tube, especially in colour TV, with its already high value of anode voltage (say 25 kV or more). The tube heater voltage can also rise if it is fed from a faulty supply line, thus reducing the life of the c.r.t.

One method of protecting against an overvoltage is to employ a network which applies a short circuit to the power supply when the voltage exceeds a predetermined value. That is, an *electronic crowbar* is applied across the supply terminals; this usually blows the fuse or operates a cut-out (this form of protection is known as *crowbar protection*). A simple circuit of this type of arrangement is shown in figure 14.17. The breakdown voltage of Zener diode D1 is chosen so that it does

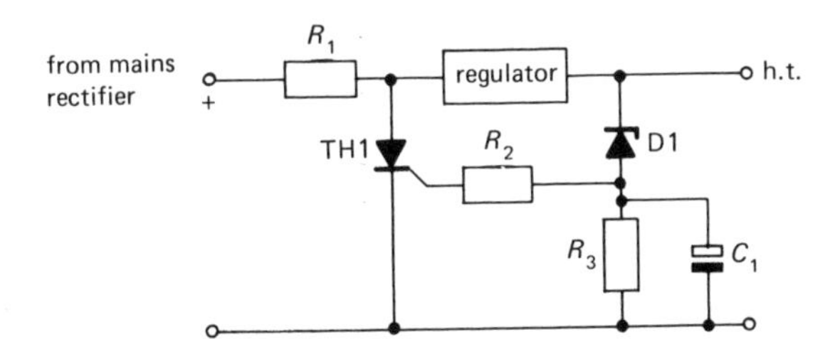

Figure 14.17 Simple electronic crowbar protection

not normally conduct; neither does thyristor TH1. If the regulator output voltage becomes excessive, say owing to a fault in the device or its control circuit, diode D1 conducts and applies a positive trigger voltage to TH1. The thyristor switches on and causes a large current to flow from the power supply, and the receiver cut-out operates. To prevent 'nuisance' tripping, capacitor C_1 absorbs momentary over-voltages produced by tube flashovers.

Another form of overvoltage protection was shown in figure 14.12. Thyristor TH2 is normally non-conducting. If the stabilised output rises above its nominal value, the voltage developed across VR2 increases above the breakdown potential of diac DA2 and the thyristor is triggered into conduction. TH2 short circuits the regulator timing capacitor C_2, and removes the firing pulses from the regulator thyristor, TH1.

The I.C. controlled power supply in figure 14.15 includes an alternative method of overvoltage protection. The magnitude of the line flyback pulses is detected by the flyback pulse rectifier and fed to pin 8 of I.C.1. The overvoltage protection block acts as a comparator which compares the potential at pin 8 with the reference voltage at pin 9. At a predetermined level the protection circuit reduces the drive waveform to zero and the sequence of events is the same as that already described in the operation of the overcurrent protection for this circuit. The advantage of using the line pulses to operate this system (instead of the more obvious source as supplied to pin 10) lies in the fact that the protection can also guard against a number of faults in the line output stage. I.C.1 includes a form of back-up protection against a loss of control voltage at pin 10. Should this happen, there would be a tendency for the comparator to induce maximum mark/space ratio and a danger of overvoltage. The 'low feedback voltage protection' block overrides the comparator and maintains a very low mark/space ratio in the drive waveform.

Some manufacturers use an overvoltage sensing circuit to shut down the line oscillator to protect the most vulnerable section of the receiver—namely the line output stage.

14.11 Power Supply Fault-finding and Safety Considerations

Statistically it appears that power supply faults are the most common source of breakdown. It is essential to establish which sections, if any, of the receiver are functioning, to see whether the 'dead' circuits share their supply rails. Where applicable, valve and tube heaters should be energised. Supplies fed from the line output stage would be immediately affected in the event of a fault in the line time base. Various protection features can also affect one circuit in the event of a fault in another.

Fault-finding in unregulated power supplies may require only the use of a voltmeter to check for the presence of a.c. input and of the various d.c. rail voltages. Usual continuity checks will then confirm the position of the faulty component.

Regulated power supplies which employ feedback present additional problems. The effect of feedback on the behaviour of the stabiliser circuit should be checked in order to decide whether the fault is in the control circuit or the regulating device itself. Varying the appropriate adjustment will disclose whether the control subcircuit responds in the manner expected.

Switched-mode power supplies can be treated in a similar manner to line time base circuits. The presence of waveforms can be detected with the aid of an oscilloscope. Unless a low capacitance probe is used, the reader should not test for the drive waveform on the base of the regulator transistor; the input capacitance of the oscilloscope can affect the timing of the switching waveform and cause damage to the output transistor. Again, the effect of feedback on the behaviour of the circuit can be observed. It must not be overlooked that a number of regulating circuits require an auxiliary supply to energise the control circuit.

Any protective features of the power supply circuit must not be immobilised for whatever reason, as damage to many components may result.

When one carries out repairs on power supplies and associated circuits, the receiver must first be switched off and unplugged from its mains socket. It is always advisable to discharge the electrolytic capacitors via a resistor of about 1 kΩ.

Before actual fault-finding anywhere in the receiver, it must be remembered that the chassis may be live at full mains potential, especially in the receivers which do not have mains transformers in the power supplies. It is necessary to ensure that the mains connections are correct to both the receiver and the plug. The standard colour code used in flexible supply leads is: *brown* for the *live* and *blue* for the *neutral.* Where a *bridge rectifier* is used in the mains power supply *without* an isolating transformer, the chassis is *always* at half the mains potential with respect to the 'true' earth. Any test equipment (oscilloscope, signal generator, etc.) used when the set is connected to the mains should be supplied via an isolating transformer. Otherwise, the test equipment must not be earthed to avoid possible damage to the circuits—but the risk of shock still remains!

Workbenches in TV repair workshops should also be supplied via isolating transformers, whose rating ought to be in the region of 500 VA per bench. This allows the TV receiver and other test equipment to be connected without causing any reduction in the supply voltage to the receiver. If the transformer is inadequately rated, it may also produce erratic operation of a thyristor regulator in a TV set.

The design *and* servicing of TV receivers in the UK must conform, by law, to strict safety standards. The quality of manufacture is governed by BS 415:1972, and the certificate of approval is given under the B.E.A.B. scheme (British Electrotechnical Approvals Board) to the appropriate make of receiver. The conditions of the approval certificate must not be altered in the process of receiver servicing. This applies to the important *safety components*, which have passed stringent tests before being adopted in the manufacture. These parts are distinguished from others in service diagrams by suitable markings. The two common methods of identification are shown in figure 14.18. In (a) the shading over R_1 and the warning sign in

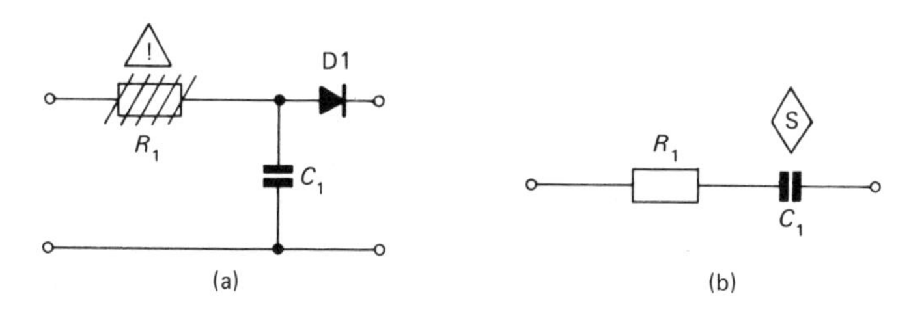

Figure 14.18 Identification markings for safety components. In (a) R_1 is a safety component, and in (b) C_1 is so designated. (Either symbol may be used in circuit diagrams)

the triangle denote that the resistor is a safety component; in (b) the letter S in a 'diamond' outline means that C_1 is a special component. Any replacement must comply with the same specifications as the original component; the method of fixing may also be critical—for example, it may be necessary to ensure correct clearances or adequate ventilation or to use low melting point solder to replace a drop-off resistor.

14.12 Mains/Battery Operation

Mains/battery receivers are capable of operating in one of several modes, including: mains operation, battery operation and recharging.

Receivers either may have an internal battery or may be supplied from an externally connected source, such as a 12 V car battery. If an internal battery is used, care should be taken in maintaining the electrolyte level in the cells. Also, if the receiver is inadequately ventilated, there is a danger of an explosion due to the accumulation of gases if a lead-acid type of battery is used.

Some receivers include facilities for *floating charge* or *trickle charge* when the receiver is connected to the mains supply. A supply circuit with this feature is shown in figure 14.19. Diode D3 is connected between the positive terminal and chassis, to protect against an accidental reversal of the battery connections. In this event, the diode would become a short circuit across the supply and cause the fuse to blow.

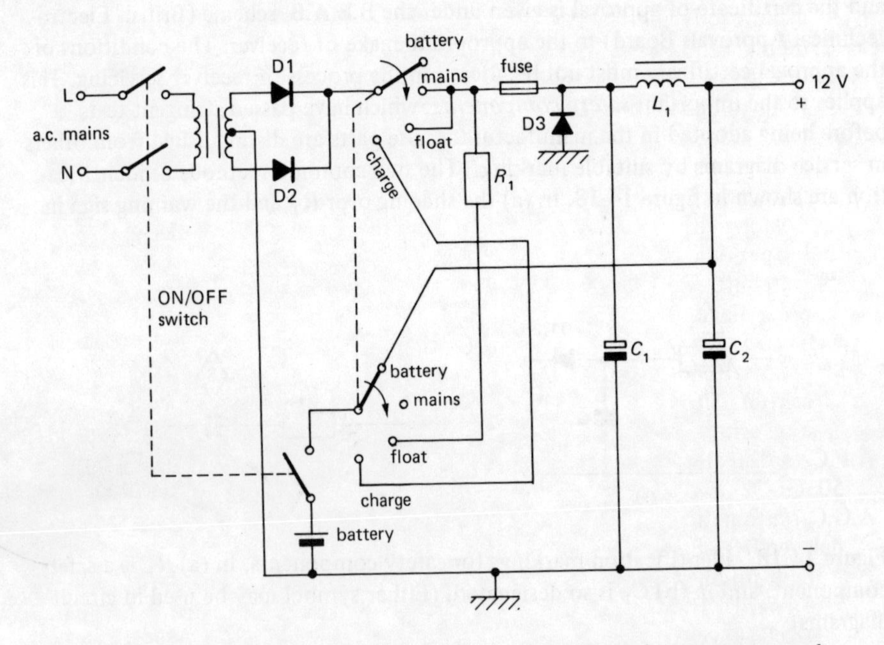

Figure 14.19 A built-in battery charging circuit in a mains/battery-operated receiver. *Note*: Diode D3 guards against the effects of incorrect polarity of the battery. This type of protection applies to most battery-operated receivers

Index